Sustainable Development Goals Series

World leaders adopted Sustainable Development Goals (SDGs) as part of the 2030 Agenda for Sustainable Development. Providing in-depth knowledge, this series fosters comprehensive research on the global targets to end poverty, fight inequality and injustice and tackle climate change.

Sustainability of Future Earth is currently a major concern for the global community and has been a central theme for a number of major global initiatives viz. Health and Well-being in Changing Urban Environment, Sendai Framework for Disaster Risk Reduction 2015–2030, COP21, Habitat III and Future Earth Initiative. Perceiving the dire need for Sustainable Development, the United Nations and world leaders formulated the SDG targets as a comprehensive framework based on the success of the Millennium Development Goals (MDGs). The goals call for action by all countries, poor, rich and middle-income, to promote prosperity while protecting the planet earth and its life support system. For sustainability to be achieved, it is important to have inputs from all sectors, societies and stakeholders. Therefore, this series on the Sustainable Development Goals aims to provide a comprehensive platform to the scientific, teaching and research communities working on various global issues in the field of geography, earth sciences, environmental science, social sciences and human geosciences, in order to contribute knowledge towards the current 17 Sustainable Development Goals.

Volumes in the Series are organized by the relevant goal, and guided by an expert international panel of advisors. Contributions are welcome from scientists, policy makers and researchers working in the field of any of the following goals:

No poverty
Zero Hunger
Good Health and Well-Being
Quality Education
Gender Equality
Clean Water and Sanitation
Affordable and Clean Energy
Decent Work and Economic Growth
Industry, Innovation and Infrastructure
Reduced Inequalities
Sustainable Cities and Communities
Responsible Consumption and Production
Climate Action
Life Below Water
Life on Land
Peace, Justice and Strong Institutions
Partnerships for the Goals

The theory, techniques and methods applied in the contributions will be benchmarks and guide researchers on the knowledge and understanding needed for future generations. The series welcomes case studies and good practices from diverse regions, and enhances the understanding at local and regional levels in order to contribute towards global sustainability.

More information about this series at http://www.springer.com/series/15486

Felix Kogan

Remote Sensing for Food Security

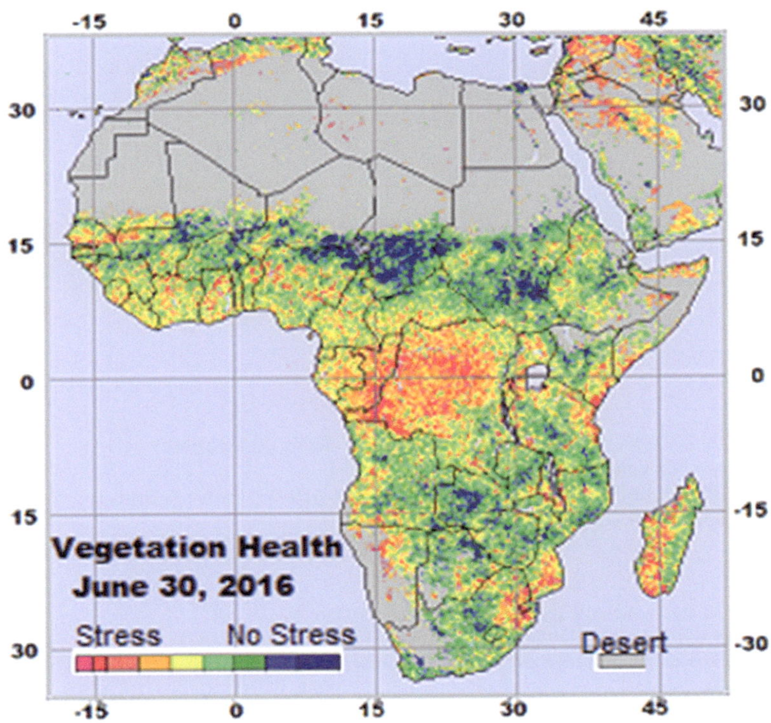

Springer

Felix Kogan
NOAA/NESDIS
College Park, MD, USA

ISSN 2523-3084 ISSN 2523-3092 (electronic)
Sustainable Development Goals Series
ISBN 978-3-030-07166-0 ISBN 978-3-319-96256-6 (eBook)
https://doi.org/10.1007/978-3-319-96256-6

Softcover re-print of the Hardcover 1st edition 2019

This Springer imprint is published by the registered company Springer Nature Switzerland AG.
The registered company address is: Gewerbestrasse 11, 6330 Cham, Switzerland

This book is dedicated to my family, who supported me at all stages of its writing and took very good care of me. I also appreciate their extreme tolerance that I could not pay too much attention to domestic issues.

Foreword

Planet Earth is dynamic, and climate change and global warming are being observed in many parts of the globe. The Earth system consists of land, ocean, biosphere, and atmosphere, and a close interaction leads to climate change and global warming. Population growth in the last three decades further complicates the interaction and crosses the carrying capacity of the planet Earth and degrades the existing environment and resources, which raises the question of sustainability of resources. The degradation of the environment affects water, ground water, monsoon, agricultural land, and agricultural productivity during drought conditions. Frequent droughts impact different parts of the world. Scientists and policy makers are concerned about food security, with increasing population especially in developing countries. Atmospheric pollution and air quality impact human health, and if proper food is not available to people it will directly impact the lives of people. About a sixth of the world population suffers from hunger, malnutrition, and lack of clean water, and the lack of food will enhance mortality. *Remote Sensing for Food Security* is an excellent book, contains ten chapters, and provides an overview of food security in the twenty-first century and productivity of crop yield. The book is written by a well-experienced scientist who has broad experience using satellite remote sensing data in monitoring vegetation and estimation of crop yield in most countries. The author has developed innovative ideas and algorithms for early detection of vegetation health and its impact on crop yield. The book has also addressed how one can predict crop yield during strong episodic climate events, such as El Niño (warm phase) and La Niño (cold phase), to prepare for the worst situation. The book also includes a discussion on the increasing CO_2 and surface temperature on vegetation and crop yield including long-term remote sensing and ground data. The author has also demonstrated how users can use satellite data and various indices to monitor vegetation growth and predict crop yield, which will be a valuable tool for readers to explore crop productivity to face food security at local, regional, and global scale.

This book by Felix Kogan provides information about the monitoring and prediction of crop yield in future scenarios (future droughts, episodic events, and global warming) so that the growing population in developing countries may not suffer from hunger.

Ramesh P. Singh, PhD

School of Life and Environmental Sciences
Schmid College of Science and Technology
Chapman University
Orange, CA, USA

Acknowledgments

I deeply appreciate the tremendous help of my daughter Maria Levinson, who has done enormous work editing this book. Being extremely busy with her work as an economist, taking care of her family, and, moreover, writing her own book, she managed to find time to improve my writing. My enormous appreciation also goes to my colleague Mr. Guo Wei, an expert in software development. I worked with Mr. Wei for many years. He developed excellent software to retrieve satellite data, process them, and convert satellite indices into numerous products based on my algorithms. He also developed an excellent website, which is regularly attended by many users, appreciating the delivered satellite data and products. My deep appreciation goes to Prof. Ramesh Singh, who found time to write a positive Foreword. I am also deeply obliged to my former colleagues Drs. Jerry Sullivan, Dan Tarpley, Doug LeComte, and many others for being very supportive in my research and development. Moreover, I appreciate the suggestions and advice from the users who communicate with me regularly. Their numerous comments helped me to improve and advance my research and development.

Contents

1 Why This Book? 1
- 1.1 Introduction 1
- 1.2 Book's Discussions 2
- 1.3 Conclusion 6
- References 7

2 Food Security: The Twenty-First Century Issue 9
- 2.1 Introduction 9
- 2.2 Food Security and Insecurity 10
  - 2.2.1 How to Measure Food Security 10
  - 2.2.2 Long-Term Food Security and Expectations 12
  - 2.2.3 Short-Term Food Security 14
- 2.3 Conclusion 20
- References 20

3 Operational Satellites for Earth Monitoring 23
- 3.1 Introduction 23
- 3.2 NOAA Polar-Orbiting Operational Environmental Satellites 24
  - 3.2.1 AVHRR Sensor 24
  - 3.2.2 AVHRR Data for Vegetation Monitoring 25
  - 3.2.3 Normalized Difference Vegetation Index 28
  - 3.2.4 Removing Noise from NDVI 29
  - 3.2.5 VIIRS Data for Vegetation Monitoring 42
  - 3.2.6 Continuity of NOAA/AVHRR, SNPP/VIIRS and J-1/VIIRS Data Records 45
- 3.3 Conclusion 48
- References 48

4 Vegetation Health Method 51
- 4.1 Introduction 51
- 4.2 Theoretical Base of Vegetation Health Method 52
  - 4.2.1 Biophysical Considerations 53
  - 4.2.2 Basic Laws for Extracting Weather Component from NDVI and BT 54

4.3 Vegetation Health Algorithm ... 54
4.4 Vegetation Health at Work ... 60
4.5 Validation ... 64
4.6 Conclusion ... 70
References ... 71

5 Monitoring Drought from Space and Food Security ... 75
5.1 Introduction ... 75
5.2 Drought as Natural Disaster ... 75
5.3 What Is Drought? ... 76
5.3.1 Drought Definition ... 76
5.3.2 Drought Features ... 78
5.3.3 Measuring Drought ... 78
5.3.4 Drought Type ... 79
5.4 Drought Detection and Monitoring Methods ... 79
5.4.1 Meteorological Methods ... 79
5.4.2 Soil Moisture and Vegetation Methods ... 81
5.4.3 Satellite-Based Methods ... 81
5.4.4 Operational Satellite-Based Vegetation Health (VH) Method and Drought Monitoring ... 84
5.5 Vegetation Health-Based Droughts from the Past to Present ... 88
5.6 Droughts at 0.5 and 1 km^2 Resolution from VIIRS, the New Operational Sensor ... 101
5.7 Devastating Droughts in 2017 and Early 2018 ... 106
5.8 Conclusion ... 109
References ... 110

6 Vegetation Health-Based Modeling Crop Yield and Food Security Prediction ... 115
6.1 Introduction ... 115
6.2 Modeling Principles ... 116
6.3 Yield-Vegetation Health Models ... 119
6.3.1 Global Grain and Food Security ... 119
6.3.2 Corn in China ... 119
6.3.3 Winter Wheat, Corn, and Sorghum in the USA ... 122
6.3.4 Winter Wheat in Ukraine ... 128
6.3.5 Corn in Argentina ... 130
6.3.6 Wheat in Australia ... 133
6.3.7 Rice in Bangladesh ... 137
6.3.8 Cereals in Russia ... 140
6.3.9 Spring Wheat in Kazakhstan ... 144
6.3.10 Corn in Zimbabwe ... 148
6.3.11 Other Countries' Crops ... 152
6.3.12 VH-Crop Modeling for Food Security: Concluding Remarks ... 158
6.4 Short Summary ... 159
References ... 159

7 Vegetation Health for Insuring Drought-Related Yield Losses and Food Security Enhancement 163
7.1 Introduction 163
7.2 Copula Approach 165
7.3 Empirical Procedure 166
7.4 Data 166
7.5 Results and Discussion 168
7.6 Conclusions 171
References 172

8 Advanced VH-Based Long-Term Drought and Food Security Prediction 175
8.1 Introduction 175
8.2 Data and Method 177
8.3 El Niño Southern Oscillation 177
8.4 Weather and Ecosystem Patterns During ENSO 177
8.5 Vegetation Health During the Strongest 2015–2016 El Niño 183
8.6 Prediction of 2015–2016 El Niño Impacts on Ecosystems During Spring and Summer 2016 185
8.7 Conclusion 186
References 187

9 Climate Change and Food Security Current and Future 191
9.1 Introduction 191
9.2 Earth Climate Change and Consequences 192
9.3 Causes of Climate Change 193
9.4 Global Temperature and CO_2 Trend 195
9.5 New Ideas About the Causes of Global Warming 197
9.6 Global Land Cover Changes During Climate Warming and Hiatus Time 201
9.6.1 World 201
9.6.2 Continents 204
9.6.3 Principle Grain-Producing Countries 204
9.6.4 Main Grain Crop Area 210
9.6.5 World Droughts 217
9.7 Conclusions 220
References 222

10 Application of Vegetation Health Data and Products for Monitoring Food Security 225
10.1 Why and How to Use Vegetation Health? 225
10.2 Vegetation Health Users 226
10.3 Vegetation Health WEB Pages 227
10.4 Vegetation Health Applications 229
10.5 Conclusion 239
References 240

Short Summary 243

Index 247

Abbreviations

BT	Brightness Temperature
C	Consumption
CFC	Chlorofluorocarbon gases
dP	Deviation of precipitation from normal
dT	Deviation of temperature from normal
dY	Yield anomaly (deviation from technological trend or from multi-year mean values)
EDF	Empirical distribution function
ENSO	El Nino Southern Oscillation
ETC	Equator crossing time
GHG	Greenhouse gas
IO	Indian Ocean
IR	Infrared
MMT	Million metric tons
NDVI	Normalized difference vegetation index
NIR	Near infrared
P	Precipitation
P	Production
PCC	Pearson correlation coefficient
R^2	Determination coefficient
RD	Relative difference
RMSE	Root mean square errors
SMN	Smoothed NDVI
SMT	Smoothed brightness temperature
SRF	Spectral response function
SST	Sea surface temperature
SSTa	Sea surface temperature anomaly
T	Temperature
TA	Temperature anomaly
TCI	Temperature condition index
t/ha	Tons per hectare
TP	Tropical Pacific
VCI	Vegetation condition index
VHI	Vegetation health index
VH	Vegetation health
VIS	Visible

1 Why This Book?

1.1 Introduction

We are living in the twenty-first century. We have reached a considerable progress in industry, economy, and finance. We have achieved great innovations in medicine, improving human health and extending life expectancy. However, in spite of this progress, today, more than one billion people, nearly a sixth of the world's population, suffer from chronic hunger and malnutrition due to a lack of food (FAO 2017; Heibuch 2011). Agricultural production is currently growing but its rate is behind the rate of population increase. Future prospects are not encouraging since we need to feed two billion people more by the middle of this century, which puts the world agriculture before a huge challenge to produce 60–70% more food than now (Charles et al. 2018; FAO 2017; Guillou and Matheron 2014; Godfray et al. 2010; Holt-Gimenez and Patel 2009; UN 2004). Unfortunately, in the twenty-first century, the continuation of previous tendencies for a rapid population growth, declining stock of natural resources, climate warming, and land cover changes has intensified world concerns about the future food supply/demand and global food security (USDA 2017; FAO 2017, 1999; Heibuch 2011). World and countries' decision makers need reliable and timely information to understand, monitor, and predict impacts of Earth's changes on global food security (FS).

The Earth changes have already affected global FS. In an effort to produce more food, agriculture put considerable constraint on the environment in overexploiting soil fertility, causing land degradation, diminishing fresh water, and a deterioration of ecosystems and climate. The current climate has started to limit staple crop production and fresh water, increased soil degradation, and slowed down the rate of agricultural output. A warmer climate is likely to constrain stronger agricultural production due to an expected increase in severity and frequency of large-scale weather extremes. This situation is further complicated by drought intensification, increasing its area, intensity, and duration, leading to a further reduction of agricultural production (Schmidhuber and Tubiello 2007; Kogan et al. 2013; Kogan 2002). If moderate-to-intensive drought covers more than 20% of the world's main agricultural area, food production becomes less than what the world needs for consumption. This situation has already deteriorated in the twenty-first century, when in the first 17 years, world grain production (the principal staple food) was below consumption in half of those years. Drought was the major environmental disaster affecting food security and world's sustainability in all these years (Kogan et al. 2015, 2017, 2018).

Unfortunately, drought is a part of Earth's climate, which occurs every year without warning, recognizing borders, political and economic differences. It has wide-ranging impacts, first on

F. Kogan, *Remote Sensing for Food Security*, Sustainable Development Goals Series,
https://doi.org/10.1007/978-3-319-96256-6_1

agriculture but also on water resources, ecosystems, energy, forestry, human health, transportation, recreation, food supply-demands, and other resources and activities (Kogan et al. 2013). Drought affects the largest number of people on Earth and is a very costly disaster. Considerable relief from drought for agriculture can be achieved if it is detected 1–3 months in advance of its start time. However, weather-based drought detection, monitoring, and prediction are currently a challenging task due first to an insufficient and sparsely distributed weather station network, especially in developing countries of Africa, South Asia, and South America. Besides, weather station measurements are not precisely characterized by crop and pasture biological requirements.

In the past 30 years, satellite technology has successfully filled weather station limitations using high-resolution and good quality data (NOAA/NESDIS 2018). Moreover, satellite products derived from the new vegetation health (VH) technology (Kogan and Guo 2014, 2017), used for drought monitoring, are based on biophysical laws, which provide cumulative approximation of the weather, soil, and hydrological impacts on vegetation and, what is the most important, the corresponding vegetation response. These features were used successfully for an advanced drought detection and estimation of agricultural losses. In addition to drought, VH products from NOAA operational polar-orbiting satellites (old and new generation) have been successfully used in recent years to model agricultural production well in advance of harvest, estimate moisture and vegetation thermal stress, monitor land degradation and climate change impacts, and predict tendencies in fish catch, fire risk, and ENSO impacts on land ecosystems. All of these new features have been successfully used for monitoring global, regional, and countries' food security situation.

1.2 Book's Discussions

This book discusses how the new vegetation health technology, designed from the NOAA operational environmental polar-orbiting satellites data, can be used to estimate vegetation health (including vegetation moisture and thermal stress) and provide advanced drought detection, monitoring drought area, intensity, duration, and impacts on agriculture, and consequently on food security (Kogan et al. 2018). Vegetation health (VH) technology includes biophysically-based VH method, operational satellite data collection and processing, VH-based indices, VH-based products (drought, fire risk, soil saturation etc.), methods for product interpretation and practical advises on data and products analysis, modeling and use (Kogan 1987, 1989, 1998, 2002; Kogan et al. 2015, 2017, 2018). Global droughts reduce agricultural production every 2–4 years and satellite-based predictions of crop and pasture losses can considerably enhance food security and improve countries' sustainability. This book consists of ten chapters. They emphasize that it is an optimal time for the book's publication because nearly one-sixth of world population suffers from a lack of food, malnutrition, and even hunger. These events are the results of a global and regional food insecurity situation. Droughts provide a very strong contribution to these events by reducing agricultural production almost every year. The current operational satellites are able to detect drought early, estimate drought intensity, area, and duration, and, what is the most important, to estimate agricultural losses in advance crops' harvest in order to provide timely food assistance to hungry people (UN 2004, USDA 2017).

Chapter 1, "Why This Book?" The book is focused on food security estimated from space data, focusing on the current situation and what to expect in the future. Environmentally based food security (FS) problems have intensified (lack of food, malnutrition, and even hunger) in the most recent 3–4 decades due to climate warming, weather change, and intensification of natural disasters, especially droughts. Therefore, this is the best time to evaluate FS from the point of view of a changing climate, droughts and their impacts on land cover changes and crop production. Besides, following lack of observations from the global weather station network and following availability of 38-year satellite-based observations and biophysical land surface models it is right time to use satellite data, which provides 10^6 times more land information (than weather stations), for evaluating environmental aspects of weather and climate changes impacts on food security. Following these considerations, the discussion will be focused on how the book can assist with global, country, and regional FS predictions using operational satellite. This chapter discusses the goals of each chapter in assessing food security, the method used for environmental predictions of such disasters as drought, crop and pasture modeling and production estimation, climate changes impacts on FS and prospects in the future food security tendencies.

Chapter 2, "Food Security: The Twenty-First Century Issue" discusses how to measure food security, analyzes a long-term disproportion between population and food production growth in the past 30–50 years, and emphasizes a very strong drought contribution to a negative annual balance between grain (major staple product) supply and demand (UN 2004). Shortages of food and the related food security issues, in addition to economic, social, environmental, and military problems, are strongly influenced by climate, whose influence is anticipated to be stronger in the future (Charles et al. 2018; USDA 2017; UN 2004). Finally, the chapter focuses on the application of satellite data for an early drought detection, monitoring drought area/intensity and two-month advanced assessment of grain (the major staple food) losses, for early prediction of food insecurity and timely planning food delivery to hungry people (Kogan and Guo 2014; Kogan et al. 2013).

Chapter 3, "Operational Satellites for Earth Monitoring" explains the principal of polar-orbiting environmental satellites (POES, old and new systems), their sensors, data and multiple products to observe land cover, to assess such natural disaster as drought, modeling agricultural production, climate change assessment and for more accurate monitoring and prediction of environmental impacts on economy, human life and consequently food security (Kogan et al. 2017; Kogan 2002). This chapter describes the operational POES data collection and processing (especially noise removal) from the two NOAA operational polar-orbiting satellite systems (initial—NOAA/AVHRR (Advanced Very High Resolution Radiometer) and new—SNPP/VIIRS and J-1/VIIRS (Visible Infrared Imaging Radiometer Suite)) and data preparation for continued applications and predictions in agriculture, human health, food security, climate and land changes, disasters detection (drought, crop losses, etc.), and other events and phenomena, affecting food security (Cracknell 1997; JPSS 2014; Kogan et al. 2018).

Chapter 4, "Vegetation Health Method." The vegetation health (VH) method was developed for space-based monitoring of moisture, thermal, and total health (combined moisture and thermal) conditions of vegetation (Kogan 1987, 1989, 1998). VH is based on biophysical laws, estimating the "carrying capacity" of the environmental resources used for crop production and humans' sustainable living. The VH data set were developed for operational purposes and investigated scientifically. The VH 38-year data set has advantages before other data sets with similar applications, being the longest, global, operational, highest spatial (0.5, 1, 4, and 16 km^2) and temporal (1 week) resolution (Kogan et al. 2017, 2015; Kogan and Guo 2017). The VH data and products are ready to be used without additional processing for monitoring, assessments, and predictions in agriculture and forestry, for climate change (forcing), vegetation health, invasive species, diseases, ecosystem addressing such topics as food security, land cover and change, climate variations impacts on agriculture, environmental security, and others (Kogan et al. 2018).

Chapter 5, "Monitoring Drought from Space and Food Security." Drought is a typical phenomenon of the Earth's climate, which reduces agricultural production, strongly affecting food security (Kogan et al. 2013; Kogan 2002). Many examples are given to explain why drought is considered as one of the main global natural disasters affecting strongly food security. Droughts are detected and monitored by the new vegetation health (VH) technology, which utilizes high- (0.5 and 1 km^2) and mid-resolution (4 and 16 km^2) 38-year NOAA operational polar-orbiting satellite data (from satellites and sensors: S-NPP/VIIRS, J-1/VIIRS and NOAA/AVHRR) for early drought detection, monitoring its features (area, intensity, duration, origination, impacts, etc.) and prediction of agricultural losses, providing an early warning for global and regional food security (Kogan et al. 2018, 2017, 2015).

Chapter 6, "Vegetation Health-Based Modeling Crop Yield and Food Security Prediction." Weather-driven crop losses have always been a major concern for farmers, traders, governments, policy makers, and international organizations for the purpose of balanced food supply/demands, trade, and distribution of aid to nations in need. They strongly affected global, country, and regional food security situation. Among many crop-weather models, the new satellite-based vegetation health (VH) method showed very good results for modeling and accurate monitoring crop yield well in advance of harvest. The VH method was used for modeling yield of agricultural crops in nearly 30 countries and for applying the models for 1–2 months advanced assessments of crop and pasture production and for 3–4 months advanced prediction of food security situation. This chapter introduces VH-based yield modeling approach and demonstrates modeling results for grain crops in major grain-producing countries such as the USA, China, Russia, Brazil, Argentina, Australia, India, Ukraine, and others (Kogan et al. 2018; Rahman et al. 2009; Salazar et al. 2007; Seiler et al. 2006; Unganai and Kogan 1998). Some examples are also shown for other crops and pasture.

Chapter 7, "Vegetation Health for Insuring Drought-Related Yield Losses and Food Security Enhancement" discusses how to use high-resolution (compared to weather data) vegetation health (VH) indices derived from operational polar-orbiting satellites for crop insurance purposes. The discussion focuses on VH-based drought predictions in combination with the Cupola method to enhance effectiveness of agricultural insurance and to protect farmers from drought-related agricultural losses. The presented experiment was performed in two main grain-producing regions of Kazakhstan, which grow spring wheat (Bokusheva et al. 2016). The results have shown that satellite-based 16 km^2 resolution of VH-based moisture assessments from vegetation condition index (VCI) and vegetation surface temperature assessments from the temperature condition index (TCI) during the critical period of crop growth are excellent predictors of farmers' wheat yields. The selected indices were used to design index-based insurance contracts by applying the copula approach. Empirical results for 47 grain-producing farms in northern Kazakhstan show that insurance contracts built on the two VH-based indices can provide substantial drought-related risk reductions for a single farm and a group of farms. For the entire sample of the experiment, risk reduction was moderate. Providing reliable VH-based farmers' insurance would simplify crop loss verification, make insurance cheaper, and provide financial security for farmers by providing compensation in case of drought-related agricultural losses (Bokusheva et al. 2016; Bokusheva and Breustedt 2012). These advantages would help farmers survive in case of weather disasters, which in turn would enhance food security situation. The effectiveness of insurance contracts can be improved using higher resolutions (4 and 1 km^2) satellite data and measuring indices at more disaggregated levels.

Chapter 8 "Advanced VH-Based Long-Term Drought and Food Security Prediction" describes drought prediction based on the interrelationship between one of the climate forcing phenomena and vegetation health indices. Among long-term predictors of world-affecting climate events such as El Nino Southern Oscillation (ENSO), inter-decadal variability of sea surface temperature (SST), sea level pressure (SLP), and others, vegetation health was found to have a strong relationship with ENSO events and might be used as several-month (4–6 months) advanced indicator of drought appearance and assessment of its impact on crop losses (Kogan and Guo 2017; Kogan 1998). It has been known that ENSO is changing global and regional temperature and precipitation patterns cyclically, every 3–7 years. El Nino, a warmer cycle of ENSO, is leading to a warmer world, while La Nina's cooler cycle is triggering a cooler world. Some regions during these cycles experience dry and warm conditions, but others, moist and cool. The ENSO-measured index was found to correlate strongly with vegetation health indices. Since ENSO events can be predicted a few months ahead of their start, they provide an advanced warning on drought quite ahead of potential crop losses. ENSO-based VH assessments can be used for 4–6-month advanced predictions of crop losses and food security situation. This chapter discusses inter-decadal variability of SST, its correlation with VH indices, and how to use these indices for advanced assessment of crop losses in advance of harvest. ENSO normally impacts weather, ecosystems, and socioeconomics (agriculture, fisheries, energy, human health, water resource, etc.) on all continents. This chapter shows how ENSO is changing world ecosystems, which areas are affected, the intensity of El Niño (warm phase) or La Niño (cold phase), and, most importantly, how to predict droughts ahead of its development from satellite data and several-month advanced crop losses. The satellite-based vegetation health (VH) method and 37-year of its data have been used as the criteria of the impact following sea surface temperature (SST) changes in Tropical Pacific. Specifically, the chapter shows VH-SST teleconnection during ENSO years, focusing on the estimation of vegetation response to the strongest El Niño, an intensity of the response and transition of the impact from boreal winter to spring and summer. Two types of ecosystem response were identified. In boreal winter, ecosystems of northern South America, southern Africa, eastern Australia, and Southeast Asia experienced strong vegetation stress during El Niño, which will negatively affect agriculture, energy, and water resources. In Argentina, south-eastern USA and the Horn of Africa ecosystem response is opposite. One of the worst disasters associated with ENSO is drought. The advantages of this study are in derivation of vegetation response to moisture, thermal, and combined conditions including an early detection of drought-related vegetation stress. For the first time, ENSO impact was evaluated based on all events with $|SSTa| > 0.5$ °C and strong events with $|SSTa| > 2.0$ °C.

Chapter 9 "Climate Change and Food Security Current and Future." Since the mid-eighteenth century, the Earth's climate has been warming. In the past 60 years, Earth was warmed up intensively leading to non-experienced before environmental, economic, and social events. One of the biggest climate-warming concerns today is how these changes have affected agricultural production and food security and what to expect in the near and distant future, considering intensive population growth and gradually leveling off trends in grain production. In the recent decades, United Nations' (WMO, UNEP) climate change actions, IPCC reports, and other scientific publications have strongly emphasized that CO_2 increase was very likely triggering global warming, which has already led to negative consequences for environment and society. The past 20 year environmental observations showed Earth changes in snow and ice areas, sea level, natural disasters, biological systems (plants, birds, etc.), and others. Some observations also indicated that climate warming has negatively affected crop yields, especially in the underdeveloped countries of Africa, Asia, and Latin America. Since climate warming anticipates drought intensification, expansion and increasing impacts on crops and pasture. One of the biggest concerns is what changes have occurred in land cover, grain production and its corresponding consequences for food security (FS) and what to expect in the near and distant future (Kogan and Guo 2014). Will FS intensify, leading to more people suffering from a lack of food and hunger, especially considering the disproportion between the Earth's population and food production increase. In addition, it is important to understand if short-term (17–35 years) climate tendencies would continue, considering that between 1998 and 2014, during the so-called "hiatus" time, global mean temperature experienced a flat trend and during the last 3 years (2015–2017), the global mean temperature was quite warmer than normal. This chapter discusses the current climate warming views, emphasizing matching upcoming trend in global CO_2 and general warming up trend in global mean temperature anomaly (TA). It also indicates that there are some mismatches between TA and CO_2 trends during 17-year from 1998 through 2014 and discusses other causes for global temperature trends. The most important part of this chapter is analysis of land surface changes in the past 4–6 decades based on remote sensing and in situ data. These observations include trend analysis of 37-year time series of land surface remote-sensing-based NDVI, brightness temperature, vegetation health indices (VCI, TCI and VHI), droughts and in situ-based precipitation, temperature, yield, and others. Based on these tendencies, the current and future food security assessments are predicted.

Chapter 10 "Application of Vegetation Health Data and Products for Monitoring Food Security" is written for users who want to apply VH data and products in their work to make analysis, monitoring and predicting weather (moisture and thermal) impacts on vegetation and help them for making decisions with their business and activities. The chapter provides vegetation health data and products' WEB address https://www.star.nesdis.noaa.gov/smcd/emb/vci/VH/index.php (NOAA/NESDIS 2018; Kogan et al. 2013; Kogan 2002), which explains how to use the information. The WEB displays real-time and historical weekly 16, 4, and 1 km^2 global and country images of moisture (VCI), thermal (TCI), and combine (VHI) conditions, drought, vegetation stress, fire risk, etc. and their changes between weeks and years (Kogan et al. 2017, 2015, 2013; Kogan and Guo 2017; Kogan 1998, 1989, 1987). Users from nearly 200 countries can access the 38-year data and product images and digital values. In addition to countries data, for the first-order divisions in each country (for example, states in the USA, oblast in Russia), VH data and products area mean values (from 4 km^2) of weekly time series can be displayed during 1981–2018 for the selected years and weeks. The users can select around 4000 administrative regions for analysis. The current users of VH-based WEB represent a wide spectrum of interests, including agriculture, traders, commodity dealers, bankers, stock Co., researchers, and others. This WEB page was developed in 2010 and attended by around 2000 users. In 2015–2017, the number of users attending the WEB page were around 64,000, 66,000, and 69,000, respectively.

1.3 Conclusion

In summary, it is important to emphasize that the Earth faces huge challenges because the population is growing much faster than agriculture can produce food (Charles et al. 2018; FAO 2017; Godfray et al. 2010). More than one billion people, nearly a sixth of the world's population, are currently suffering from chronic hunger and malnutrition due to a lack of food (Guillou and Matheron 2014; Heibuch 2011). The efforts of agriculture to produce more food put considerable constraints on the environment, which is gradually deteriorating; demands for more food puts constraints on many resources, water and soil fertility, which have been overexploited. A warmer climate supposed to change drought dynamics and its impacts on land degradation and crop production. Feeding nine billion people by 2050 is expected to be a challenging task because it requires world agriculture to produce nearly 70% more food in the future compared to the current level (Charles et al. 2018). Intensification of agriculture put strong pressure on the fragile world ecosystems, considering the current trends in turning marginal lands to agriculture, exhaustion of water resources and ongoing climate and land cover changes. These global changes put additional strain on food security. Our Earth is very big. We need strong technology to monitor environment and its impacts on food production to satisfy all nations. Satellite technology has already proved to be successful in this endeavor, which is presented to help food security enhancement. This chapter explains the principles of this book, focusing on prediction of food security (FS) based on monitoring environmental factors, deteriorating FS, from operational satellites.

The material presented in the book is unique because it shows that (1) operational satellites proved to be a very good single source of information (without weather) for environmental monitoring, advanced agricultural assessments, and food security predictions; (2) vegetation health (VH) method is a new powerful tool and, what is the most important, was comprehensively validated in different world climates and ecosystems; (3) many developed VH products were successfully used for detection and monitoring natural disasters (drought, malaria, etc.) helping to improve food security and sustainability (Kogan et al. 2015); (4) VH method is used in nearly 30 countries to model many crops and pasture production (Kogan and Guo 2014); (5) VH data and products are extremely popular since during 2015, 2016, and 2017, 64,000, 66,000, and 69,000 users, respectively, accessed the WEB to apply this information in their work; (6) users of VH data and products represent a wide spectrum of interests, including agriculture, traders, commodity dealers, bankers, stock Co., researchers, and others; (7) the first time all of the VH-based application are presented in one publication.

References

Bokusheva, R., and G. Breustedt. 2012. The effectiveness of weather-based index insurance and area-yield crop insurance: how reliable are post predictions for yield risk reduction? *Quarterly Journal of International Agriculture* 51 (2): 135–156.

Bokusheva, R., F. Kogan, I. Vitkovskaya, S. Conradt, and M. Batyrbayeva. 2016. Satellite-based vegetation health indices as criteria for insuring against drought-related yield losses. *Journal of Agricultural and Forest Meteorology* 220: 200–206.

Charles H., J. Godfray, J.R. Beddington, R. Crute, L. Haddad, D. Lawrence, J.F. Muir, J. Pretty, S. Robinson, S.M. Thomas, and C. Toulmin. 2018. Food security: The challenge of feeding 9 billion people. *Food security*. https://waterlandandecosystems.wikispaces.com/file/view/Component2_CIAT.pdf.

Cracknell, A.P. 1997. *The advanced very high resolution radiometer*. USA: Taylor & Francis 534 p.

FAO. 1999. Food insecurity: When people must live with hunger and fear of starvation. *The State of Food Insecurity in the World Report*, Rome, 76.

———. 2017. How close we are to zero hunger. http://www.fao.org/state-of-food-security-nutrition/en/.

Godfray, H.C.J., J.R. Beddington, I.R. Crute, L. Haddad, D. Lawrence, J.F. Muir, J. Pretty, S. Robinson, S.M. Thomas, and C. Toulmin. 2010. Food security: The challenge of feeding 9 billion people. *Science* 327: 812–818. https://doi.org/10.1126/science.1185383.

Guillou, M., and G. Matheron. 2014. *The world challenge to feed 9 billion people*. New York: Springer http://www.springer.com/us/book/9789401785686.

Heibuch, J. 2011. Food insecurity on the rise. *Treehugger*, October 31. https://www.treehugger.com/health/food-insecurity-rise-infographic.html

Holt-Gimenez, E., and R. Patel. 2009. *Crisis and the hunger for justice*. Boston: Pambazuka Press https://books.google.com/books.

JPSS. 2014. Joint polar satellite system. http://www.jpss.noaa.gov

Kogan, F.N. 1987. Vegetation health index for areal analysis of NDVI in monitoring crop conditions. *Preprint 18th Conference on Agricultural and Forest Meteorology AMS*, Boston, 103–114.

Kogan, F. 1989. Remote sensing of weather impacts on vegetation in non-homogeneous areas. *International Journal of Remote Sensing* 11 (8): 1405–1419.

Kogan, F.N. 1998. A typical pattern of vegetation conditions in southern Africa during El Nino years detected from AVHRR data using three-channel numerical index. *International Journal of Remote Sensing* 19 (18): 3689–3695.

———. 2002. World droughts in the new millennium from AVHRR-based vegetation health indices. *Eos Transaction of American Geophysical Union* 83 (48): 562–563.

Kogan, F., and W. Guo. 2014. Early twenty-first-century droughts during the warmest climate. *Geomatics, Natural Hazards and Risk* 7: 127–137. https://doi.org/10.1080/19475705.2013.878399.

———. 2017. Strong 2015-2016 El Niño and implications to global ecosystems from space. *International Journal of Remote Sensing* 38 (1): 161–178. https://doi.org/10.1080/01431161.2016.1259679.

Kogan, F., T. Adamenko, and W. Guo. 2013. Global and regional drought dynamics in the climate warming era. *Remote Sensing Letters* 4: 364–372. https://doi.org/10.1080/2150704X.2012.736033.

Kogan, F., M. Goldberg, T. Schott, and W. Guo. 2015. SUOMI NPP/VIIRS: Improve drought watch, crop losses prediction and food security. *International Journal Remote Sensing* 36 (21): 5373–5383. https://doi.org/10.1080/01431161.2015.1095370.

Kogan, F., W. Guo, and W. Yang. 2017. SNPP/VIIRS vegetation health to assess 500 California drought. *Geomatics, Natural Hazards and Risk* 8 (2): 1383–1395. https://doi.org/10.1080/19475705.2017.1337654.

Kogan, F., W. Guo, W. Yang, and S. Harlan. 2018. Space-based vegetation health for wheat yield modeling and prediction in Australia. *Journal of Applied Remote Sensing* 12 (2): 026002. https://doi.org/10.1117/1.JRS.12.026002.

NOAA/NESDIS 2018. Vegetation health. https://www.star.nesdis.noaa.gov/smcd/emb/vci/VH/vh_browse.php

Rahman, A., F. Kogan, L. Roytman, N.Y. Krakauer, M. Nizamuddin, and M. Goldberg. 2009. Use of vegetation health data for estimation of aus rice yield in bangladesh. *Sensors* 9: 2968–2975.

Salazar, L., F. Kogan, and L. Roytman. 2007. Use of remote sensing data for estimation of winter wheat yield in the United States. *International Journal of Remote Sensing* 28: 3795–3811.

Schmidhuber, J., and F.N. Tubiello. 2007. Global food security under climate change. *Proceedings of the National Academy of Sciences of the United States of America* 104: 19703–19708.

Seiler, R.A., F.N. Kogan, G. Wei, and M. Vinocur. 2006. Seasonal and inter-annual responses the vegetation and production of crops in Cordoba Argentina assessed by AVHRR derived vegetation indices. *Advances in Space Research*. https://doi.org/10.1016/j.asr.2006.05.024.

UN. 2004. World Population Prospects. https://www.google.com/search? tbm=isch&sa=1&ei=PMBsWrSRBYTbzwLW26foDA&q=global+population+from+1960+through2017

Unganai, L.S., and F.N. Kogan. 1998. Drought monitoring and corn yield estimation in southern Africa from AVHRR data. *Remote Sensing of Environment* 63: 219–232.

USDA. 2017. Food security. https://www.usda.gov/topics/food-and-nutrition/food-security.

2 Food Security: The Twenty-First Century Issue

2.1 Introduction

At the end of the twentieth century, around one-quarter of the world population has not had enough food for normal living and nearly one billion people became hungry every year. In addition, 800 million people are undernourished globally (Charles et al. 2018; Godfray et al. 2010; USDA 2017, 2018). Africa, South America, and South Asia face the greatest hunger burden. During 1999–2009, food insecurity rose from 10.1% to 14.7%. This means that the number of households that had trouble getting enough food for the family rose by 4.6%. While food insecurity took a slight dip to 14.5% in 2010, it is still far too high (Heibuch 2011; WGS 2013). Nearly 280 million people in Asia are undernourished, and in Sub-Saharan Africa, the current rate of undernourishment is around 23% (Charles et al. 2018). According to WFP (2014), famines have been declared previously in southern areas of Sudan in 2008, in Ethiopia in 1984–1985 and 2000, in the Democratic People's Republic of Korea (DPRK) in 1996, and in Somalia in 1991–1992 (WFP 2014). Poor nutrition causes nearly half the deaths in children under five, and one in four children suffer stunted growth. Moreover, malnutrition, including undernutrition and micronutrient (nearly two billion people) deficiencies, is the top contributor to global disease burden. In addition to malnutrition, nearly 700 million people in 76 countries are food insecure (USDA 2017). The vast majority of the world's hungry live in developing countries (Gottlieb and Joshi 2010; Holt-Gimenez and Patel 2009).

In the recent 20–40 years, global food security has become an extremely important issue. International organizations and developed countries have worked together to diminish the number of hungry and malnutrition people (USDA 2017; WFP 2014; FAO 1999, 2018). Since food security is controlled by financial, economic, political, and environmental factors and also the number of people consuming the food, it should be considered from global and regional aspects. From a regional point of view, developed countries with a strong political system do not have major problems with a lack of food, although there are some indicators about undernutrition households (USDA 2017). However, developing countries have become affected by a lack of food to feed their population in the recent 30 years. Since many developing countries have presently had this problem, food security has become a global issue, especially in the twenty-first century (Charles et al. 2018; USDA 2017; Gottlieb and Joshi 2010). In 2011, at the Camp David Summit, G-8 and African leaders committed to the *New Alliance for Food Security and Nutrition* (WH 2012; Heibuch 2011) to achieving global food security. Goals were set to increase responsible domestic and foreign private investments in African agriculture, take innovations to enhance

F. Kogan, *Remote Sensing for Food Security*, Sustainable Development Goals Series,
https://doi.org/10.1007/978-3-319-96256-6_2

agricultural productivity, and reduce the risk of food insecurity in African communities. Such important organizations as World Bank, African Development Bank, the United Nations' World Food Program, the International Fund for Agricultural Development, and Food and Agriculture Organization have joined the *New Alliance* (WH 2012). Some of the far-reaching goals were set up to take innovations that can enhance agricultural productivity and reducing losses from natural disaster, especially drought (Heibuch 2011; WGS 2013).

This chapter discusses how to measure food security (FS), global and regional FS performance in the past 30–50 years in relation to population and agricultural production growth, how FS depends on earth resources, environmental conditions, natural disasters, specifically drought, the long-term aspects of FS in relation to world population and agricultural production dynamics, water limitation and climate-related land cover change. Finally, discussion will focus on application of satellite data for assessment of FS from satellite-based models, frequent monitoring FS situation and advance FS predictions.

2.2 Food Security and Insecurity

According to the United Nations (FAO 2017), food security is strong when people in a region have access to sufficient food to meet their dietary needs. No or limited amount of food leads to malnutrition, hunger, famine, and development of food insecurity situation (Coleman-Jensen et al. 2011; Holt-Gimenez and Patel 2009). Although it is difficult to measure food security/insecurity, the number of hungry and/or malnourished people can determine the existence of an FS problem. Following FAO (1999), 1.2 billion people globally have been chronically food insecure in the late 1990s due to undernourishment. One half of these people (642 millions) were in India. Over 15 million of the undernourished were even in developed countries. Famine, which is the worst insecurity, amounts to when at least 20% of households in an area face extreme food shortages leading to acute malnutrition with rates exceeding 30% and the death rate of more than two persons per day per 10,000 persons (FAO 1999). A few large-area famines have been recorded in the past (Holt-Gimenez and Patel 2009). For example, in the late nineteenth century, India and China experienced famine. Fortunately, in the recent 50 years, famine has not been declared, except for small regions (FAO 1999, 2017; USDA 2017; GRACE 2014). However, the food insecurity situation has intensified in the recent two decades, especially in African and Asian countries (Charles et al. 2018; Godfray et al. 2010) and future FS assessments are not encouraging (Charles et al. 2018; USDA 2017; FAO 2017, 2018).

2.2.1 How to Measure Food Security

Among economic, political, financial, population, agricultural, and environmental factors affecting food security (FS), the world population consuming food and annual agricultural production determines food supplies and demand and results in a FS problems or no problems. These two principal factors: annual food supply or production from agriculture and annual food demand or consumption can be considered as the main first-order contributors to food security/insecurity assessments. At the first level of assessment, global food demand (consumption) is determined by the number of world people and global food supply which is estimated by the amount of food produced by the world agriculture. Agricultural production is multidimensional, which includes crops, vegetables, fruits, animals, etc. From all these varieties, grain is the principal product used by the entire world for both food and feed. Therefore, on a global, continental, and country scale, the amount of grain produced annually in the fields can be considered as an indicator of global food supply. World population and global grain production can be assessed for FS purposes from two prospective: long-term (multiyear) changes and short-term (annual) variations. The long-term component is estimated statistically by multiyear (more than three decades) dynamic (or trend) in global population and grain production. The short-term component can be estimated by the annual variations in the same factors or their combination, expressed as the amount of grain per capita. Summarizing, it is important to emphasize that the amount of grain produced by the world agriculture can be used as the principal indicator characterizing food availability and the amount of grain consumed annually and the number of people who need this grain can serve as indicators of food demand (Godfray et al. 2010, 2018).

The multiyear dynamics of world population and total grain (cereal) production shown in Fig. 2.1 were used as an indicator of long-term supply-demand situation. Trend analysis of these characteristics indicates some similarities and differences. Both parameters are similar in their multiyear increasing dynamics since 1960s (Fig. 2.1a, b). Global population more than doubled, from around 3 billion in 1960 to nearly 7.6 billion by January 2018 (WP 2018; WM 2018; UN 2004) and world grain production increased almost four times (following the "Green Revolution" based agricultural technology improvement) from 0.70 billion tons in 1960 to nearly 2.75 billion tons by 2017 (WP 2018; WB 2018). However, the most important are differences between population and cereal production dynamics, estimated by the quadratic equations of each factor against the year number (the equation is in Fig. 2.1a, b). The rate of their increase, estimated by the coefficient with the quadratic term of equation is absolutely different. Population accelerated during 1960–2015 (the rate is positive, +0.108 per year), while grain production increase is decelerating or leveling off (the rate is negative, −0.019 per year) over 55 years (Fig. 2.1a, b). As the result of these differences, grain production per capita during 1960–2015 experienced a flat trend (Fig. 2.1c) from the early 1980s (prior to that time the trend was increasing). However, two multiyear variations in the dynamics of grain production per capita after the early 1980s is noticeable, especially some increases from 2005. Therefore, farther dynamics should be reevaluated in 5–7 years.

In addition to total world cereal production, which includes wheat, corn, rice, coarse grain, and sorghum, it is important to estimate long-term trends in major grains (wheat, corn, and

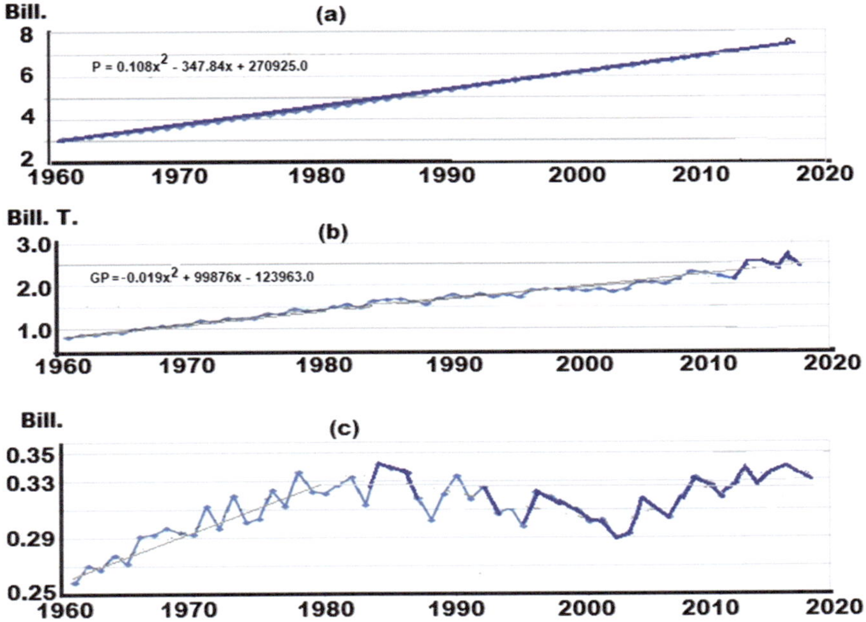

Fig. 2.1 World (**a**) population, (**b**) cereal (grain) production, and (**c**) cereal production per capita (USDA 2018; WB 2018; FAO 2018; Statistica 2017; WP 2018; Brown 2012)

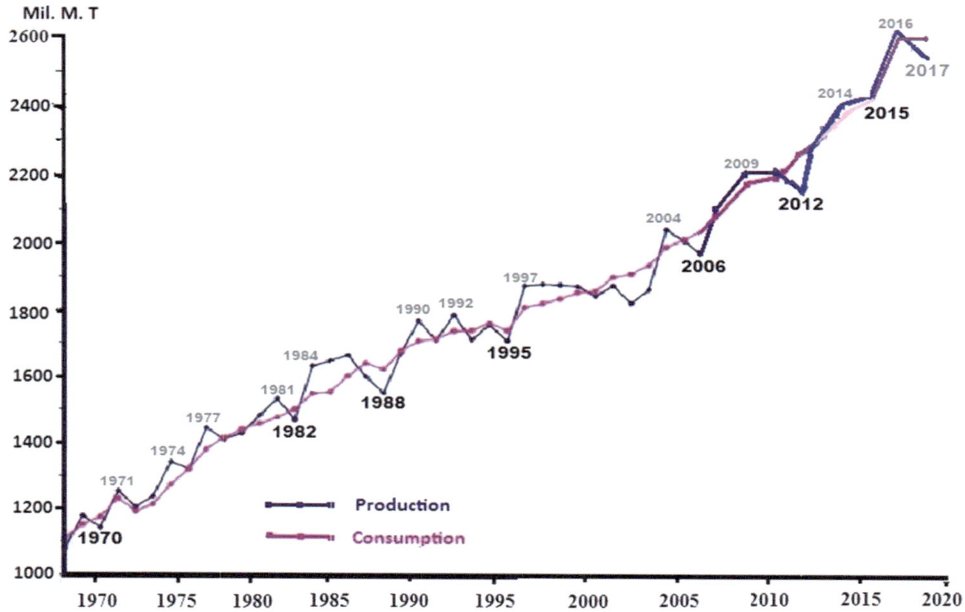

Fig. 2.2 World major grain (wheat, corn, rice) production versus consumption [2017 is based on preliminary estimates (USDA 2018)]

rice), which are mostly consumed by humans (USDA 2017; Brown 2012). These grain productions are normally compared to consumption. In general, when global annual major grain production is below consumption, this negative balance indicates a food security problem. Statistical analysis of 1968–2017 total major grains production compared to its consumption, shown in Fig. 2.2, indicates that in the first 17 years of the twenty-first century, food security has become a serious problem, since major grain production has become below the amount of consumed grains in half of those years [USDA 2017; WB 2018; FAO 2017)]. Moreover, if before the mid-1980s major grain consumption slightly exceeded the production (less than 20 million metric tons, MMT), after that period, the negative balance gradually increased to nearly 100 MMT and repeated more frequently. An increase in the difference values and annual repetition indicates an intensification of food insecurity in the world from the mid-1980s. Similar results were obtained for coarse grain, which is used primarily for feed and which contributed to nearly 40% of total global cereal production (USDA 2017). At the background of coarse grain increase over the past 37 years, in almost half of them, production was below consumption. In summary, we emphasize that FS measures can be considered from long- and short-term prospective. The long-term (multiyear) measures of FS are global population and grain production dynamics. From the short-term (annual) prospective, FS can be estimated by an annual balance between grain production and consumption, when a negative balance can be considered as food insecurity situation.

2.2.2 Long-Term Food Security and Expectations

Strong, exponential population growth compared to linear food production increase has been observed by economist Malthus in the late eighteenth century (Malthus 2015; Handy 2009). Malthus warned that global population would exceed the earth's capacity to grow food. Despite having been largely debunked, this theory has remained prominent in discourse regarding hunger, the world's population of carrying capac-

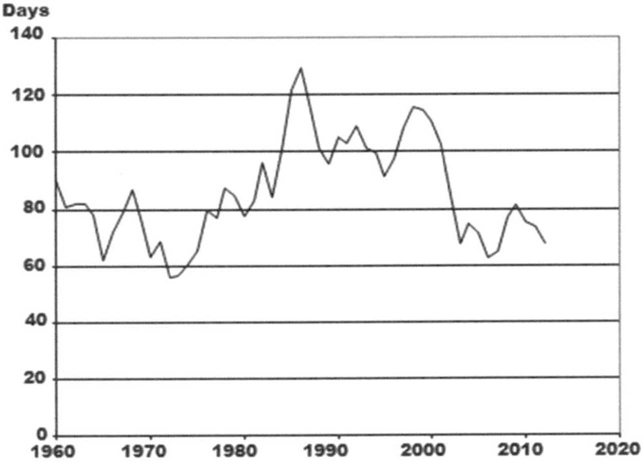

Fig. 2.3 World grain stocks as days of consumption by global population 1960–2012 (*Source USDA, USDA/ERS 2018*)

ity and the need for improved agricultural technology to increase production faster. In 2011, world population exceeded 7 billion thresholds, and by the early 2018, earth population increased to 7.6 billion (WP 2018). With nearly 100 million people added every year, by the mid-twenty-first century, global population is expected to increase nearly 30%, reaching 9 billion (UN 2004). The estimates have shown that such a strong population increase would require a 60–70% more food compared to what is produced now (Charles et al. 2018; FAO 2017; Brown 2012; Gottlieb and Joshi 2010; Handy 2009; Schmidhuber and Tubiello 2007). However, the best forecasts for such grain production increase are not optimistic, since as the new agricultural technology (called "Second Green Revolution" [Bennett et al. 2012; Pretty et al. 2006)], which is supposed to stimulate long-term grain increase, is improving slower, due to Earth resource exhaustion, climate change, land degradation (and some of them removing out of agriculture for real estate), water limitation, poor soils, inefficient water use, and intensification of natural disasters, especially drought (WB 2018; Statistica 2017; McKenna 2017; PotashCorp 2013; Bennett et al. 2012; Schmidhuber and Tubiello 2007; Solomon et al. 2007).

It is known that an annual shortage of grain production is normally compensated by consuming the amount of grain in stock [accumulated grains, which were not used in the years when grain production exceeded consumption (USDA 2017)]. However, statistical analysis of long-term prospects (1960–2012) with grain stocks does not show encouraging results. Following Fig. 2.3 (Brown 2012; WGS 2013; WFP 2014; USDA 2017), if the amount of grain in stocks from 1960s through 1979s was enough to feed the world population for 60–70 days and this number of days almost doubled increasing to 120 days by 1990s, after 1990s the trend has gone down to the same 60–70 days, which was at the very beginning. Other economic indicators showed some long-term problems with agriculture, even in developed countries. For example, following Fig. 2.4, the US farm income, which has almost tripled, increasing from $40–50 billion in the early 2000s to $110–120 billion in 2013, dropped back to $60 billion in 2015 (Newman 2018). The estimates for 2016–2017 and predictions for 2018 are not encouraging. Moreover, farmers' spending has increased requiring them to borrow more money from banks (Newman 2018). There are other important limitations such as 9–11% decrease in the sown cereal area of major world grain producers of North America and Europe (WB 2018).

An intensification of long-term food insecurity is not only a global problem. It is much more of a continental problem, specifically in Africa and South Asia, where most of developing countries are struggling with a lack of food, malnutrition, and hunger. Major grains are the most important commodity there. Table 2.1 shows total major grain (wheat, corn and rice) produc-

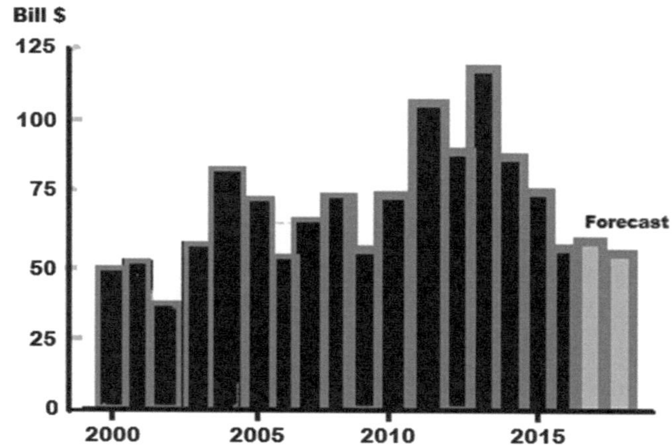

Fig. 2.4 Farm income ($ Billion) in the USA from 2000 (Newman 2018)

Table 2.1 Grain production (P), consumption (C), and (P−C) difference in million metric ton (MMT) during 2012–2017 in Sub-Sahara Africa and South Asia

	Sub-Sahara AFRICA			Southern ASIA		
Year	Production (P)	Consumption (C)	P−C	Production (P)	Consumption (C)	P−C
2012	80.4	108.6	−28.2	308.2	294.8	13.4
2013	84.8	115.0	−30.2	314.2	305.1	9.1
2014	85.2	117.8	−32.0	317.4	308.3	9.1
2015	78.5	118.9	−40.4	304.9	302.8	2.1
2016	86.9	120.9	−34.0	310.8	317.5	−6.7
2017	84.8	122.9	−38.1	323.8	324.9	−1.1

tion and consumption during 2012–2017 in Sub-Sahara Africa and South Asia (USDA/FAS 2017a, b; USDA 2018). Although South Asia is producing 3.5-time and is consuming almost 3-time more grain than Sub-Sahara Africa, in both subcontinents grain production and consumption were experiencing some increasing tendencies over 6-year period, which is a positive fact. However, on a negative side, grain consumption was growing faster during 2012–2017: in Sub-Sahara grain consumption has increased 10% compared to 6% production growth and in South Asia 10% versus 5%, respectively. But more important, a long-term grain conclusion can be drawn from the last 6-year production-consumption balance. Per Table 2.1, Sub-Sahara Africa was consuming 35–40% more grain than it was producing. Moreover, over the last 6 years, P−C balance was widening: in Sub-Sahara, negative P−C difference is increasing and in South Asia was changing from positive to slightly negative, which are not favorable tendencies.

2.2.3 Short-Term Food Security

Short-term food security/insecurity are normally considered from their annual changes. In addition to be influenced by economic, political, population, agricultural, and technological factors, the annual FS situation is strongly controlled by weather variations, which affect the amount of grain produced annually and consequently their contribution to a P−C balance. Following much research, one of the biggest FS problems is developed after large weather-related annual grain losses, which are normally associated with droughts (FAO 2017; USDA 2017; PotashCorpo 2013; Meroni et al. 2013; Brown 2012; Bennett et al. 2012; Godfray et al. 2010; Schmidhuber and Tubiello 2007; Kogan et al. 2013, 2015, 2017; Kogan 1995, 1997, 2002). Figure 2.5 presents global differences between annual major grain production and consumption (USDA/FAS 2017a, b; USDA 2018), which characterizes shortages of grain and potential for development of the food insecurity situation. Analysis of

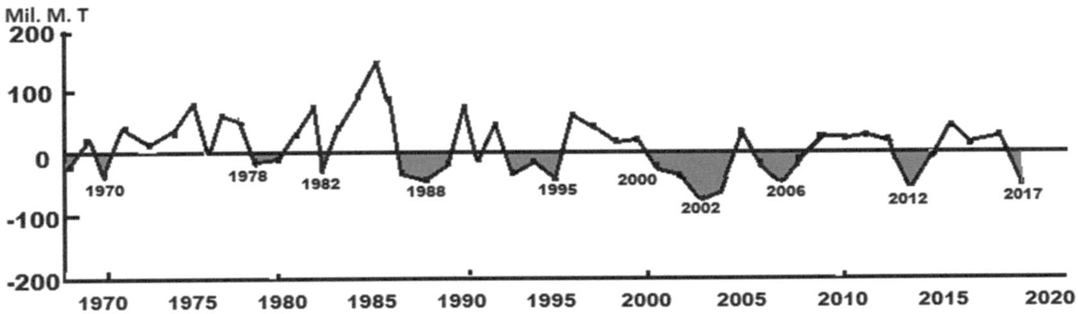

Fig. 2.5 Difference between total global major grain (wheat, corn, rice) production and consumption

Table 2.2 Countries and their parts affected by droughts identified from (NOAA/NESDIS 2018) in the years of negative balance between global major grain production and consumption shown in Fig. 2.4 (gray-shaded area)

Year	Negative P–C balance (MMT)	Countries, their part, and drought intensity
1982	15	Mexico$^{3-1}$, Brazil$^{3-1}$ (east), India$^{2-1}$, Kazakhstan$^{2-1}$ (north), Australia$^{2-1}$ (south), Asia^{1} (s. east)
1987	22	India^{3}, Africa$^{3-1}$ (Sub-Sahara), Europe$^{2-1}$ (south), Brazil$^{2-1}$
1988	43	USA$^{3-2}$, Argentina$^{3-2}$ (central), Europe^{1} (south), India$^{2-1}$, China$^{2-1}$ (s. east), Kazakhstan$^{2-1}$
1989	8	USA$^{3-2}$ (north), Canada$^{3-2}$ (s. east), Mexico$^{2-1}$, Argentina$^{2-1}$
1993	24	Australia$^{3-1}$ (north), Europe$^{2-1}$ (south), Brazi$^{2-1}$, Africa^{1} (central), India^{1}, China^{1} (east)
1994	10	Europe$^{3-1}$ (east), India^{1} (south), China^{1} (south)
1995	20	Argentina$^{3-2}$, Kazakhstan$^{3-2}$ (north), Europe$^{3-1}$ (west), Africa$^{3-1}$ (e. central), China$^{3-1}$ (west)
2000	15	Europe$^{3-2}$ (s. east), USA^{1} (west), Africa^{1} (Sub-Sahara), India^{1}
2001	31	USA$^{3-2}$, Canada$^{3-2}$ (south), Europe$^{3-1}$ (west), Australia$^{3-2}$, Mexico$^{3-2}$, Middle-East$^{2-1}$, Africa$^{2-1}$ (north),
2002	94	USA^{3} (central), India$^{3-2}$, Europe$^{3-2}$ (east), European Russia$^{3-2}$, Australia$^{3-2}$, Africa$^{3-2}$ (south and Sub-Sahara)
2003	73	Canada^{3}, Europe$^{3-2}$, USA$^{3-2}$ (north), Argentina$^{3-2}$ (central), Kazakhstan^{3} (north), Asia$^{3-2}$ (s. east)
2005	10	Africa$^{2-1}$ (north and Sub-Sahara), Europe$^{2-1}$ (west), Mexico^{1}, Brazil^{1}, Australia^{1}
2006	42	Australia^{3}, Kazakhstan^{3} (north), USA$^{3-2}$ (n. central), Argentina$^{3-1}$ (central), Europe$^{3-1}$ (west), Brazil^{1}
2007	5	Ukraine^{3} (south), USA^{1} (west), Brazil^{1} (east), Australia$^{1-2}$ (center)
2012	54	USA^{3}, Europe^{3} (south), Brazil^{3} (east), Kazakhstan^{3}, Argentina^{2}, Middle-East^{1}, Africa$^{1-2}$ (north and Sub-Sahara), Ausralia^{1}
2017	55	Mexico^{3}, USA^{2} (west), Africa^{2} (central), Europe$^{2-1}$ (west), Kazakhstan^{1}, Australia^{1} (central), Brazil^{1} (east)

The numbers indicate drought strength: 1—moderate, 2—severe, 3—extreme

Fig. 2.5 and Table 2.2 indicates that negative production-consumption (P–C) balance (production below consumption) occurred in drought years. If the negative balance exceeds 15 million metric tons (MMT), then drought-related grain losses occurred in quite a few countries, including some of the largest grain producers; for the negative balance greater than 40 MMT, drought is normally very intensive and affects major grain areas.

Presented in Table 2.2, drought area and intensity were estimated from satellite-based vegetation health (VH) indices described in detail in Chaps. 3 and 4. Table 2.2 indicates that in 2002, 2003, 2006, and 2012 (2017 provide preliminary analysis since P and C numbers have been

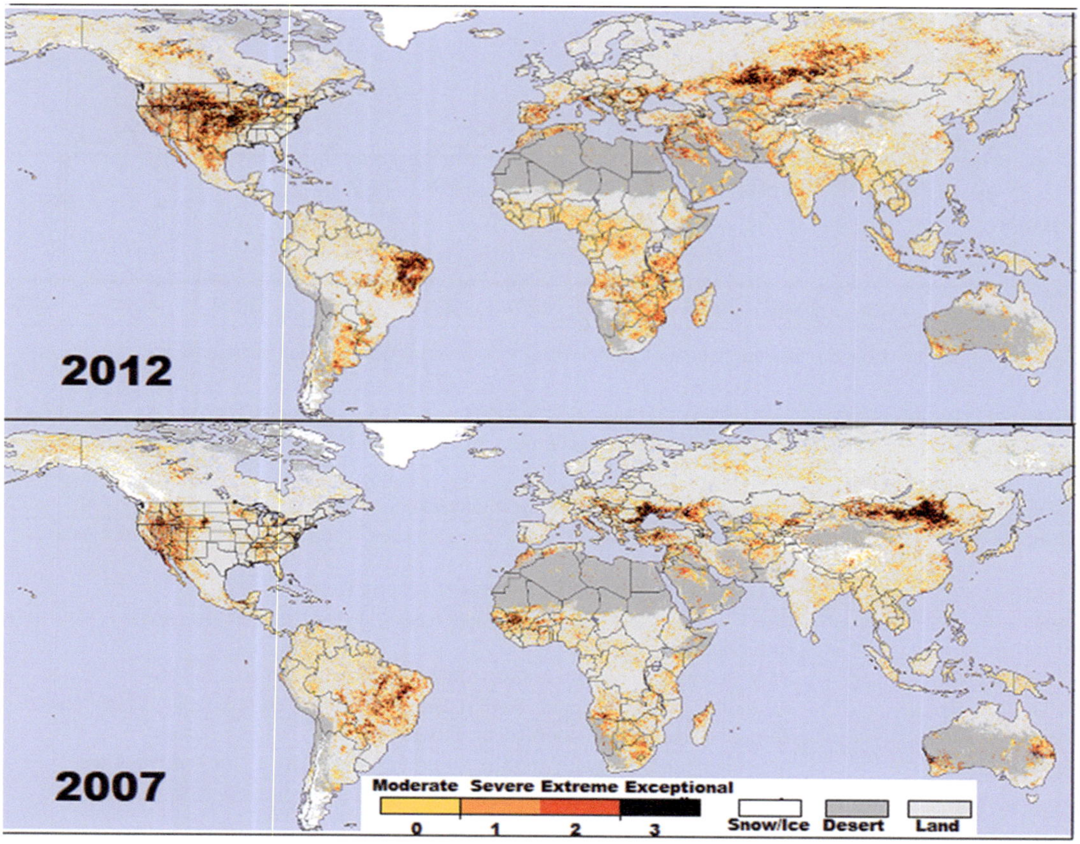

Fig. 2.6 Drought area and intensity in July 2012 and 2007 (NOAA/NESDIS 2018), the years with large (54) and small (5), correspondingly, grain production-consumption negative balance from Table 2.2

estimated), a negative balance was greater than 40 MMT because drought intensity was the strongest [extreme (3) and severe (2)], covering the entire country or its principal grain areas. For example, in 2012, such major grain producers as the USA [contributed 15.7% to 2014 total global grain production (WB 2018)], Europe (11.9%), Brazil (3.6%), and Kazakhstan (0.6%) were affected by extreme (level 3) drought and Argentina (1.8%) by severe (level 2) drought. Following the 2012 drought-related losses of grain, its production was below the consumption by 54 MMT. The 2007 drought also affected such principal grain producers as the USA, Brazil, and Australia. However, the drought was much lighter [level (1)] and affected partially grain areas. Only Ukraine experienced considerable losses of grain from extreme drought (level 3) in 2007. However, Ukraine grain production accounts only 2.2% from global. Therefore, a combination of a lighter 2007 drought, which affected small areas of major grain producers (the USA, Brazil, and Australia), and small Ukrainian contribution to the collected global grain in 2007, resulted in a very negligible (5 MMT) global grain P–C negative balance. The described differences between 2012 and 2007 in drought area, intensity and affected areas are well seen on the VH-based drought images of Fig. 2.6. Comparison of other years (not shown) with large and small negative difference between grain production and consumption (2006 versus 2005 or 2003 versus 2000 and others) showed similar drought location and intensity results to 2012 versus 2007.

From the three major grain crops, wheat, corn, and rice, negative global P–C difference values

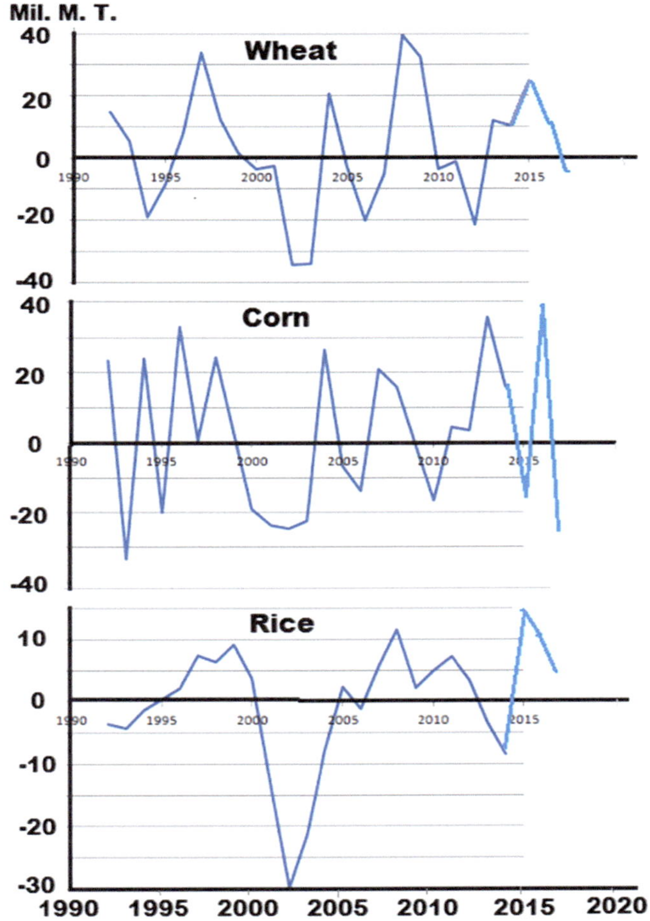

Fig. 2.7 Difference between global wheat, corn, and rice individual production and consumption (MMT)

might occur in the same year for all crops or only for some of the crops (two or even one), while others might not experience shortage of specific grains compared to the demands (consumption). This is demonstrated in Fig. 2.7, showing a P–C difference for global wheat, corn, and rice during 1991–2017. The analysis indicates some specific P–C features. First, from the past 27 years, shown in Fig. 2.7, in half of the years, global wheat and corn production was below consumption (negative P–C difference), but only 25% of such years were similar for rice. A fewer years for rice are due to rice is irrigated, helping to avoid some minor droughts. Second, coincidence of large negative P–C balance for all three crops during 2001–2003 (less affected in 2001 due to negligible wheat contribution). Third, a negative P–C balance for two crops, wheat and corn in 2005, 2006, and 1995 and for corn and rice in 1993. Fourth, all years with negative global total grain P–C balance (Fig. 2.6) coincides with the years of identical balance of individual and groups of crops in Fig. 2.7. For example, a very large shortage of total major grains produced in 2002 and 2003 compared to consumption (94 and 73 MMT, respectively) were because all three crops contributed to a negative P–C global grain balance: wheat contributed 43 and 42 MMT in 2002 and 2003, respectively, corn—23 and 22 MMT, and rice—30 and 10 MMT. In 1993 and 1995, two crops contributed to a total grain negative balance: large from corn (33 MMT) and small from rice (8 MMT) in 1993 and, wheat and corn (10 and 20 MMT) in 1995. In all these years, drought

Table 2.3 Global production (P), consumption (C), and P–C difference (MMT) for wheat (a), corn (b), and rice (c) in Sub-Sahara Africa and South Asia

	Sub-Sahara AFRICA			South ASIA		
Year	Production (P)	Consumption (C)	P–C	Production (P)	Consumption (C)	P–C
(a) Wheat						
2012	6.7	24.0	−17.3	126.4	131.0	−4.6
2013	7.2	26.4	−19.2	125.9	132.0	−6.1
2014	7.5	27.2	−19.7	130.1	132.3	−2.2
2015	6.4	28.6	−22.2	119.7	128.9	−9.2
2016	7.2	29.2	−22.0	120.1	139.1	−19.0
2017	7.1	30.3	−23.2	132.9	143.1	−10.2
(b) Corn						
2012	60.5	59.9	0.6	31.2	26.5	4.7
2013	63.5	62.4	1.1	34.1	29.8	4.3
2014	63.1	64.0	−0.9	34.1	32.8	1.3
2015	57.3	63.4	−6.1	32.7	35.1	−2.4
2016	64.4	64.1	0.3	36.8	36.9	−0.1
2017	62.5	64.8	−2.3	36.3	38.6	−2.3
(c) Rice						
2012	13.2	24.7	−11.5	150.6	137.3	13.3
2013	14.1	26.2	−12.1	154.2	143.3	10.9
2014	14.6	26.6	−12.0	153.2	143.2	10.0
2015	14.8	26.9	−12.1	152.3	138.8	13.5
2016	15.3	27.6	−12.3	153.9	141.5	12.4
2017	15.2	27.8	−12.6	154.6	143.2	11.4

was the principal disaster affecting wheat, corn, and even rice, although it is irrigated.

Two subcontinents, Sub-Sahara Africa and South Asia, which suffer the most from food insecurity, have developed a certain tendency over the years in the type of the grain crops they prefer. However, an intensive population increase, slow improvement in agricultural technology and the resulted lack of food, especially in a frequent drought years, has led to some changes in traditions. Following Table 2.3a, in recent 6 years (2012–2017), South Asia's (traditionally rice consuming region) wheat consumption exceeded production and the negative P–C gap expanded from 2–6 to 10–19 MMT. In Sub-Sahara Africa, wheat production is still much below consumption, since corn was traditionally the accepted crop. Meanwhile, in the recent two decades, wheat has become a widely grown staple food in Sub-Sahara (Table 2.3a) due to enormous population increase (WG 2017). Future prospects with population growth in those subcontinents are not encouraging since 450 million people are expected to be added by the 2030s to the current population of 1.2 billion in Africa (WG 2017). An intensified population growth is also expected in South Asia, where the current population of 1.883 billion is anticipated to increase nearly 1% in almost two decades (WM 2018). Meanwhile, the 2012–2017 data (Table 2.3a) indicates that wheat consumption on these subcontinents is gradually increasing (in Sub-Sahara from 26 to 30 MMT and in South Asia from 132 to 143 MMT). However, FS is gradually intensifying since a P–C negative wheat gap is growing (Table 2.3a). Besides, in Sub-Sahara Africa, the 2012–2017 wheat production is still three to four times below consumption. Therefore, Africa is the largest world importer (27%) of the global wheat market (WG 2017). Following the current trend, the widened P–C gap is expected to grow in the next one to two decades.

Per data in Table 2.3b, among the major grain crops, corn is number one in Sub-Sahara Africa, which is produced and consumed. Similar to wheat, corn consumption increased nearly 8%

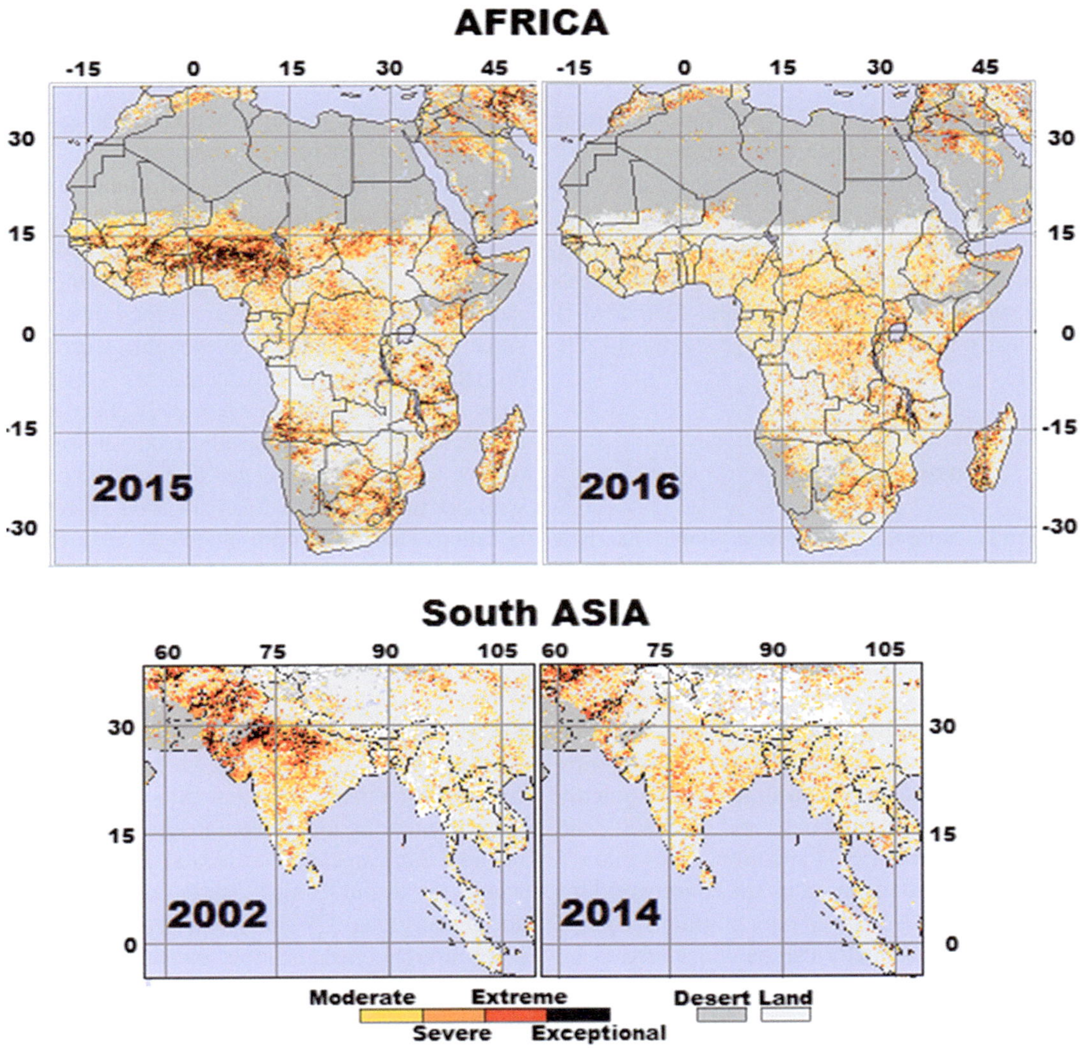

Fig. 2.8 Vegetation health-based drought maps (NOAA/NESDIS 2018) during the years of small (Africa 2016 and South Asia 2014) and large (Africa 2015 and South Asia 2002) grain production minus consumption (P−C) balance

during 2012–2017, while production has remained relatively stable with some annual variations. Moreover, in half of the years, the P−C balance was slightly negative due to a lower corn production. Rice consumption in Sub-Sahara (Table 2.3c) was also increasing with similar speed to production increase (7–8%). However, it is important to emphasize that rice consumption exceeds by nearly 80% production in Sub-Sahara, requiring considerable rice imports. South Asia is traditionally a rice-producing region. The 2012–2017 data (Table 2.3c) indicates a 1–2% annual rice production increase, which stably exceeds consumption of nearly 10–13 MMT per year. Therefore, although there is some progress in wheat, corn, and rice production increase in the recent 6-year in Sub-Sahara Africa and South Asia, the long-term food security situation has remained tense since consumption of these crops is increasing faster and stronger, following the demands from the growing population. Besides, agricultural technology is progressing slow in these subcontinents. One of the biggest problems is drought, which reduces grain production. An example in Fig. 2.8 compares VH-derived droughts, which affected these subcontinents in years of large and small P–C balance. As seen, the 2015 drought in Africa was much stronger (level 2 and 3) compared to lighter (level 1) drought in 2016. As the result, total three-crop grain production in 2015 (78.5 MMT) was 10% lower than in 2016 (88.4 MMT). In South Asia, the most intensive drought (level 2 and 3) was in 2002, when total grain production (270 MMT) was nearly 14% below the 2014 production (317.1 MMT).

2.3 Conclusion

Currently, more than one billion people (nearly a sixth of the world's population), mostly from developing countries of Africa and Asia, suffer from chronic hunger and malnutrition due to a lack of food, which creates problem with food security. Among economic, political, population, agriculture, and environmental factors affecting food security, the world population consuming food and annual agricultural production mainly determines food supplies and demand and through their balance FS situation. Since grain is the principal product used by the entire world for both food and feed, the amount of grain produced by agriculture annually can be considered as an indicator of global food supply. Therefore, world population, grain production and their balance can be used for food security assessment. Food security/insecurity can be considered as a long-term and short-term issue. Long-term global FS is strongly controlled by an intensity of increase in the number of world people and the amount of produced grain. Both of these parameters are experiencing an upward trend in the recent 50 years. However, multiyear dynamics indicates that the world population is growing much faster than grain production is increasing. By the mid-twenty-first century, the world population may increase 26%, adding 2 billion to the current 7.6 billion. The estimates show that in order to feed 9.5 billion people, agriculture would require to produce 60–70% more grain. Such a huge grain increase by 2050s is a very big challenge since the current grain production multiyear trend is leveling off because the second Green Revolution, which is applying currently to advance agriculture quickly, is progressing extremely slow. Therefore, future prospects for elimination of long-term food insecurity are not encouraging.

There are better ways of controlling the FS situation annually through the prediction of grain production and early assessments if the predicted grain amount is sufficient for consumption. Annual grain losses, which create food insecurity situation, are associated with droughts. Currently, drought-related grain losses can be predicted early (2 months in advance of harvest) from operational satellites, providing advanced assessments of how much grain is needed to cover the deficit with consumption and to avoid food insecurity situation. This chapter presented evidences that if drought-related grain losses, detected 2 months in advance of harvest, occurred in many countries, including largest grain producers, negative global P–C balance might exceed 20 and in case of the strongest drought might exceed 40–50 MMT. Finally, it is important to emphasize that food insecurity is intensified in the twenty-first century, triggering by population increase and frequent annual drought-related losses in grain production. Such an early assessment of grain losses would be useful for FS prediction. Currently, additional global FS concern arises from anticipated drought intensification, expansion, and penetration to the new areas following climate change (Gillins 2014; Solomon et al. 2007). Operational satellite can detect, monitor drought (area, duration, intensity), and predict losses of grain production on every 0.5, 1, and 4 km^2 of land surface, addressing annual food security situation, which are presented in the next chapters.

References

Bennett, A., G. Bending, D. Chandler, S. Hilton, and P. Mills. 2012. Meeting the demand for crop production: The challenge of yield declines in crops grown in short rotations. *Biological Reviews* 87: 52–71.

Brown, L.R. 2012. *Full planet, empty plates: The new geopolitics the food scarcity*, 79. Washington: Earth Policy Institute www.earth-policy.org.

Charles, H., J. Godfray, J.R. Beddington, R. Crute, L. Haddad, D. Lawrence, J.F. Muir, J. Pretty, S. Robinson, S.M. Thomas, and C. Toulmin. 2018.

Food security: The challenge of feeding 9 billion people. *Food Security* https://waterlandandecosystems.wikispaces.com/file/view/Component2_CIAT.pdf.

Coleman-Jensen, Alisha, Mark Nord, Margaret S. Andrews, and Steven Carlson. 2011. Household food security in the United States in 2010, Economic Research Report 118021. United States Department of Agriculture, Economic Research Service. https://ideas.repec.org/p/ags/uersrr/118021.html.

FAO. 1999. Food insecurity: When people must live with hunger and fear of starvation. *The State of Food Insecurity in the World Report*, Rome, 76.

———. 2017. How close we are to zero hunger. http://www.fao.org/state-of-food-security-nutrition/en/.

———. 2018. FAO cereal supply and demand brief. *World food situation*. http://www.fao.org/worldfoodsituation/csdb/en/.

Gillins, J. 2014. U.S.A. Climate Has Already Changed, Study Find, Citing Heat and Floods. *The New York Times*, May 6.

Godfray, H.C.J., J.R. Beddington, I.R. Crute, L. Haddad, D. Lawrence, J.F. Muir, J. Pretty, S. Robinson, S.M. Thomas, and C. Toulmin. 2010. Food security: The challenge of feeding 9 billion people. *Science* 327: 812–818. https://doi.org/10.1126/science.1185383.

Godfray, H.C.J., P. Aveyard, T. Garnett, J.W. Hall, T.J. Key, J. Lorimar, R.T. Pirrehumbert, P. Scarborough, M. Springmann and S.A. Jebb. 2018. Meat consumption health and the environment. Science 361 (5324): 1–8.

Gottlieb, R., and A. Joshi. 2010. *Food justice*, 320. Cambridge: MIT Press.

GRACE. 2014. Food security food access. *Food program*. http://www.sustainabletable.org/280/food-security-food-access.

Guillou, M., and G. Matheron. 2014. *The world challenge to feed 9 billion people*. New York: Springer http://www.springer.com/us/book/9789401785686.

Handy, J. 2009. 'Almost idiotic wretchedness': A long history of blaming peasants. *Journal of Peasant Studies* 36: 325–344.

Heibuch, J. 2011. Food insecurity on the rise. *Treehugger*, October 31. https://www.treehugger.com/health/food-insecurity-rise-infographic.html.

Holt-Gimenez, E., and R. Patel. 2009. *Crisis and the hunger for justice*. Boston: Pambazuka Press https://books.google.com/books.

Kogan, F.N. 1995. Droughts of the late 1980s in the United States as derived from NOAA polar-orbiting satellite data. *Bulletin of the American Meteorological Society* 76 (5): 655–667.

———. 1997. Global drought watch from space. *Bulletin of the American Meteorological Society* 78: 621–636.

———. 2002. World droughts in the new millennium from AVHRR-based vegetation health indices. *Eos Transaction of American Geophysical Union* 83 (48): 562–563.

Kogan, F., T. Adamenko, and W. Guo. 2013. Global and regional drought dynamics in the climate warming era. *Remote Sensing Letters* 4: 364–372. https://doi.org/10.1080/2150704X.2012.736033.

Kogan, F., M. Goldberg, T. Schott, and W. Guo. 2015. SUOMI NPP/VIIRS: Improve drought watch, crop losses prediction and food security. *International Journal of Remote Sensing*. https://doi.org/10.1080/01431161.2015.1095370.

Kogan, F., W. Guo, and W. Yang. 2017. SNPP/VIIRS vegetation health to assess 500 California drought. *Geomatics, Natural Hazards and Risk*. https://doi.org/10.1080/19475705.2017.1337654.

Malthus, T.R. 2015. *Essay on the principal of population and other writings*, 352. London: Penguin https://books.google.com/books/about/An Essay on the Principle of Population.html?id= Z0eBgAAQBAJ.

McKenna, M. 2017. Revived climate changed forum focused on threats to human health. *EOS*, 98(4), April 7.

Meroni, M., F. Kayitakire, and M.E. Brown. 2013. Remote sensing of vegetation: Potential application for index insurance. In *The challenges of index-based insurance for food security in developing countries*, ed. R. Gommes and F. Kayitakire, 238–245. Sevilla: European Commission/Joint Research Center.

Newman, J. 2018. Farm belt braces for steep income drop, trade spats. *The Wall Street Journal*, February 9, 3A.

NOAA/NESDIS. 2018. Vegetation Health Data and Products. https://www.star.nesdis.noaa.gov/smcd/emb/vci/VH/vh_browse.php.

PotashCorpo. 2013. Agriculture: Crop overview. http://www.potashcorp.com/industry_overview/2011/agriculture/16.

Pretty, J.N., A.D. Noble, D. Bossio, J. Dixon, R.E. Hine, F.W.T. Penning de Vries, and J.I.L. Morison. 2006. Resource-conserving agriculture increases yields in developing countries. *Environmental Science and Technology* 40: 1114–1119.

Schmidhuber, J., and F.N. Tubiello. 2007. Global food security under climate change. *Proceedings of the National Academy of Sciences of the United States of America* 104: 19703–19708.

Solomon, S., D. Qin, M. Manning, Z. Chen, M. Marquis, K.B. Averyt, M. Tignor, and H.L. Miller, eds. 2007. *Contribution of Working Group I to the Fourth Assessment Report of the Intergovernmental Panel on Climate Change*, 996. Cambridge: Cambridge University Press.

Statistica. 2017. Total global grain production from 2008/09 to 2017/18. *The Statistics Portal*. https://www.statista.com/statistics/271943/total-world-grain-production-since-2008-2009/.

UN. 2004. World population prospects. https://www.google.com/search? tbm=isch&sa=1&ei=PMBsWrSRBYTbzwLW26foDA&q=global+population+from+1960+through2017.

USDA. 2017. Food security. https://www.usda.gov/topics/food-and-nutrition/food-security.

———. 2018. Grain world market and trade. https://apps.fas.usda.gov/psdonline/circulars/grain.pdf.

USDA/ERS. 2018. Highlight from February 2018 farm income forecast, Feb 07. https://www.ers.usda.gov/topics/farm-economy/farm-sector-income-finances/highlights-from-the-farm-income-forecast/.

USDA/FAS. 2017a. Grain: World market and trade. May. https://gain.fas.usda.gov/psdonline/circulars/grain.pdf.

———. 2017b. South Africa – Republic of Grain and Feed Annual. *Gain Report* (Prepared by D. Esterhuizen), March 15. https://gain.fas.usda.gov/Recent%20GAIN%20Publications/Grain%20and%20Feed%20Annual_Pretoria_South%20Africa%20-%20Republic%20of_3-16-2017.pdf.

WB. 2018. Agriculture & rural development. https://data.worldbank.org/topic/agriculture-and-rural-development.

WFP. 2014. https://www.google.com/search?q=world+food+program&oq=world+food+program&aqs=chrome..69i57j015.6319j0j7&sourceid=chrome&ie=UTF.

WG. 2017. World Grain. http://www.world-grain.com/.

WGS (World Grain Stocks). 2013. World grain stocks as days of consumption 1960-2012. http://www.treehugger.com/sustainable-agriculture/global-grain-stocks-drop-dangerously-low2012-consumption-exceeded-production.html.

WH (The White House). 2012. Fact sheet: G-8 action on food security and nutrition. May 18. https://obamawhitehouse.archives.gov/the-press-office/2012/05/18/fact-sheet-g-8-action-food-security-and-nutrition.

WM. 2018. Southern Asia population *Worldometer.* http://www.worldometers.info/world-population/southern-asia-population/.

WP (World population). 2018. http://www.worldometers.info/world-population/.

3 Operational Satellites for Earth Monitoring

3.1 Introduction

In the second half of the twentieth century, following the world recovery after World War II, through its intensive economic and cultural development, it became clear that this recovery could be done effectively with more accurate environmental monitoring of ocean atmosphere and land. That stimulated the development of satellite systems to observe the Earth and to use the data for modeling and prediction of the impacts of environment on economy and human life. Two operational satellite systems have been developed: geostationary (GEO) and polar-orbiting (POES). The GEO satellites are set in space to frequently monitor (every 5–15 min) weather over the same limited latitudinal area (between two poles) of the Earth. Global weather monitoring by GEO satellites was achieved through international cooperation using GEO satellites from several countries. Currently, global coverage of GEO satellites is achieved through several systems developed in the USA (GOES and GOES-R) to observe North and South America; a European satellite (METEOSAT, MSG) to observe Europe and Africa; and a Japanese (GMC, MTSAT) system to cover Asia. There are also geostationary satellites in other countries (Russia (METEOR) and others). GEO satellite data are used frequently for weather monitoring and prediction.

The POES (polar-orbiting environmental satellites) produce less-frequent, daily observations but cover the entire globe. They encircle the Earth in a north-south (from pole to pole) in sun synchronous orbit. Each new satellite circles around the Earth between the two poles rotating around Earth shows the new area. Therefore, after 24-h the entire globe is covered with daily environmental measurements, which are used for monitoring land, atmosphere, and ocean and their impacts on economy and human life. In the most recent two to three decades, POES data were used intensively to detect and monitor drought (start, area, intensity, duration, impacts, etc.), soil moisture (both shortage and saturation), fire risk, vegetation stress, and other environmental parameters and phenomena. These POES measurements and developed products were especially successful in providing assessments of drought impact on agriculture and prediction of crops and pasture losses. These predictions were applied to estimation of food shortages and analysis of the food security situation. Further discussions will be focused on the NOAA POES operational meteorological satellites and their applications.

F. Kogan, *Remote Sensing for Food Security*, Sustainable Development Goals Series,
https://doi.org/10.1007/978-3-319-96256-6_3

3.2 NOAA Polar-Orbiting Operational Environmental Satellites

Two NOAA POES satellite systems are available: morning and afternoon. They have several instruments on board to measure many parameters of the Earth's ocean, atmosphere (clouds, aerosols, temperature, precipitation, etc.), land cover, incoming solar radiation, the energy spectrum, and other parameters. For an assessment of environmental impacts on agriculture and food security, two of the most important instruments have been used: the Advanced Very High Resolution Radiometer (AVHRR) on the original, National Oceanic and Atmospheric Administration (NOAA) system, servicing operationally for the global community since the early 1980s (Cracknell 1997) and the Visible Infrared Imaging Radiometer Suite (VIIRS) onboard the new operational system, called Suomi National Polar-Orbiting Operational Environmental Satellite System Preparatory Project (Suomi-NPP or S-NPP), which has been in service since 2011, which is followed by the most advanced J-1 system, launched in November 2017 (JPSS 2014; NOAA 2017), which will continued until the mid of the current century.

Prior to the first US operational POES satellites, a few experimental meteorological satellites have been launched to investigate their performance and applications. The first such satellite, called TIROS-1, was launched in 1960. In the next 15 years, TIROS-2 to 10 and NOAA-1 to 5 were launched to test satellites and sensors performance and to develop applications. Following these tests, the first operational POES satellite in the USA, NOAA-6, was launched on June 27, 1979. That day, the operational POES era began (Cracknell 1997). One of the most successful NOAA polar-orbiting operational satellite sensors, which observations have been used for monitoring the Earth and providing environmental monitoring and prediction, from the very beginning, have been the Advanced Very High Resolution Radiometer (AVHRR). The AVHRR sensor has five channels to observe the Earth: visible, near-infrared, and three infrared. Since the early 1980s, 14 polar-orbiting satellites were launched with the AVHRR instrument on board to be used for global numerical measurements of environmental parameters, data accumulation, product development, and to monitor environmental impacts on Earth. Currently, the 37-year (1981–2017) of NOAA/AVHRR operational polar-orbiting satellite data and products characterizing land, at 4 and 16 km^2 resolution, are available for the entire world. These data and products are providing numerical assessment of land cover (greenness), temperature, vegetation health, droughts and their impacts, soil moisture, prediction of crop and pasture production, evaluation of vegetation stress (moisture and thermal), fire risk, climate trend, and some other parameters, which are used for numerical assessments and prediction of food security.

3.2.1 AVHRR Sensor

The NOAA/AVHRR is a cross-track scanning system (Kidwell 1995, 1997; Cracknell 1997). The system's AVHRR instrument has been observing the Earth continuously throughout its 37-year history (from 1981 to 2017) in the following wavelengths of the solar spectrum: visible [VIS, 0.58–0.68 μm, channel 1 (Ch1)], near-infrared [NIR, 0.725–1.1 μm, channel 2 (Ch2)], and three infrareds [IR, 3.5–3.9 μm, channel 3; 10.3–11.3 μm, channel 4 (Ch4); and 11.5–12.5 μm, channel 5 (Ch5)]. The NOAA/AVHRR scans the Earth continuously at medium, 4 km^2 [Global Area Coverage (GAC)] and at high, 1.1-km (Local Area Coverage (LAC)) resolution from selected portions of each orbit (Cracknell 1997).

From the 14th NOAA morning and afternoon operational satellites, flying in sun-synchronous orbit and carrying the AVHRR instruments, data sets for vegetation monitoring were developed from seven afternoon satellites: NOAA-7, 9, 11, 14, 16, 18, and 19. They were launched, correspondingly, on June 23, 1981 (local day time at launch 14:30 pm), December 12, 1984 (14:20 pm), September 24, 1988 (13:30 pm), December 30, 1994 (13:30 pm), September 21,

2000 (13:44 pm), May 20, 2005 (13:50 pm), and June 2, 2009 (13:44 pm). These satellites operated during the following years: 1981–1985, 1986–1989, 1989–1994, 1995–2000, 2001–2005, 2005-present, and 2009-present, respectively (Kogan et al. 2017; Kidwell 1997; Cracknell 1997). From September 1994 through January 1995, no afternoon operational observations were produced since NOAA-11 satellite malfunctioned and the new NOAA-13 satellite failed soon after launch. Also, between January and June 2005, NOAA-16 malfunctioned, from time to time, and its data was replaced with NOAA-17 (morning satellite) preliminary calibrated to the NOAA-16 afternoon data. From the indicated satellites, NOAA-7 and 9 carried AVHRR-1 instrument, NOAA-11 and 14—AVHRR-2, and the rest—AVHRR-3. All of them have identical design but slightly different response functions (Kidwell 1997; Kogan 2006).

3.2.2 AVHRR Data for Vegetation Monitoring

Several global data sets were developed from the AVHRR records since the early 1980s. Those oriented toward application were NOAA's global vegetation index (GVI and GVI-2) and NASA's Pathfinder and GIMMS (Tarpley et al. 1984; James and Kalluri 1994; Kidwell 1997; Tucker et al. 2004). These data were focused only on two channels (VIS and NIR), ignoring infrared measurements, which are very important for monitoring thermal-caused vegetation stress, production losses, and food security. As a result, NOAA developed a new data set, entitled the global vegetation health (VH, Kogan 1989; LeComte and Kogan 1988). The NOAA operational VH has many advantages over the other nonoperational global data sets (Pathfinder, GIMMS, and others). The most important are (a) application of infrared channels in addition to VIS and NIR, (b) 38-year longevity, (c) 0.5, 1, 4, and 16 km^2 specical and 1-week temporal resolution, (d) strong theoretical background based on bio-physical laws (Law of Minimum, Law of Tolerance, and Principal of Carrying Capacity), (e) numerous applications, (f) climate/land cover change monitoring, (g) comprehensive validation , and others. Following these advantages, the global VH was the most accurate in assessment of droughts impacts on crop losses and prediction of the food security situation. Currently, the AVHRR-based VH data set is global, the longest (38 years), has the highest spatial (0.5, 1, 4, 16 km^2) and temporal (one week) resolution, contains originally observed reflectance and emission, and has many indices, including those with suppressed noise, biophysical climatology, and, what is the most important, products used for monitoring the environmental and socioeconomic activities (Kogan et al. 2017, 2018; Kogan 1995a, b, 1997, 1989). This subchapter describes the new, considerably improved and currently available (to users since 1981 AVHRR-based operational global vegetation health) data set at 4 and 16 km^2 (0.036° and 0.144°) resolution.

The VH system algorithm starts form data extraction from the NOAA AVHRR/CLAVR-x processing system (Jacobowitz et al. 2003; Heidinger and Pavolonis 2005) and collating the data from 1.1 km^2 onto a global 4 km^2 VH grid. This grid is based on the Plate Carree map projection. The global data spans from 75.024° (north edge) to −55.152° (south edge) in the latitudinal and from −180° (west) to 180° (east) in the longitude directions. This processing supports nominal grid cell size of 4 by 4 km resolution with the 3616*10,000 grid elements for the entire world. The VH input includes the CLAVR-x navigation (NAV), observation (OBS), and geolocation (GEO) files for each Global Area Coverage (GAC) Level 1b orbit. The VH is using three AVHRR channels, VIS (Ch1), NIR (Ch2), and IR-4 (Ch4).

One of the important steps in primary data processing is radiometric calibration of VIS, NIR and correction of IR-4. Visible channels' calibration consists of generally two steps: pre- and post-launch calibration. Based on Kidwell (1995, 1997), the following pre-launch linear formula ($A = S*C + I$) is applied, where (A) is albedo, (S)—slope, and (I)—intercept. Since the instrument output does not remain the same after launch, post-launch calibration was applied

($R = S\times(C-cd)$) to AVHRR on NOAA-7 to 14 satellites, where C is 10-bit radiance count and cd—dark count) following Rao and Chen (1995, 1996, 1999). For NOAA-16 through 19, a dual slope calibration method was applied. For the best thermal characteristics of the environment, Ch4 data (from two IR channels) were used since they are less affected by moisture in the atmosphere compared to Ch5 (Kidwell 1997; Cracknell 1997). One of the most important conditions was the collection of thermal channel data from afternoon satellites in order to monitor the highest temperature during each day, which is crucial for detecting thermal stress in vegetation, estimation of agricultural production loss, and applying more accurate approach to food security.

The VIS and NIR channels have been used by scientists since the very beginning of the satellite era, although there was very limited biophysical explanation of the principles. As it is known from plants, biophysics, due to the availability of the chlorophyll-a and carotenoids (Fig. 3.1a), vegetation absorbs the second highest amount of solar energy in the VIS (0.58–0.68 μm) range of solar spectrum for its biological processes. High solar energy absorption stimulates the highest photosynthetic rate and green mass accumulation (Fig. 3.1b). However, it is important to note that the highest solar energy absorption and the highest photosynthetic rate, as seen in Fig. 3.1, are in the 0.40–0.55 μm range of solar spectrum (Myers 1970; Gates 1970, Gray and McCrary 1981; James and Kalluri 1994). Unfortunately, both operational AVHRR and VIIRS sensors, as well as scientific MODIS sensor, were not designed to produce measurements in that range because the main goals of the sensors during the period of their development were to observe and measure environmental parameters controlling weather (Cracknell 1997).

Meanwhile, land and ecosystem scientists found ways of using the selected bands to characterize land cover with specific focus on vegetation. Using Landsat 1 satellite VIS and NIR data, scientists examined vegetation green-up during the growing season (Gray and McCrary 1981; ASP 1975). Following their analysis, the VIS and NIR bands provided useful information about

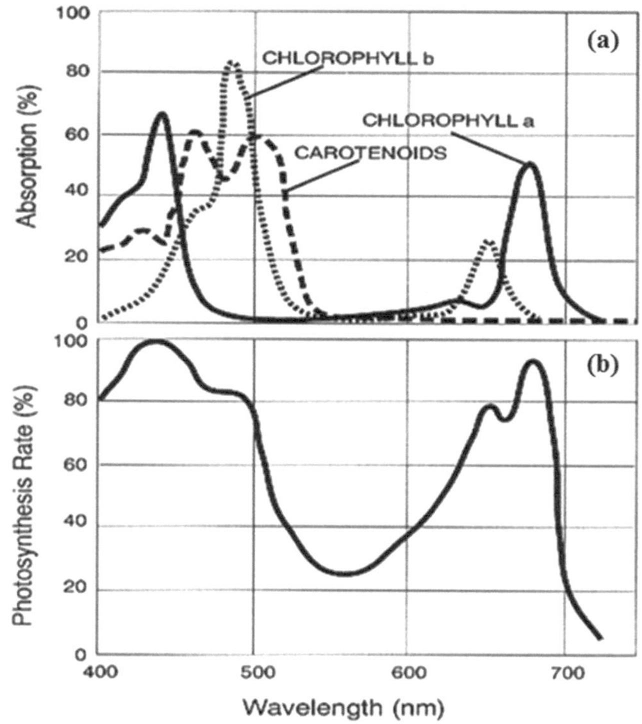

Fig. 3.1 Absorption of chlorophyll-*a*, -*b* and carotenoids (**a**) and photosynthesis rate (**b**) at different wavelength

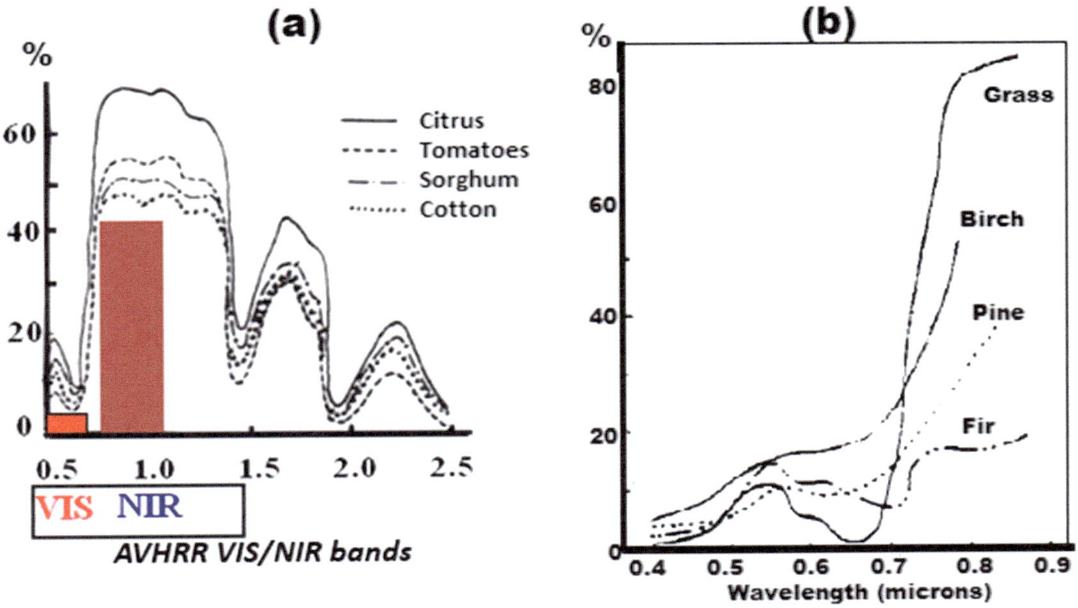

Fig. 3.2 Solar radiation reflected by vegetation (**a**) and reflectance (%) of fir, pine, birch, grass (**b**) in the VIS and NIR bands of AVHRR sensor

ecosystems, both wild and cultivated. Figure 3.2 shows these differences, comparing the amount of solar radiation reflected by the vegetation in these two bands. First, what is noticeable from the figure is that in all ecosystems very little radiation is reflected (<20%) in the VIS band and very high (more than 20% for trees and 40% for crops) in the NIR. Second, different ecosystems reflect (and consequently absorb) different amounts of solar radiation in the two bands. Third, and most important, changes between VIS and NIR bands for crops and wild vegetation are not proportional. For example, grass has the highest reflection (90%) in NIR and the lowest (13%) in VIS, while fir trees have the lowest reflection (20%) in NIR and the highest (17%) in VIS (Fig. 3.2b). For crops, cotton has lower NIR but higher VIS refection than tomatoes (Fig. 3.2a). These inconsistencies indicate that reflectance in VIS and NIR bands can't be used individually for effective characterization of vegetation. Moreover, as seen in Fig. 3.3, when wheat crop is under environmental stress, reflectance in VIS and NIR is

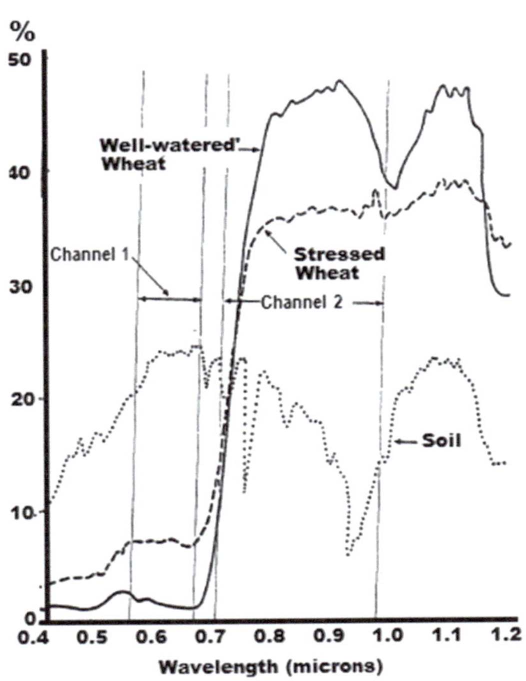

Fig. 3.3 Wheat and bare soil reflectance (%) across NOAA/AVHRR VIS (Ch 1) and NIR (Ch 2) bands

changing in an opposite direction, creating another disproportional problem for using these two bands separately (Gray and McCrary 1981). Therefore, the ratio of the difference of the red (VIS) and near-infrared (NIR) radiances over their sum with the name vegetation index was introduced to characterize vegetation performance in response to the sun's radiation and ambient conditions (Cracknell 1997).

3.2.3 Normalized Difference Vegetation Index

The introduced vegetation index has been very popular and has been widely used for monitoring the Earth's vegetation. In later scientific publications, the index was renamed to the normalized difference vegetation index (NDVI). The NDVI has been investigated thoroughly, providing excellent characterization of vegetation external and internal features. Thus, following VIS channel performance (Fig. 3.1), NDVI has showed strong relationship with chlorophyll concentration on land and in water. As seen in Fig. 3.4a, their relationship for land vegetation is very strong [$R^2 = 0.75$, (c)] and the equation shown on the diagram can be used to derive the amount of this important physiological component, controlling vegetation health. Following Kushnir et al. (2009), NDVI of water near the shore (derived from SeeWiFS data) is increasing with an increase in chlorophyll concentration (Fig. 3.4b). This relationship can be used to monitor water contamination in the coastal zone. NDVI has also been used to estimate the ecosystem's primary production, biomass, vegetation condition, and other parameters of land surface, atmosphere near the ground, and even the concentration of mineral suspension in the water near the shore (Gray and McCrary 1981; Tarpley et al. 1984; Tucker et al. 1983; Tucker and Seiler 1986; Tucker et al. 2004; James and Kalluri 1994; Kogan 1987, 1989; Kogan et al. 2003; Hashemi and Chenani 2004; Saleous 2005; Kushnir et al. 2009).

In addition to chlorophyll monitoring, there are two other very important NDVI properties. First, its values for vegetative ecosystems are quite different from values of other big targets on Earth's intercepting solar lights, such as bare soil, water, and clouds; second, NDVI values can be used to distinguish vegetation density. Table 3.1 shows typical channels and no noise NDVI values for the three levels of vegetation density and for the indicated big targets. Following these properties, it is easy to distinguish NDVI of vegetative ecosystems from desert, snow/ice, water, and clouds. This is important when NDVI is processing to remove the noise and for the development

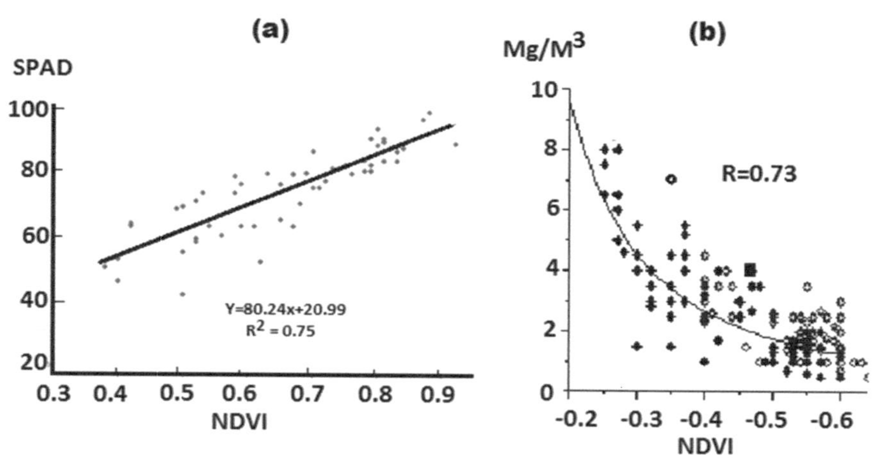

Fig. 3.4 The correlation between NDVI and chlorophyll content: (**a**) chlorophyll measured with SPAD-502 (Chlorophyll Meter form Minolta SPAD Ltd) in northern Iran; (**b**) water chlorophyll measured with SeaWiFS 90 m from the shore of Black Sea

Table 3.1 Typical AVHRR-based Reflectance and NDVI in the USA's northern Great Plains, southern Canada, European Russia (northern and central)

Earth Big Cover Type	VIS Channel (CH 1)	NIR Channel (CH2)	NDVI
Vegetation: Dense	0.050	0.150	0.500
Medium	0.080	0.110	0.140
Light	0.100	0.120	0.090
Bare soil	0.269	0.283	0.025
Clouds and snow/ice	0.227	0.228	0.002
Water	0.022	0.013	−0.260

of NDVI products characterizing vegetation cover, drought, soil moisture, fire risk, and many others shown on the following WEB site: (https://www.star.nesdis.noaa.gov/smcd/emb/vci/VH/index.php). These techniques will be discussed in the next subchapters.

Another important NDVI property is its change with weather variation. This is demonstrated by AVHRR data collected from NOAA-7 and 9 satellites, and also precipitation and temperature from weather stations in central Ethiopia (Horn of Africa). As seen in Fig. 3.5, the growing season (June–October), no noise weekly NDVI time series during 1984–1987 versus the same years' amount of total July–September precipitation anomaly (deviation from multiyear climatological normal, collected from meteorological observations). This example clearly indicates that during 1984, the lowest (from all shown years) NDVI values (0.10–0.12 in August–September) during the entire growing season were associated with severe shortages of precipitation (51% normal), while in 1987, higher than normal precipitation (115%) created good moisture conditions for the 4-year highest NDVI (0.3 during the middle of the season in August–September). It is interesting that near-normal (95%) precipitation in 1985 produced similar to above normal precipitation impacts in 1987 on NDVI values. In 1986, 73% of the normal growing season rainfall provided slower NDVI growth in the first part of the growing season (June–August) and intensified growth in the second part (September–October). However, the highest NDVI values

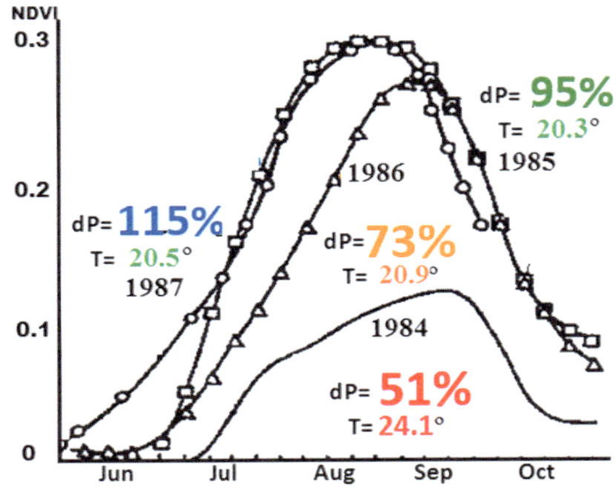

Fig. 3.5 Weekly growing season (June–October) NDVI during 1984–1987, total July–September precipitation anomaly (dP, deviation from multiyear climatology) and maximum mid-season temperature (T) from weather station in central Ethiopia

became lower in 1986 compared to 1985 and 1987. This example indicates that NDVI, in addition to characterization of the earth surface, is a very good indicator of rainfall and can serve as a criterion of moisture availability for vegetation. Figure 3.5 shows the highest daily air temperature in the middle of the growing season. Only 1984, a dry year, had elevated daily air temperature (24.1°). Other years did not show strong temperature correspondence to NDVI changes.

3.2.4 Removing Noise from NDVI

Over the years of NDVI's processing, research, and development of applications, it was established that noise in AVHRR data creates fundamental constraints to the remote sensing of the Earth's vegetation, atmosphere near the ground and water. The multiple noise sources are physical, geometrical, mechanical, instrumental, mapping, environmental, and random. Some of them are long-term and some are short-term and some are both (Kogan 1987, 2001; Kogan et al. 1996; Rao and Chen 1995, 1999; Cracknell 1997; Kidwell 1997). Some sources of noise, for example, cloud creates the biggest and continual problem in using satellite data for Earth, and especially vegetation monitoring (Tarpley et al. 1984;

Cracknel 1997). As was seen in Table 3.1, clouds reflection equals amount of sunlight in both VIS and NIR channels, making NDVI values near zero. If not for such considerations, the situation of the earth's ground, in the presence of clouds in vegetative areas, can be assessed erroneously as being without vegetation.

Therefore, the weekly composite technique has been proposed and successfully applied over the years. The essence of this simple method was to use, instead of daily NDVI, several-day composite NDVI values considering that during a few-day period 1 or 2 days might be found without or with less clouds. Besides clouds, in daily data, some pixel measurements might be missing due to problems with the position of the satellite and sensor temporary malfunctioning. NDVI compositing technique replaced pixels with no NDVI data with the pixels with the data from other days. From the standpoint of physiological aspects of vegetation monitoring, such an approach is fairly grounded since vegetation greenness changes very little during a few-day period. Seven-day composite NDVI values (saving the pixels for the day with the highest NDVI during 7-day) were selected as an optimal time-based approach. Figure 3.6, showing sixteen 4 by 4 km resolution pixels, demonstrates the work of this technique. As seen on day one (upper row) 5 out of 16 pixels are without data (white pixels) and the rest have much lower NDVI (probably due to some clouds and aerosols) compared to it and should be in the middle of the growing season in Illinois. The first day comes to 1-day composite NDVI (the lower row), as is. After the second day was overlaid, the first-day's missing values were filled and higher second-day NDVI values replaced those in the first-day with the 2-day composite. By the end of the seventh day, all missing pixels had some values and NDVI was the highest in the 7-day composite (lower row). The same NDVI-based 7-day composite procedure was also applied to brightness temperature data. The compositing procedure begins on the first day of a year and each 7-day period must have at least 4 days in the same week and year (if less than a week, data will be considered as missing). The output is a single file for each processing period containing metadata for each output variable, sensor and solar zenith angle, relative azimuth angle and Ch1, Ch2, Ch4 and Ch5 counts in the Hierarchical Data Format (HDF) similar to CLAVR-x output.

Figure 3.7 demonstrates Illinois's (USA) area mean NDVI and BT weekly composite time series during 2014. As seen, both indices clearly show the Northern Hemisphere seasonal signal when NDVI and BT are low in winter (January and December and around) and high in summer (June–August). At the same time, there are a lot of short-time (from week to week) variations in both indices, especially in BT. For example, in July 2014, NDVI for week 28 was higher (NDVI = 0.48) than for weeks 27 (0.43) and 29

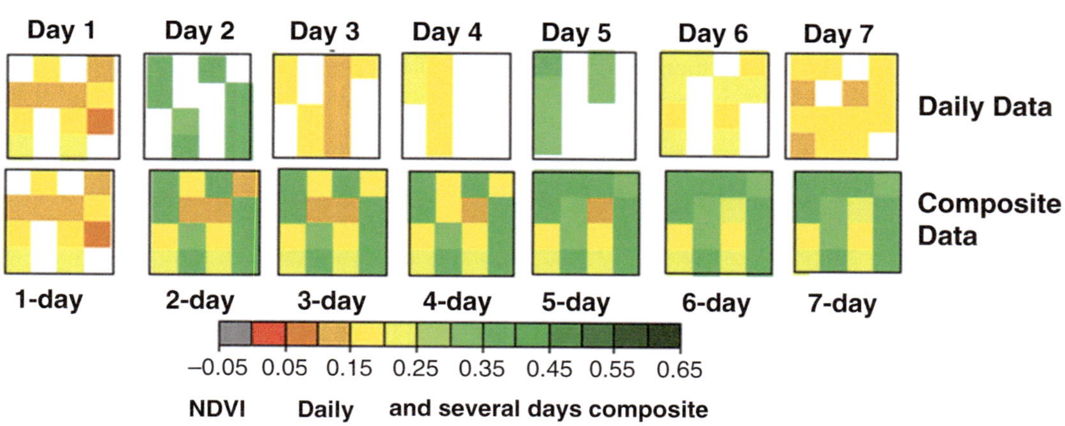

Fig. 3.6 NDVI daily and 7-day composite in July 21–27, 1987 (central Illinois, USA, 4 by 4 km pixels)

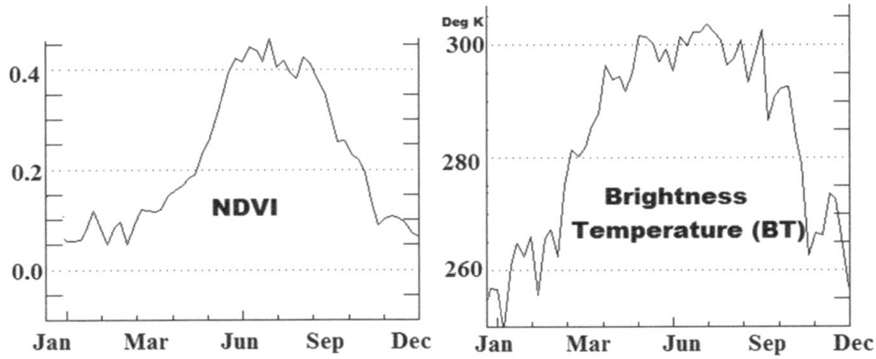

Fig. 3.7 Weekly composite NDVI and brightness temperature (BT) weekly time series during 2014 (central Illinois, USA, 4 by 4 km pixel)

(0.41). If not to consider these fluctuations as noise, vegetation conditions might be assessed erroneously as an improvement from week 27 to 28 and deterioration from week 28 to 29. A similar situation is observed with BT in August, when in a 3-week period, BT jumped from 293°K in week 34 to 306°K in 35 and dropped to 287°K in week 36. Such variations in NDVI and BT must be removed in order to use these indices for monitoring the vegetation state and weather impacts on its changes.

Unfortunately, at the background of good cloud detection methods (Heidinger and Pavolonis 2005), there are no methods of complete cloud clearing. The indicated problems definitely affect the indices, which must be corrected in order to use them for Earth monitoring. Meanwhile, there are no techniques to correct these and other noise problems, as shown in Fig. 3.7. Therefore, vegetation-oriented method for a comprehensive noise reduction was developed using a statistical approximation of the vegetation and temperature annual time series. The idea was to (1) single out the seasonal cycle, (2) suppress high-frequency noise (weekly fluctuations), and (3) enhance medium- and low-frequency variations related to persistent weather fluctuations. This technique considers smoothing the weekly time series with a combination of a compound median filter and the least squares technique (Kogan 1987, 1989; Kogan and Sullivan 1993). Numerous tests showed that this smoothing completely eliminated the high-frequency outliers, including random, approximated accurately the annual NDVI and BT cycles, and, more importantly, singled out medium-frequency weather-related fluctuations (valleys and hills in the NDVI and BT time series) during the annual cycle (Kogan 1995a, b).

The compound median filter and the least squares statistical technique (Kogan and Sullivan 1993) is based on such assumptions: (1) NDVI and BT time series clearly show seasonal cycle, (2) small and large jumps in NDVI between neighboring data points indicate unrealistic values, and (3) existence of persistent weather fluctuations. A short essence of the technique is the following: if the measured, raw data are labeled as $R_1, R_2, R_3, \ldots, R_N$ and the smoothed as $S_1, S_2, S_3, \ldots, S_N$, a simple measure for the cumulative difference in height between two consecutive smoothed data points is the following SUM:

$$\sum(S_{n+1} - S_n)^2, \text{ where } n \text{ changes from 1 to } N \quad (3.1)$$

The difference between smoothed and R data values can be estimated as:

$$\sum(S_n - R_n)^2, \text{ where } n \text{ changes from 1 to } N. \quad (3.2)$$

These SUMs act in opposite directions. If Sum (3.1) is small, then $S_n \approx S_{n+1}$ and Sum (3.2) is large. If Sum (3.2) is small, then $S_n \approx R_n$ and Sum (3.1) is large.

A good compromise is to make both sums as small as possible, at the same time. In other words, $S_1, S_2, S_3, \ldots, S_N$ are selected in a way that the compound SUM is minimal:

$$\mathrm{SUM} = \sum (S_{n+1} - S_n)^2 + W \times \sum (S_n - R_n)^2 = \text{minimum}$$

The relative weight W is a positive number that the data user chooses. If smoothness ($S_n \approx S_{n+1}$) is the most important, which is the main condition for NDVI and BT filtering, then W is near 0.

Figure 3.8 demonstrates Illinois's areal mean raw (original) and smooth NDVI and BT derived from AVHRR and VIIRS during 2012–2014. Smoothed data are called SMN (Smoothed NDVI) and SMT (Smoothed temperature, although it would be more correct to name brightness temperature). Following the mentioned assumptions, the produced SMN and SMT clearly emphasize a seasonal cycle (compatible from year-to-year), remove noise-related high-frequency jumps [very strong for BT (April 2013)], and enhance weather-related changes (at the background of jumps from week-to-week). The last statement is supported by example of March 2012 AVHRR-based BT (Fig. 3.8) showing that after a strong BT drop and its following weather-related increase, SMT shows an increase. From other examples, it is important to indicate smooth winter temperature changes in NDVI and BT, enhanced NDVI in November 2014 and others.

In addition to high-frequency cloud noise, a very strong long-term NDVI and BT distortions are developed due to heavy aerosol in the atmosphere (both troposphere and stratosphere) after volcano eruptions. Such a situation, during the 38-year period, has developed following eruption of El Chichon (in Mexico, 17.36° N, 93.22° W) in April 1982 and Pinatubo (in the Philippines, 15.13° N, 120.35° E) in mid-June 1991 (Kogan et al. 1996; Vargas et al. 2009). The heaviest aerosol concentration in the stratosphere is normally developed in a few weeks after the eruption, when the injected staff encircled the Earth with the major airflow. As seen in Fig. 3.9, the Pinatubo aerosol was distributed between 30–20° N and 20–25° S and El Chichon—between 30–20° N and 5–15° S. Both volcanoes emitted massive clouds of ash and sulfur dioxide (SO_2) into the atmosphere. The SO_2 reacted with water vapor in the atmosphere to form sulfuric acid particles, a known form of aerosol. The aerosols were rapidly distributed around the globe by stratospheric winds, but at different times depending on the seasonal circulation patterns. The maximum concentration of aerosol particles was confined to tropical latitudes causing an increase in the aerosol optical thickness by more than one order of magnitude (Fig. 3.9). The aerosol cloud from these two volcanic eruptions remained in the atmosphere for several years (Kogan et al. 1996).

Compared to El Chichon, the area of the Pinatubo aerosol was larger and the aerosol was heavier. Besides, the aerosol stayed in those areas for several years reducing NDVI and BT values. Pinatubo example in Fig. 3.10 clearly indicates that in October 1991, equatorial Africa has reduced NDVI (less greenness) compared to the

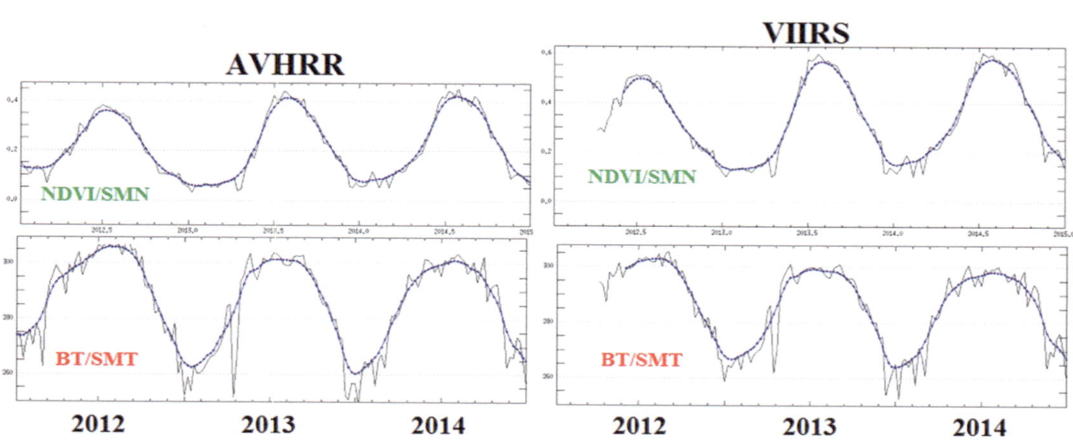

Fig. 3.8 Illinois's areal mean raw (original) and smooth NDVI and BT derived from AVHRR and VIIRS during 2012–2014

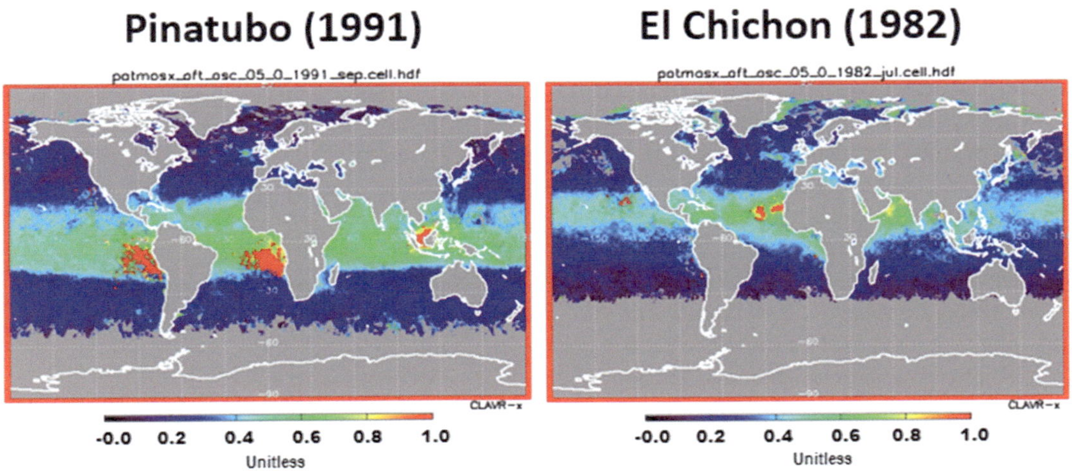

Fig. 3.9 Stratospheric aerosol from volcanoes: Mount Pinatubo, 1991 and El Chichon, 1982

Fig. 3.10 Equatorial Africa NDVI in October 1989, 1990, 1991, and 1992

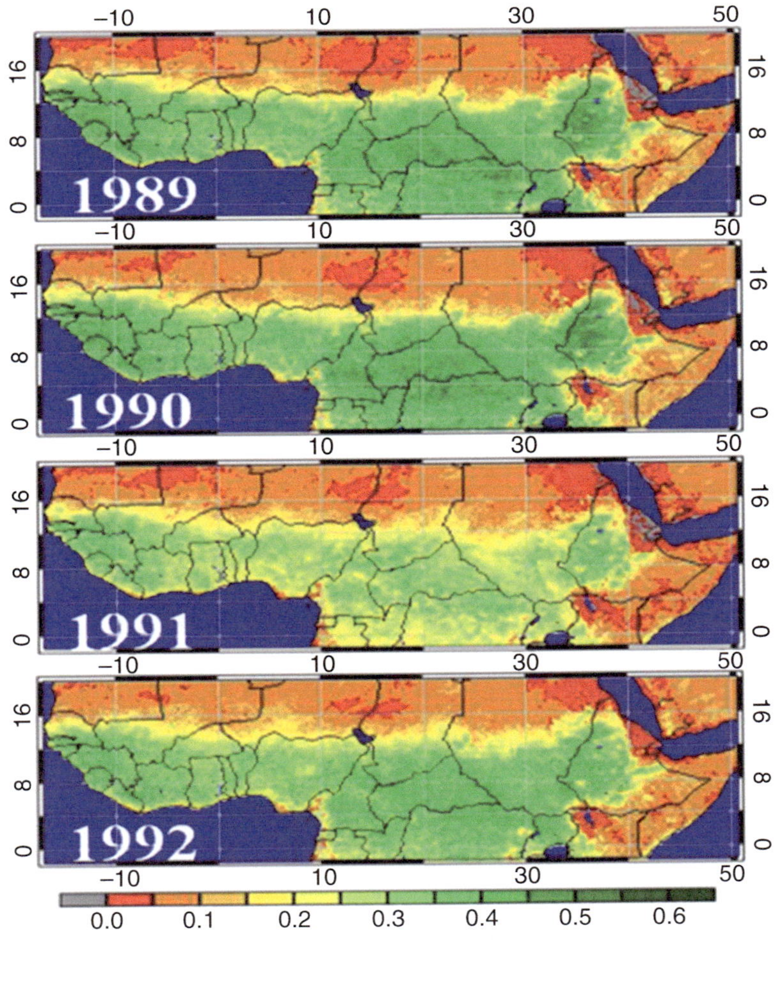

NDVI in 1990 and 1989 (before eruption) and even in 1992, 1.5 year after Pinatubo eruption, although the recovery was not complete (greenness slightly less than in pre-Pinatubo years). NDVI and BT attenuation by the volcanoes' aerosol must be corrected before their use. Post-factum estimates indicated that NDVI in the aerosol-affected area decreased by half during the second part of 1991 and through 1992 (Kogan et al. 1996; Vargas et al. 2009). Similarly, NDVI was reduced in the world's tropical latitudes from April 1982 for 2–5 years after El Chichon eruption in Mexico (Vargas et al. 2009).

A few attempts were made in the past to correct the effect of volcanic aerosols on AVHRR-derived NDVI. However, in situ data has not confirmed the restoration of NDVI. Due to the absence of validation, considerable difficulties with adequate aerosol observations, quick aerosol change, no method for BT restoration and low accuracy of its assessment, some researchers simply removed the pixels that were most likely contaminated by volcanic aerosol. Therefore, we were looking for a reliable method of NDVI and BT correction.

Such method is needed because besides clouds and heavy aerosol problem, NDVI and BT corrections are required due to many other noise-created problems, such as satellite orbital drift, sensor degradation, satellite change, difference in response function, delays in equator crossing time, differences between AVHRR sensors, sensor change, and others (Kogan et al. 2003, 2017). These problems are indicated in Table 3.2. Satellite orbital drift and AVHRR sensor degradation create long-term noise in channels and two NDVI and BT indices. As known, operational satellites are developed to work in space for 5–6 years each (Kidwell 1995, 1997; Cracknell 1997). According to Fig. 3.11, the afternoon satellites have been normally launched to cross the equator around 14:00 (afternoon) time. However, after 2–3 years of each satellite in space, its orbit drifts resulted in later equator crossing time (ECT), which is continuing with different speed for the operational afternoon satellites (Fig. 3.11a). Following the drift, the sensor observes the Earth later and later making observation under lower sun angel compare to the time of their launch. This will result in NDVI and BT reduction in late afternoon and their values to be incompatible with the values in earlier time.

Table 3.2 Noise in NOAA/AVHRR operational satellite data (Gitelson et al. 1998; James and Kalluri 1994; Kogan et al. 1996; Vargas et al. 2009)

Frequency	Noise type
Short-term	Clouds
	Water vapor
	Aerosol
	Sun glint
	Sun-surface-sensor geometry
	Atmosphere path radiance
	Bidirectional reflectance
	Rayleigh and ozone absorption
	Atmospheric resistance
	Open canopy background
	Human errors
Long-term	Volcanoes' aerosol
	Sensor change
	Satellite change
	Satellite orbit degradation
	Satellite malfunctioning
	Response function differences
	Sensor deterioration
	Calibration changes

For example, to investigate a lower NDVI level for NOAA-7 and 9 (1981–1989) compared to the later satellites, a half an hour data were collected from the Spinning Enhanced Visible and Infrared Imager (SEVIRI) on the Meteosat Second Generation (MSG-2) European satellites to investigate NDVI dynamics in relation to the local time of observations. The results indicated that the bias developed because the satellite's equator crossing time was one and a half hour later than for the following satellites. SEVIRI's NDVI analysis for the 2009 indicated that NDVI drops between 6% and 11% if observation local time moves from 13:00 to 14:30. This problem creates some difference in channels, NDVI and BT values not only with earlier observations but also with observations with other NOAA operational satellites, affecting strongly NDVI and BT time series and creating problems for their use in climate and land cover change studies and applications. Following the ECT problem NDVI and BT reduction were observed on the data collected from NOAA-9 during 1987 and 1988, NOAA-

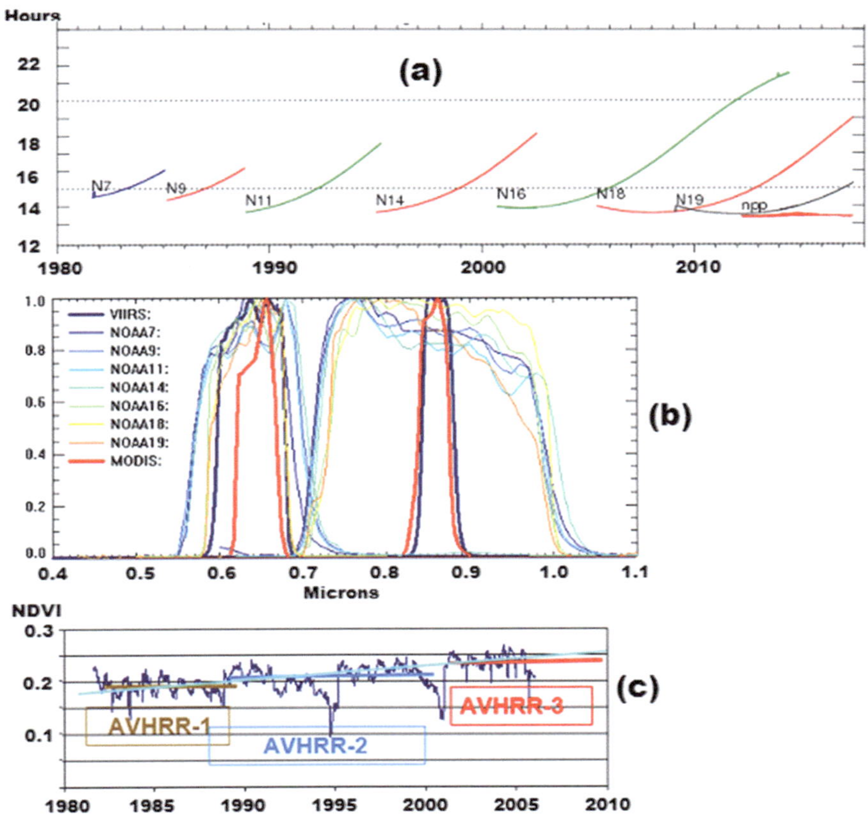

Fig. 3.11 NOAA afternoon operational satellites: (**a**) equator crossing local time, (**b**) satellite spectral response function for VIS and NIR, and (**c**) NDVI time series during 1980–2006 for three AVHRR sensors

11 in 1993 and 1994, and NOAA-14 in 1999 and 2000 (Fig. 3.11a). This problem strongly affects Ch4 measurements and calculated BT, especially in late afternoon due to a reduction in the maximum daily temperature. Following the ETC problem, we limited the use of NOAA-7 through NOAA-19 satellite data by the 15:00 Equator crossing time.

The other two examples in Fig. 3.11b, c show some differences in spectral response function (SRF) of VIS and NIR channels between the same AVHRR sensors on different NOAA operational satellites (Fig. 3.11b). SRF indicates the area under the curves covering the range of micrometer measurements for each channel (VIS, NIR and IR) of the sensor (Trishchenko et al. 2002). Although most of SRF's areas are close to each other, some differences are well seen. For example, SRF from AVHRR sensor on NOAA-18 and -19 (yellow) are narrower than on previous satellites. This results in slightly higher NDVI values for the last two NOAA afternoon satellites compared to others. The larges differences in NDVI is observed between old, AVHRR and new, VIIRS (blue) generation of the sensors [including scientific MODIS (red)]. Developing the first generation of NOAA afternoon polar-orbiting satellites for the period 1980–2020, three AVHRR sensors were constructed, AVHRR-1, AVHRR-2, and AVHRR-3. Although the sensors were the same, NDVI time series values, as seen from Fig. 3.11c, are increasing from AVHRR-1 to AVHRR-3. In addition to examples in Fig. 3.11, some difference between NDVI calculated from NOAA-16 and -18 and from NOAA-14 and -11 satellites reflectance existed during 2000–2010. This occurred because the AVHRR-3 instrument has slightly different characteristics than AVHRR-2 and AVHRR-1. Specifically, the area under SRF curve for the AVHRR-3 was 8–12%

narrower than for the AVHRR-1 and -2. As the result, the NDVI from NOAA-16 and 18 turned out to be higher than for the previous satellites. Moreover, in addition to NDVI reduction for NOAA-7 and 9 due to stratospheric aerosol and sever orbit degradation, general level of NDVI for the 1981–1988 period was 10–15% smaller than for the next two satellites because the time of the equator crossing for the first two satellites was almost 1 h later. As the result of these problems and also orbit-degradation errors, large jumps in NDVI were observed between the end of previous and the beginning of the next satellites. The IR channels are also affected by stratospheric aerosol and orbit degradation (although less than the VIS) and must be corrected.

Since there is currently no technique to completely remove noise from volcanoes' heavy stratospheric aerosol and other noise shown in Table 3.2 and Fig. 3.11, a numerical investigation was performed to estimate NDVI distortion. The idea was to bypass possible corrections to VIS and NIR channels and correct NDVI directly. A simple approach consists of (1) developing a benchmark NDVI for the years not affected by heavy volcanos' aerosol and other noise described in Table 3.2, Figs. 3.9 and 3.10 (especially, satellite orbital drift) during the work of NOAA-9, -11, and -14 satellites and (2) compare distorted with the benchmark NDVI. The benchmark NDVI statistical sample was composited from the following 5 years of NDVI data: 1989, 1990, 1995, 1996, and 1997. Mean 5-year value of NDVI was calculated for each week and each 4 km^2 resolution global land pixel. An 188,032 sample of benchmark NDVI were developed, one for each of the 3616 latitude lines and each of the 52 weeks of the year. A comparison of noise-affected NDVI (NDVI_a) with a benchmark NDVI (NDVI_b) was performed by taking a simple difference between them. Since noise affects NDVI during the nearly 30 years, the (NDVI_b – NDVI_a) was derived for 1981–2007 for each week and 4 km^2 land pixel of latitude–longitude directions.

Figure 3.12 presents a numerical investigation of changes in pixel-level NDVI_a for each year compared to the benchmark NDVI_b (without the years of heavy aerosol and other noise). The figure shows a numerical image of pixel-level

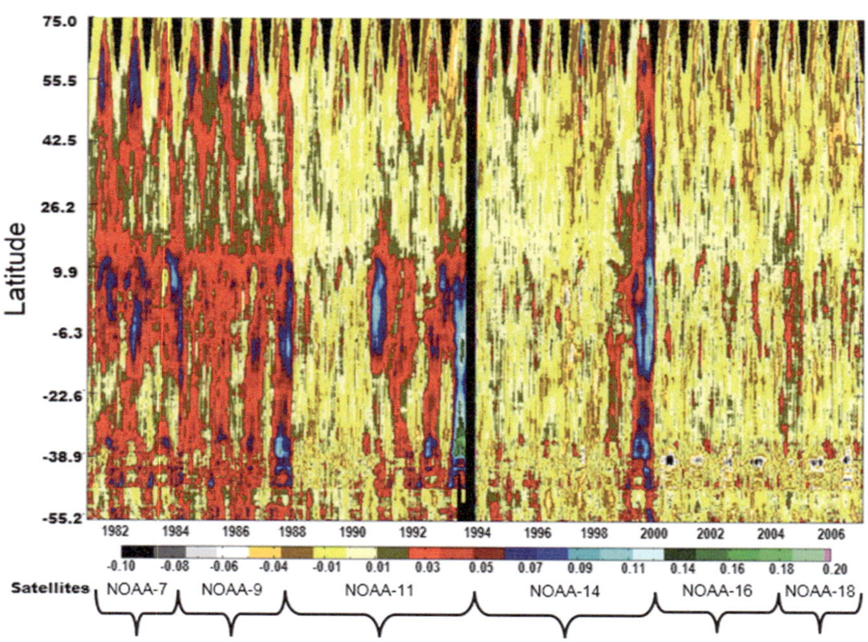

Fig. 3.12 Pixel-level (4 km^2 resolution) difference between distorted NDVI (NDVI_a) and the benchmark NDVI (NDVI_b) during 1982–2007

difference between these weekly values (NDVI_b – NDVI_a) for the world land. The analysis showed that if NDVI_b exceeds NDVI_a more than 0.01, then the pixel value is contaminated with aerosol and other sources of noise. The (NDVI_b – NDVI_a) for each week and 4 km^2 land pixel shown in Fig. 3.12 confirms that a very strong NDVI distortion is observed in equatorial zone (17° N - 15° S) during 1982–1983 and 1991–1992 (red-blue). Aerosol from El Chichon distorted NDVI for 3 to 4 years longer than from Pinatubo (only 2-year). Moreover, during a 6-year period, the El Chichon NDVI distortion was observed farther north, up to 56° N and up to 55° S, although, north of 20° N and south of 15° S, some areas were not distorted. As seen in Fig. 3.12, besides aerosol-based distortion, NDVI is distorted ((NDVI_a – NDVI_b) < 0) due to a satellite change from NOAA-7 to -9 (1985), from NOAA-9 to -11 (1988), and from NOAA-14 to -16, due to the loss of a new satellite in 1993 and data substitute with morning satellite. As seen in the figure, there are a few other unexplained NDVI distortions. The Fig. 3.12 analysis indicates that multiple corrections are needed in order to use the indices for vegetation monitoring. Since the corrections for all noises are not available, the Empirical Distribution Function (EDF), a normalization statistical approach, was applied to weekly 4 km^2 global land's NDVI (Crosby et al. 1996; Kogan et al. 1996; Vargas et al. 2009).

The EDF technique was used to statistically remove the negative bias from NDVI caused by multiple noises. This technique has been applied before for calibration of Geostationary Operational Environmental Satellites (GOES) imager channels, to remove striping and banding from satellite imagery, to perform inter-satellite calibration and to adjust for long trends in multi-year satellite data (Crosby et al. 1996). An EDF is a cumulative distribution function that gives the probability that a random variable X is less than a given value x *or* $[F(x) = Pr\{X < x\}]$. The application of the EDF technique to reduce the negative effect of volcanic aerosols and other noises on NDVI is based on the assumption that for large areas, the NDVI reduction due to technical and external forces (orbital drift, volcanic eruptions, etc.) is larger than the weather-related NDVI changes from year to year. Following this assumption, large changes in NDVI can signal unexpected disturbances due to non-weather noises. This negative effect can be reduced by adjusting the affected NDVI's EDF with a benchmark NDVI's EDF from the years, which were not affected by external noises.

Brightness temperature (BT) is also reduced (not shown) in the areas of volcano-based aerosol and from other noises (Table 3.2), especially such as satellite orbital drift, sensor degradation and change, land anisotropy, and others. Unfortunately, in addition to noise-related errors, BT also has frequent weather-related fluctuations (more than NDVI), which can be erroneously interpreted as noise. Therefore, noise-detected and noise-applied corrections developed for NDVI were also used to correct BT. Finally, following NDVI noise analysis and correction development, several conclusions can be made: (1) in 2–4 years, after the two volcanoes' eruption, NDVI and BT were considerably reduced; (2) in the equatorial zone, this reduction was the strongest; (3) in addition to the equatorial zone, some areas to the north and south of that zone had reduced NDVI and BT values as well, indicating aerosol penetration and noise development, and (4) some areas outside of aerosol influence have also had reduced NDVI and BT (1999, 1985–1988 and other), indicating other causes of indices distortion.

The NDVI's EDF-based normalization procedure was applied to correct all noise-related NDVI reduction. Figure 3.13a displays the normalization procedure. In this example, for the NDVI-affected pixel value of 0.2 (X_1) the corresponding EDF-corrected value $P(X_1) = 0.6$ was found. The intersection of line $P(X_1) = P(X_2)$ with the benchmark NDVI's EDF provides a normalized NDVI pixel value of 0.29 (X_2) for the pixels of the affected NDVI. This normalization procedure was applied to each pixel in the latitude line being processed. If the difference between the corresponding actual NDVI (NDVI_a) and benchmark NDVI (NDVI_b) is greater than the preestablished NDVI threshold of 0.01, then the

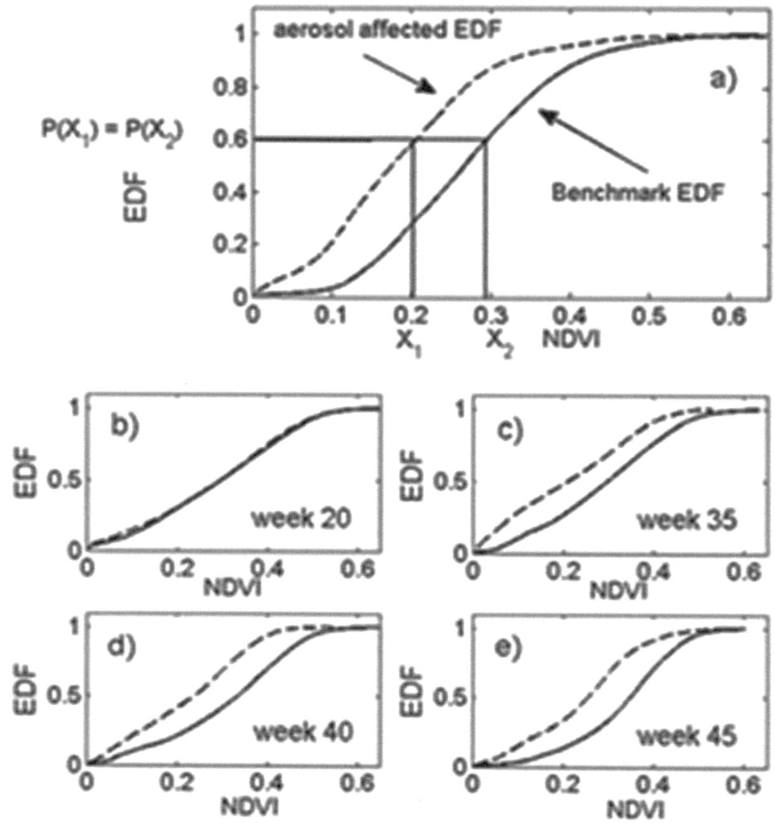

Fig. 3.13 Aerosol-affected (dashed line) and benchmark (solid line) NDVI's EDF: Principles of NDVI EDF normalization (**a**); Real Pinatubo1991 NDVI EDF data for latitude band 7° N (**b**, **c**): week 20 (May) pre-eruption time (**b**), weeks 35 (August) post-eruption (**c**), weeks 40 (October) post-eruption (**d**), and week 45 (November) post-eruption (**e**)

corresponding correction is made. Figure 3.13b–e provides real data during the pre- and post-Pinatubo weeks of 1991. In pre-Pinatubo time (week 20, Fig. 3.13b) no difference was observed between the two EDFs. This fact supports the assumption about statistical stability of the non-disturbed land surface vegetation. After the eruption, the affected NDVI's EDF shifted to the left of the benchmark NDVI's EDF (toward lower NDVI values). This difference gradually increased over time as the volcanic aerosols developed and encircled the Earth. The NDVI reduction was approximately 0.1–0.15 for the NDVI benchmark range between 0.2 and 0.45. For El Chichon aerosols the normalization was done from week 13 (April), 1982 to week 52 (December) 1984 and for Mount Pinatubo from week 26 (June) 1991 to week 52 1993.

Figure 3.14 shows latitudinal profiles of aerosol-affected and normalized NDVI of Pinatubo. As seen, for latitudes 20° N – 20° S in pre-eruption time (week 20, May) the two lines coincide. In post-eruption time (weeks 35, 40, and 45), the affected NDVI shifted left to the lower NDVI values. In central latitudes, the normalized NDVI during the build-up of Pinatubo aerosols increased up to 0.15. Approaching the north and south borders of the affected area (±20° latitude), these values decreased considerably. Similar results were obtained for El Chichon. Finally, Fig. 3.15 shows NDVI for week 40 (October) over part of the aerosol-affected South America and central Africa in pre-eruption (1989), post-eruption (1991) time and the normalized 1991 NDVI. As seen, the NDVI is higher (greener) in 1989 (Fig. 3.15a) without stratospheric aerosols than in 1991 (Fig. 3.15b) three to four months after the Mount Pinatubo eruption and aerosol has developed. The normalization procedure returns NDVI back to pre-eruption

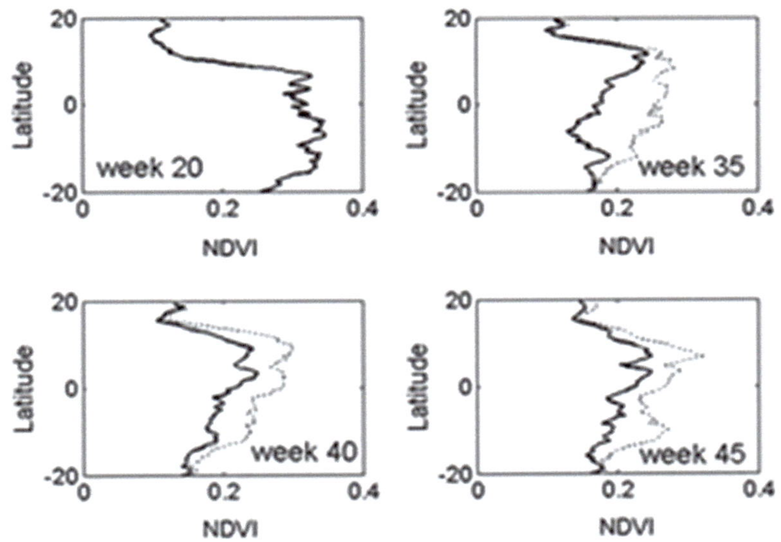

Fig. 3.14 NDVI's latitudinal (from 20° N to 20° S or −20) profiles (actual–solid line and EDF normalized—dotted line) for year 1991, week 20 pre-Pinatubo and weeks 35, 40, and 45 post-Pinatubo

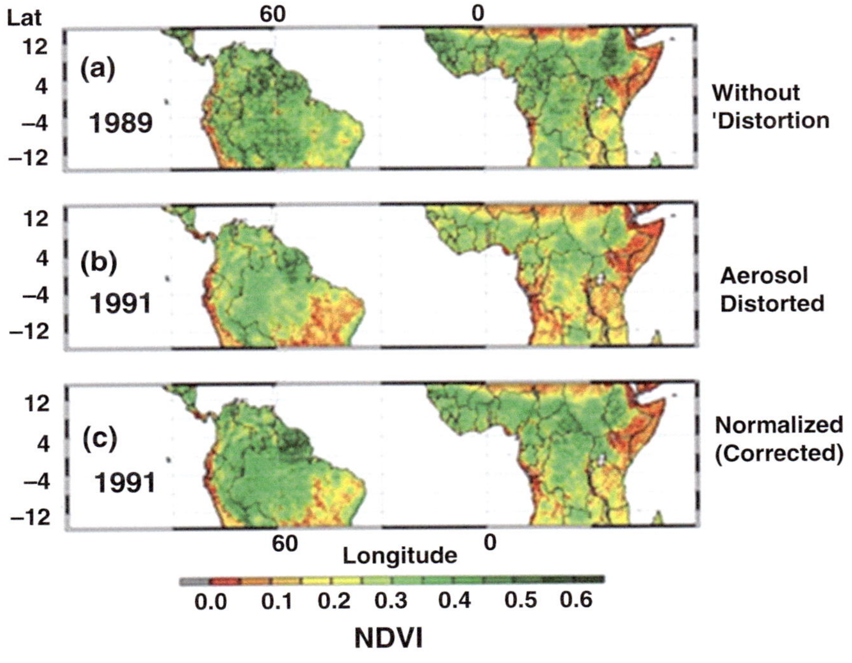

Fig. 3.15 AVHRR-based NDVI week 40 for South America and Africa in October 1989 without volcanic aerosol distortion (**a**), 1991 aerosol distorted after Mt. Pinatubo eruption in June 1991 (**b**), and 1991 normalized by EDF (**c**)

period. The normalization (Fig. 3.15c) increased NDVI considerably (by 0.03 to 0.18) and the image looks in general as green as in pre-Pinatubo time in October 1989.

Following the applied procedure, considerable improvements in the corrected NDVI are seen in Figs. 3.16 and 3.17. The first figure compares raw with the corrected long-term noise NDVI and the second one displays NDVI without short-term noise versus completely corrected (both short/long noise removed) SMN. The result of more than a dozen corrections, the noise

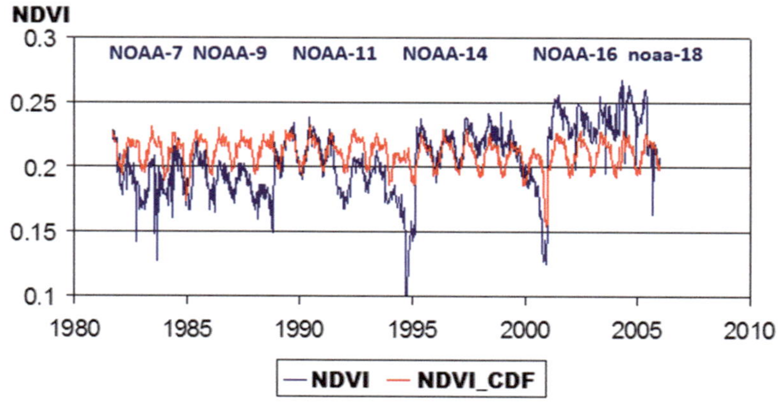

Fig. 3.16 Average NDVI (40° N to 40° S): raw [not corrected, (NDVI)] and CDF-Corrected NDVI (NDVI_CDF) during 1981–2005

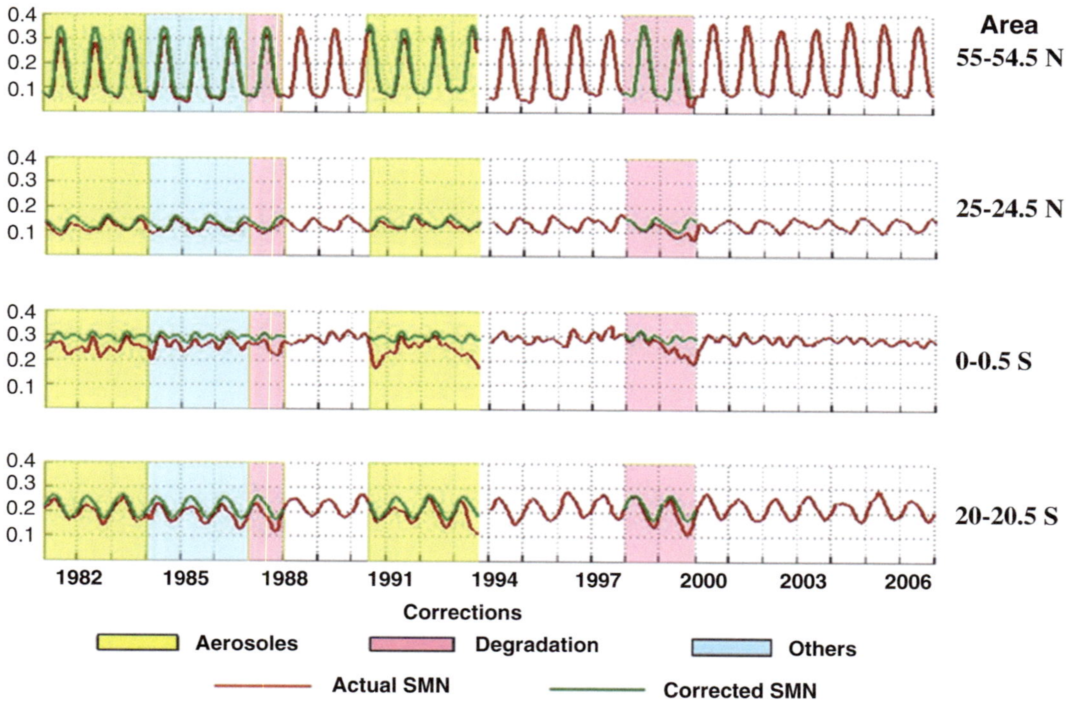

Fig. 3.17 Smoothed NDVI (SMN): not corrected SMN (red) and completely corrected SMN (green) time series for several global ecosystems during 1981–2007

was definitely removed, especially such as clouds, heavy aerosol (from the recent two strong volcanoes), satellite orbit and sensor degradation over time, difference between the same and different sensors (spectral response function and technical), satellite/sensor malfunctioning, human errors (there were found such), random and some others shown in Fig. 3.17. It is interesting that different ecosystems with high vegetation in the north as well as medium and low vegetation in grass land, tropical forest and semiarid areas shows similar results of SMN improvements. Since BT has faster and stronger fluctuations in response to weather changes, the NDVI-developed procedure was applied to remove similar BT distortions for each 4 km^2 (in both latitude/longitude directions) of the 37-year of the NOAA/AVHRR operational data.

Finally, the image of completely corrected NOAA/AVHRR smoothed NDVI (SMN) and smoothed BT (SMT) for July 2017 is shown in Fig. 3.18. First, what is seen is right global ecosystem distribution for both parameters resulting from all corrections. In Africa, it is low SMN (0.01–0.1) and high SMT (40–50 °C) in desert, clear transition to semi-desert/grassland and tropical forest (SMN = 0.20–0.35 and 0.35–0.55, respectively), especially with temperature in grassland (25–30 °C) and tropical forest (15–20 °C). Moreover, since Africa has a food security problem and the world is concerned about this situation, it is important to emphasize that the major areas of agriculture (Sub-Sahara and southern Africa) have reduced SMN (0.12–0.17) and elevated SMT (25–35) compared to other world agricultural areas. This also indicates that weather might produce more severe and reoccurring drought conditions reducing agricultural production. Therefore, operational satellite efforts in monitoring droughts and food security in Africa should be more intensive. In addition, the figure also shows that the highest world SMN (0.50–0.70) is observed in the best agricultural areas of eastern USA, southern Canada, the black forest zone of Russia, and the tropical zone of South America. In addition to a very clear identification of ecosystems, it is also important to notice a very smooth SMN and SMT change over space, which is not typical for raw NDVI and BT (Kogan 1997, 2001).

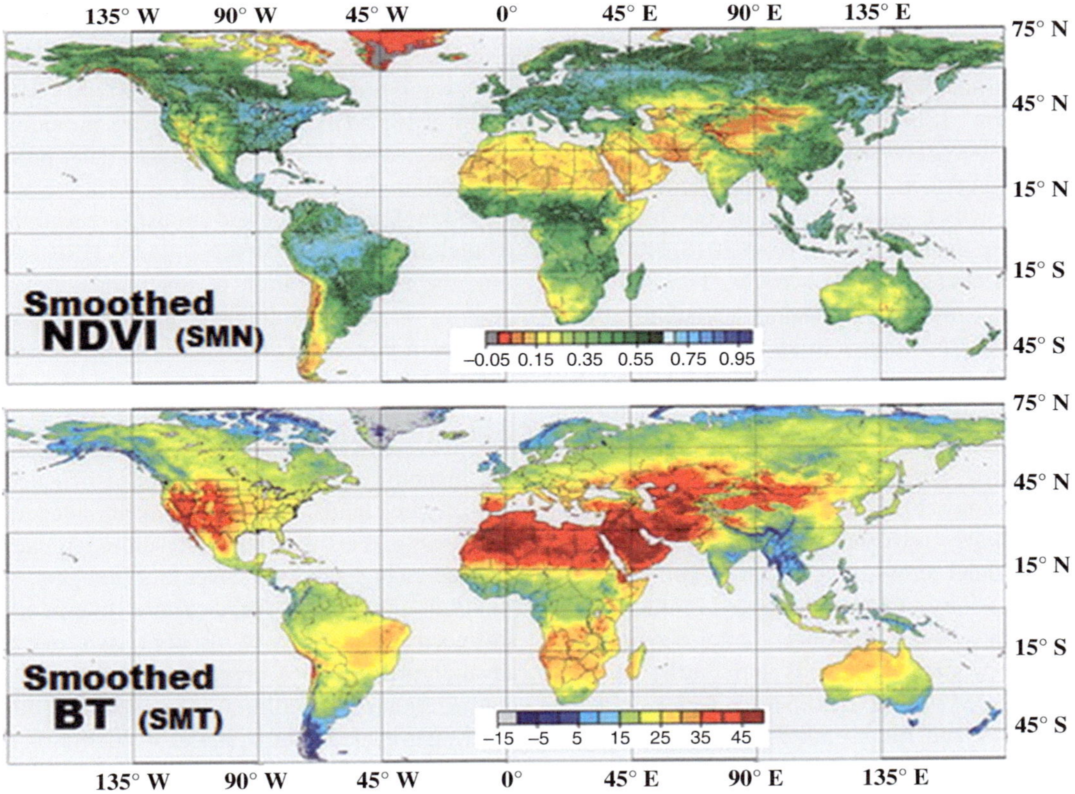

Fig. 3.18 Global image of complete noise-corrected NOAA/AVHRR-based smoothed NDVI (SMN) and smoothed BT (SMT, °C) for 8–14 July 2013

3.2.5 VIIRS Data for Vegetation Monitoring

Declining stock of natural resources, climate warming, intensification of weather disasters, and land cover changes, in combination with rapid population growth, have raised concerns about environmental degradation, future food supply/demand, and global food security. In order to address these issues, the world community needs reliable, real-time climate, land and weather monitoring system, provided by both high-resolution satellite and low-resolution in situ observations and in a timely manner. In the recent two decades, the role of satellite observations for climate services increased considerably since they provided more comprehensive, reliable, and high-resolution information, especially with the introduction of the new generation of operational satellites. The new operational polar-orbiting satellite system, the Suomi NPOESS (National Polar-orbiting Operational Satellite System) Preparatory Platform (SNPP) SNPP was launched in November 2011 and has five sensors on board, signifying a new era of environmental space crafts, which will continue as the Joint Polar Satellite System (JPSS), from 2018, NOAA-20, for the next two-three decades. The SNPP five sensors on board include Visible-Infrared Imaging Radiometer Suite (VIIRS) to observe ocean, atmosphere, and land, Cross-track Infrared Sounder (CrIS) for three-dimensional measurements of atmospheric temperature, air pressure, and moisture profiles, Advanced Technology Microwave Sounder (ATMS) for microwave retrieval of atmospheric events, 22-channel cross-track scanner for sounding observations; Ozone Mapper Profiler Suite (OMPS), to measure total-ozone and ozone-profile and Clouds and Earth's Radiant Energy System (CERES) to measure solar-reflected and Earth-emitted radiation from the top of the atmosphere to the Earth's surface. The JPSS-1 launched in 2017 with VIIRS, CrIS, ATMS, OMPS, CERES, and JPSS-2 will be launched during 2023–2025 with VIIRS, CrIS, ATMS, and OMPS (JPSS 2014).

The Visible-Infrared Imaging Radiometer Suite (VIIRS) designed for observing the Earth environment operationally includes vegetation monitoring. VIIRS accommodates the best technical and scientific features of its predecessors (including AVHRR) and has new features such as a wider swath, higher resolution, a sharper view at the swath's edge, faster data processing and availability, advanced climate/weather impacts monitoring, and other attributes. The new features will provide advanced land and economic applications following the experience with NOAA/AVHRR performance, applications and the new contemporary requirements from climate, economy and humanity (JPSS 2014). VIIRS will considerably improve environmental and socioeconomic services provided to global and regional communities. Specifically, S-NPP and JPSS will address the impacts of the changing climate and weather extremes on industries, water, energy, population health, and other resources and activities.

Currently, the S-NPP/VIIRS and J-1/VIIRS (from 2018, NOAA-20) systems are providing an advanced observation, processed data, products and predictions for each 4, 1, 0.75, 0.5 and 0.375 km^2 Earth surface and atmosphere near the ground. Similar to the first NOAA/AVHRR system, the new generation of operational polar-orbiting satellites are focused on predictions of vegetation condition and health, droughts, soil moisture, crop/pasture production, food security, fires, mosquitoes-borne diseases, etc. However, the new applications are currently developing and planning such as assessments of irrigated/non-irrigated lands, micro-assessment of vegetation stress and extreme fires, landslides, climate-induced changes in land cover and droughts, El Nino/La Nina-based predictions, one-month advanced predictions of drought, two-month advance prediction of crop/pasture production, optimization of agricultural insurance, and others (Kogan et al. 2013a, b, 2015a, b, 2016; Kogan and Guo 2014, 2015, 2016; Hoerling et al. 2014; Kogan 1995a, b, 1987, 1989, 2001, 2002, 2006; Gitelson et al. 1998; Salazar et al. 2008; Unganai and Kogan 1998). All these VIIRS products have been widely applied by global users for high resolution monitoring vegetation health, estimation of agricultural production losses, and even assessment and prediction of food security situation.

VIIRS provides observations in 22 reflective and emissive bands, with five of them, designed as imagery bands (I-band), having the highest 375 m spatial resolution, and the rest, moderate bands (M-bands), with 750 m resolution. The M-bands are designed to produce such observations as aerosol, cloud, ocean color, snow, sea surface temperature, fire detection, and others. I-band are used to observe land surface reflectance/emission, cloud, snow/ice, normalized difference vegetation index (NDVI), vegetation health (VH) indices, and a few others (JPSS 2014; Kogan et al. 2017, 2018). For vegetation monitoring, VIIRS observes the world's land, between 180° W and 180° E and between 55° S to 75° N, every day. Since the data volume is massive, it is collected and processed by subregions (granules, total 64 for the world) with the size 45° W–E by 16°30′ N–S each. VIIRS acquires 330 Mb/min data and 2.3×10^9 grids/day. The total number of optimally processed world grid cells is approximately 36×10^6 at 500 m^2 spatial resolution. Similar to AVHRR operational sensor, the I-band data are global, weekly, and applied to each 16, 4, 1, and 0.5 km^2 of Earth land and atmosphere near the ground. One of the primary components of these services is drought detection, monitoring, prediction of drought-related agricultural losses, and assessment of food security situation. VIIRS accommodated the best technical and scientific features of its US predecessors: the three and a half decades of AVHRR sensor on NOAA polar-orbiting operational satellites and the nearly 15 years of Moderate Resolution Imaging Spectra-Radiometer (MODIS) on NASA's MODIS/AQUA scientific satellite. Similar to its predecessors, VIIRS flies in a polar, sun-synchronous orbit, collecting afternoon data to measure vegetation health, condition, greenness, and the highest daily temperature, which are very important parameters for monitoring drought and its impacts. VIIRS will also provide multiyear services vital for land cover and climate change detection. Significant VIIRS observation improvements compared to the predecessors (Table 3.2) include a 1.3 times wider swath, a sharper view of the swath edge, multiple spectral channels, higher spatial resolution, excellent radiometric features, faster data processing, and the availability of new products and services (JPSS 2014; Kuciauskas et al. 2013; Solbrig and Lee 2013). Among the numerous VIIRS improved and new products (daytime and night-time low clouds, light emission, sea surface temperature, aerosols, elevated dust, land cover, ocean color, chlorophyll concentration, moonlight, volcanic activities and ash dispersion, wildfires, snow–cloud discrimination, etc.), enhanced high-resolution drought detection and monitoring will improve the prediction of its detrimental consequences for agriculture, economics, energy, human health, recreation, transportation, and water resources management.

A brief algorithm from data collecting/reading to product development consists of the following important steps: retrieval of orbital granule data, projecting each 375 m^2 granule pixel data to a 500 m^2 grid's (set up as an optimal standard) orbit map, selecting pixels close to the center of the grid and close to nadir, filling gaps, producing daily maps for I1, I2, and I5, radiance calibration, calculation of daily normalized difference vegetation index (NDVI = $(I_2 - I_1)/(I_2 + I_1)$), conversion of I5 emission to brightness temperature (BT), weekly compositing pixel level NDVI and BT (following NDVI's maximum value composite technique), suppression of high-frequency noise described in the Sect. 3.2.1, and mapping the data. Due to only 7-year SNPP/VIIRS services, the data has not deteriorated yet. Therefore, a complete procedure for noise removal from the indices was not required, except for correction of cloud/aerosol-related NDVI and BT distortions. Following this correction SNPP/VIIRS SMN and SMT indices were calculated.

In addition to NOAA/AVHRR, NOAA's S-NPP/VIIRS, and J-1/VIIRS satellites (currently NOAA-20), the data from NASA's scientific MODIS/AQUA were also investigated since this sensor was generally similar to S-NPP/VIIRS and has a nearly 15-year history. Figure 3.19 shows a

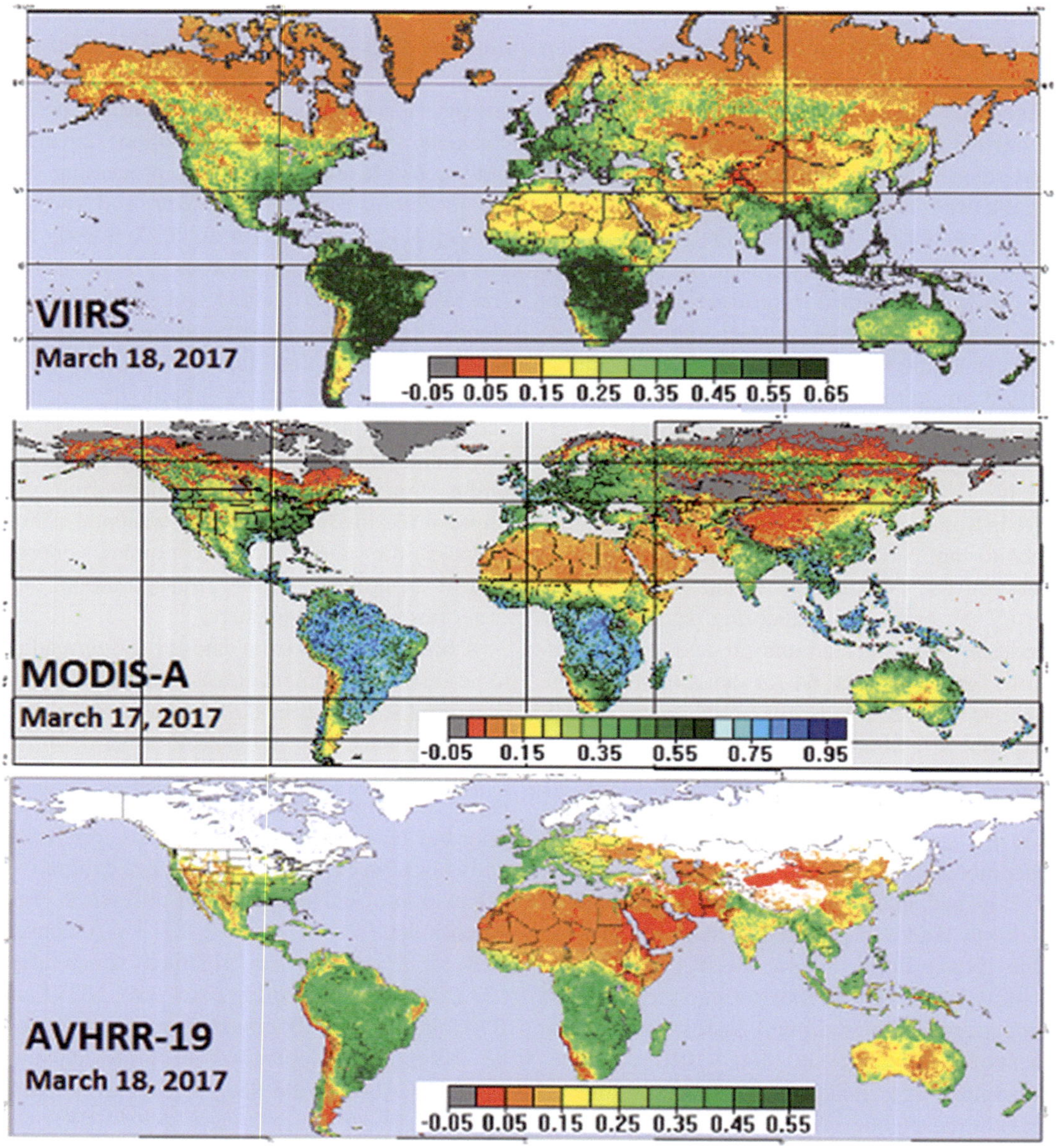

Fig. 3.19 Global Weekly-composite NDVI (with noise) for NOAA/AVHRR, SNPP/VIIRS, and MODIS/AQUA for March 18, 2017

global image of NDVI (without noise removal) from these three afternoon satellites. They show a similar vegetation distribution such as a tropical zone of Africa and South America, deserts of Africa, North America, Australia, grassland of Asia, European and North American and other ecosystems. This NDVI image clearly shows the difference between the satellite systems. AVHRR-based NDVI has the smallest values and MODIS NDVI has the largest, with VIIRS NDVI in the middle. Although the NDVIs from these threes sensors correlate with each other, combining them together might create additional error in the data resulting in erroneous environmental assessments. Therefore, following operational goals of satellite applications and also a requirement to finish NOAA/AVHRR data use transferring to S-NPP/VIIRS and J-1/VIIRS data system for the next

20–30 years, the decision was made to use a combination of operational NOAA/AVHRR with operational SNPP/VIIRS and NOAA-20/VIIRS observations and data. Since NASA's data are not operational and quite different from NOAA observations, they were not used.

3.2.6 Continuity of NOAA/AVHRR, SNPP/VIIRS and J-1/VIIRS Data Records

As mentioned above, the longest, 38-year, 4 and 16 km^2 spatial and 7-day composite temporal data records were developed from NOAA/AVHRR polar-orbiting operational satellites, S-NPP/VIIRS and NOAA-20/VIIRS. From the current satellites of this series in space, the NOAA-19 still provided observations of reflectance/emission. However, NOAA-19, launched in 2009, has almost reached its life expectancy and its AVHRR data has begun to deteriorate, requiring a complete switch to S-NPP/VIIRS, in space since 2012, which is followed by J-1/VIIRS (NOAA-20) launched in November 2017 and be followed by the next J-x/VIIRS series until the mid-2040s. Currently, blended AVHRR and VIIRS data are used to calculate regularly VH data, indices (NDVI, BT, SMN, SMT, VCI, TCI, VHI), and land-atmosphere-climate products, available at NOAA WEB (https://www.star.nesdis.noaa.gov/smcd/emb/vci/VH/vh_browse.php). The 32-year (1981-2012) NOAA/AVHRR data have been used successfully by thousands of world users. One of the principle questions for the switch to new system (considering the sensors' differences) is whether S-NPP/VIIRS 6-year and NOAA-20/VIIRS's 2018 weekly-composite NDVI and BT (main indices) data records are at the same level as NOAA/AVHRR 32-year weekly-composite records in order to continue time series of VH indices and products for correct assessment and prediction of land, air near the ground, climate conditions and food security.

Our investigation comparing simultaneous data records showed that NOAA/AVHRR NDVI and BT indices are not compatible with S-NPP/VIIRS- and NOAA-20/VIIRS-derived similar indices. As seen in Fig. 3.20a, while the spectral response function (SRF) of VIS channels is almost the same for the two sensors, NIR's SRF for AVHRR is wider compared to a very narrow SRF for VIIRS. This indicates that VIIRS is measuring a much higher reflectance in NIR, which results in a higher NDVI. The relationship shown in Fig. 3.20b indicates that the line of the strong correlation between global VIIRS/NDVI and AVHRR/NDVI is shifted up from the 1-by-1 line. Following this shift, VIIRS/NDVI is nearly 30% higher than AVHRR/NDVI. Similar results were obtained for the other narrow bandwidth sensors (Gonsamo and Chen 2013; Trishchenko et al. 2002). Following this investigation, VIIRS NDVI should be adjusted to AVHRR NDVI to keep them at the same general level and also keep continuity of the available 38-year AVHRR and VIIRS time series. In contrast to NDVI, global AVHRR/BT versus VIIRS/BT correlation line almost coincides with the 1-by-1 line, although for the 310–340°K interval, the line is slightly shifted to the higher AVHRR-measured BT values (Fig. 3.20d). These differences are also related to a dissimilarity in IR's SRF (Fig. 3.20c). The AVHRR sensor has been collecting data from two IR bandwidths: 10–11 μm (Ch 4) and 11–12 μm (Ch 5), from which only Ch 4 was used for modeling the vegetation canopy's BT, as being less contaminated by atmospheric moisture. VIIRS collects high-resolution IR data from the entire 10–12 μm spectral range, providing slightly smaller BT values because atmospheric moisture in the 11–12 μm diapason slightly attenuates NIR and BT values. To answer the question of whether VIIRS-BT should be adjusted to AVHRR-BT for maintaining continuity of data records, additional investigation confirms the need for such adjustments since based on Fig. 3.21 in variety of ecosystems (desert, crops, forest, such as broadleaf, coniferous, tropical, and combination), during 2013-2018, when correction was done, the difference between these values in summer could be as large as 1.5–4°K (difference between green (not corrected) and pink (corrected) lines); the exception is deciduous forests with a smaller difference (0–1.5°K).

NDVI and BT adjustments are also warranted by the fact that during September 2012–November 2013 period of simultaneous two-sensor observa-

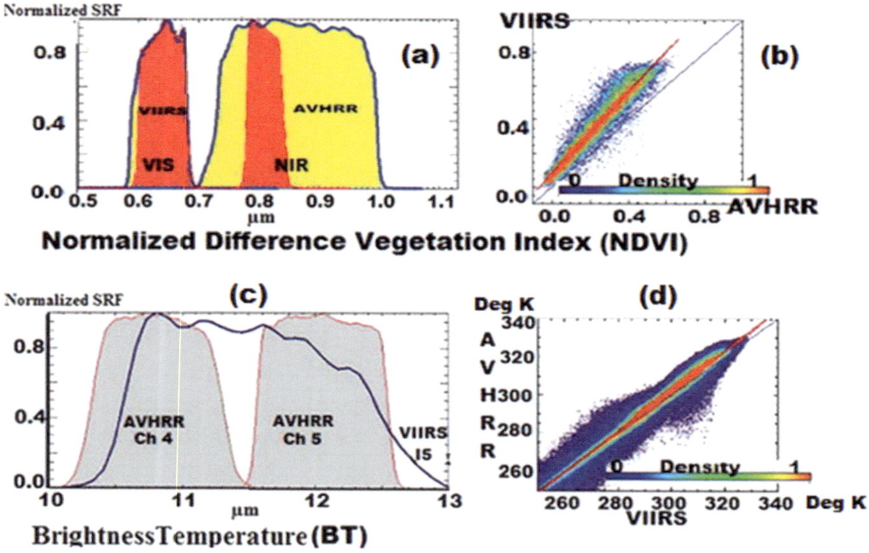

Fig. 3.20 VIIRS and AVHRR Normalized Spectral response functions (SRF) for visible (VIS), near-infrared (NIR), and infrared (IR) channels (**a** and **c**) and correlation between VIIRS and AVHRR NDVI (**b**) and brightness temperature (**d**)

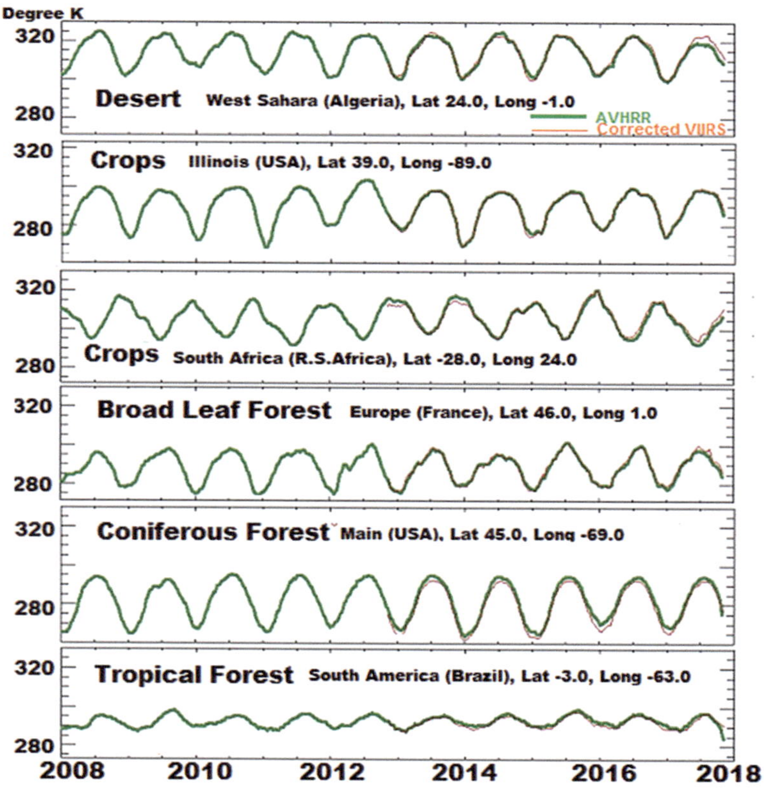

Fig. 3.21 Difference between AVHRR BT and VIIRS BT (in °K) for different ecosystems (desert, crops, and forests) during September 2008 through September 2017

tions, the correlation between VIIRS/BT with AVHRR/BT as well as for VIIRS/NDVI and AVHRR/NDVI in the principal global vegetation area (between longitude 180° W and 180° E and latitude 40° S and 40° N, containing nearly 1,000,000 of 4 km^2 pixels) was very high and stable over each week (Fig. 3.22, Pearson correlation coefficient is 0.96 for NDVI and 0.95 for BT). This is confirmed by the slope of linear regression trend, which is stable with a mean value of 0.702 for NDVI and 1.047 for BT (the intercept fluctuates depending on the season). The VIIRS to AVHRR adjustments would require a reduction in VIIRS NDVI by around 30% and an increase in VIIRS BT by approximately 0.5–4°K depending on ecosystem type.

Since the correlation shown in Fig. 3.22 is strong and stable over weeks, VIIRS indices were adjusted to the corresponding AVHRR ones' in order to keep a continuity of vegetation health time series since 2013. The results of this procedure in the form of the indices' time series are shown in Fig. 3.23. The figure demonstrates 2010–2016 global primary agricultural area mean SMN and SMT time series combining original AVHRR 2010–2012 (up to week 24) weekly records with adjusted VIIRS records from week 25, 2012 to week 49, 2016. The figure confirms the previous statement that global VIIRS/SMN (red line on Fig. 3.23a) is much higher, nearly 45% (SMN = 0.36 for the middle of the growing season) compared to AVHRR/

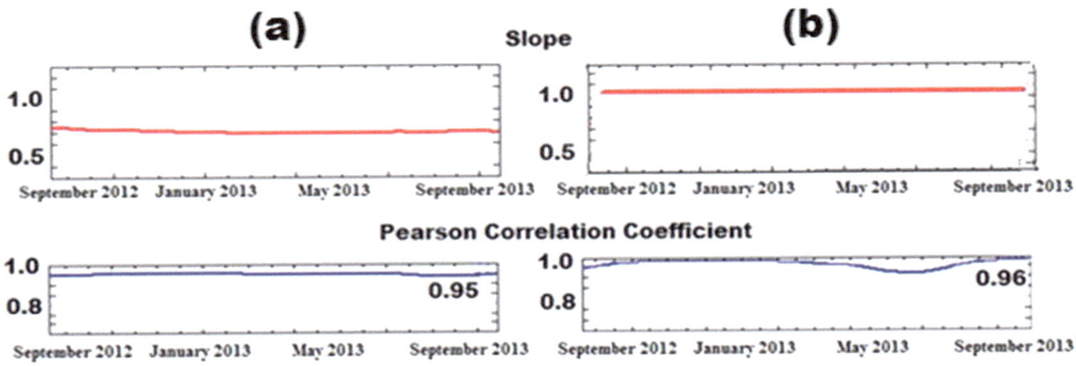

Fig. 3.22 Dynamics of slopes and Pearson correlation coefficients during the period September (week 35) 2012 to November (week 47) 2013 for linear regression of (**a**) weekly AVHRR/NDVI versus VIIRS/NDVI and (**b**) AVHRR/BT versus VIIRS/BT

Fig. 3.23 Global (180° W – 180° E and 40° S – 40° N) mean weekly SMN (**a**) and SMT (**b**) time series during 2010–2016 from AVHRR overlaid with 3.5-year (from week 21, 2012 through week 48, 2016) of VIIRS indices original and adjusted to AVHRR values

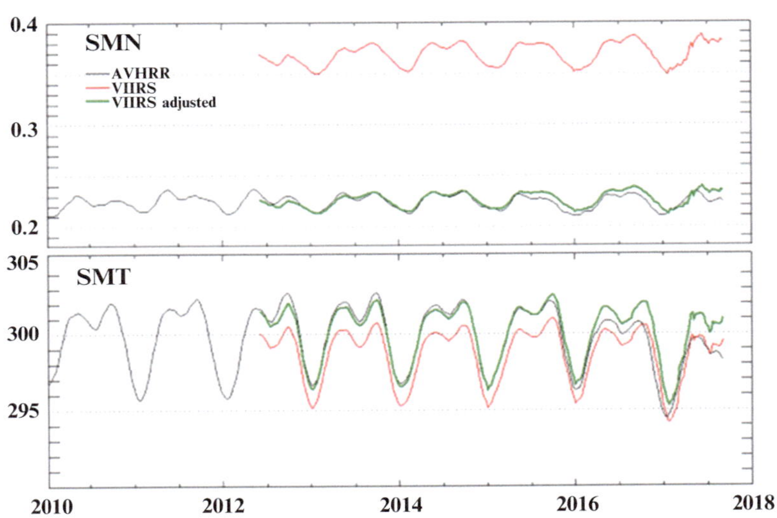

SMN (SMN = 0.25). For thermal estimates, SMT values also confirm (Fig. 3.23b) that VIIRS/SMT (red line, SMT = 300° K) is nearly 2° K lower than AVHRR (black line, SMT = 302° K). Both VIIRS parameters have been adjusted to AVHRR ones, which is shown in green on Fig. 3.23. The estimates show that the adjustments are fairly reasonable since the correlation coefficient for both indices is 0.97–0.98.

3.3 Conclusion

Since 1981, NOAA has started regular monitoring of land surface, ocean, and atmosphere from NOAA-type **operational polar-orbiting afternoon environmental satellites**, serving global community. These services for land monitoring have been continuing for the past 38 years, including present time. In 2011, the new, considerably improved satellite system, S-NPP type with VIIRS instrument on board and from the end of 2017, J-1/VIIRS system (currently NOAA-20) were launched to continue monitoring the Earth land, environment, and its impact on global economy and human activities until the mid-twenty-first century. From many sensors on these two systems, the AVHRR on NOAA satellites and VIIRS on S-NPP and on just launched NOAA-20 satellites have been used to design a considerably improved technology, called vegetation health (VH), with many new products. The VH was designed to monitor, from space, weather impacts, especially drought-related vegetation stress, to predict crop and pasture production, climate warming impacts on land cover and, based on these characteristics, to assess and even predict food security situation in the world. Vegetation health-based global data sets was developed based on these two sensors from the early 1980s and used operationally since then. However, a development of correct data sets was not a straightforward process since atmosphere is on a way of satellite beam, contaminating its signal (for example, clouds or aerosol) and climate, land cover plus weather are experiencing considerable changes over space and time, complicating the development. Besides, unexpected events (sensor malfunctioning), environmental phenomenon (volcanoes), satellite/sensor deterioration, technical (problem with computers), human, and random errors provide more problems for the data development and application. As the result of these issues, satellite data are extremely contaminated with noise, which prevent using VH data successfully for Earth monitoring and food security prediction. Therefore, this chapter describes, in detail, many problems with data and methods used for noise detection, correction, elimination of data distortion, and, what is the most important, verification if the data have been improved and are reliable to be applied. Among a very important method used to eliminate noise from satellite data, this chapter describes statistical procedures used for removing high-frequency noise related mostly to clouds, regular aerosol, volcano-based aerosol, satellites' orbit degradation, sensor deterioration and change, satellite malfunctioning and changes, etc. An important consideration is given to combining afternoon brightness temperature with NDVI and also adjusting the newer sensor data to the older sensor data in order to keep continuity of the data for climate and land cover changes prediction. In addition, this chapter explains theoretical background for using operational NOAA, S-NPP and J-1 satellite systems for vegetation monitoring from the point of view of vegetation physiology. All of these detailed descriptions have been provided that the users were absolutely sure that the data are reliable to be used for humanity.

References

ASP (American Society of Photogrammetry). 1975. *Manual of remote sensing*, 47. Falls Church VA: ASP (American Society of Photogrammetry).

Cracknell, A.P. 1997. *The advanced very high resolution radiometer*, 534. Boca Raton: Taylor & Francis.

Crosby, D.S., M.D. Goldberg, and W. Chung. 1996, Inter-satellite calibration using empirical distribution functions. *8th conference on satellite meteorology and ocean*, Atlanta, GA, American Meteorological Society, 188–190.

Gates, D.M. 1970. Physical and physiological properties of plants. In *Remote sensing with Specific Reference to Agriculture and Forestry*, 224–252. Washington: National Academy of Sciences.

Gitelson, A.A., F. Kogan, E. Zakarin, L. Spivak, and L. Lebed. 1998. Using AVHRR data for quantitative estimation of vegetation conditions: Calibration and validation. *Advances in Space Research* 22: 673–676.

Gonsamo, A., and J.M. Chen. 2013. Spectral response function comparability among 21 satellite sensors for vegetation monitoring. *Geosciences and Remote Sensing* 51: 1319–1335. https://doi.org/10.1109/TGRS.2012.2198828.

Gray, T.T., and D.G. McCrary. 1981. The environmental vegetative index: The tool potentially useful for arid land management. *Proc. 5th Conf. on Biometeorology*, Anaheim CA, 205–209.

Hashemi, S.A, and S.K. Chenani. 2004. Investigation of NDVI index in relation to chlorophyll content change and phenological event. *Recent Advances in Environment, Energy Systems and Naval Science*. http://www.wseas.us/e-library/conferences/2011/Barcelona/MNICEG/MNICEG-02.pdf.

Heidinger, A.K., and M.J. Pavolonis. 2005. Global daytime distribution of overlapping cirrus cloud from NOAA's advanced very high resolution radiometer. *Journal of Climate* 18 (22): 4772–4784.

Hoerling, M., J. Eischeid, A. Kumar, R. Leung, A. Mariotti, K. Mo, S. Schubert, and R. Seager. 2014. Causes and predictability of the 2012 great plains drought. *Bulletin of the American Meteorological Society* 95: 269–282. https://doi.org/10.1175/BAMS-D-13-00055.1.

Jacobowitz, H., L.L. Stow, G. Ohring, A. Heidinger, K. Knapp, N. Nalli. 2003. The advanced very high resolution radiometer PATHFINDER atmosphere (PATMOS) climate data set: A resource for climate research. *Bulletin of the American Meteorological Society* 785–793.

James, M.E., and S.N. Kalluri. 1994. The Pathfinder AVHRR land data set: An improved coarse resolution data set for terrestrial monitoring. *International Journal of Remote Sensing* 15: 3347–3363.

JPSS. 2014. Joint polar satellite system. http://www.jpss.noaa.gov.

Kidwell, K.B., ed. 1995. *NOAA polar orbiter data users guide*. Camp Springs, MD: National Oceanic and Atmospheric Administration, National Environmental Satellite Data and Information Services, National Climatic Data Center.

———., ed. 1997. *Global vegetation index user's guide*. Camp Springs, MD: National Oceanic and Atmospheric Administration, National Environmental Satellite Data and Information Services, National Climatic Data Center.

Kogan, F.N. 1987. Vegetation health index for areal analysis of NDVI in monitoring crop conditions. *Preprint 18th Conference on Agricultural and Forest Meteorology* AMS, Boston, 103–114.

Kogan, F. 1989. Remote sensing of weather impacts on vegetation in non-homogeneous areas. *International Journal of Remote Sensing* 11 (8): 1405–1419.

Kogan, F.N. 1995a. Droughts of the late 1980s in the United States as derived from NOAA polar-orbiting satellite data. *Bulletin of the American Meteorological Society* 76 (5): 655–667.

———. 1995b. Application of vegetation index and brightness temperature for drought detection. *Advances in Space Research* 15 (11): 91–100. https://doi.org/10.1016/0273-1177(95)00079-T.

———. 1997. Global drought watch from space. *Bulletin of the American Meteorological Society* 78: 621–636.

———. 2001. Operational space technology for global vegetation assessment. *Bulletin of the American Meteorological Society* 82 (9): 1949–1964.

———. 2002. World droughts in the new millennium from AVHRR-based vegetation health indices. *Eos Transaction of American Geophysical Union* 83 (48): 562–563.

Kogan, F. 2006. Early drought detection, monitoring and assessment of crop losses from space: Global approach. In *Disaster forewarning diagnostic methods and management*, ed. F. Kogan, S. Habib, and M. Matsuoka, 91–95. Bellingham: SPIE.

Kogan, F., and W. Guo. 2014. Early twenty-first-century droughts during the warmest climate. *Geomatics, Natural Hazards and Risk*: 1–11. https://doi.org/10.1080/19475705.2013.878399.

———. 2015. 2006–2015 mega-drought in the western USA and its monitoring from space data. *Geomatic, Natural Hazards and Risk*. https://doi.org/10.1080/19475705.2015.1079265.

———. 2016. Strong 2015–2016 El Niño and implication to global ecosystems from space data. *International Journal of Remote Sensing* 38 (1): 161–178. https://doi.org/10.1080/01431161.2016.1259679.

Kogan, F., and J. Sullivan. 1993. Development of global drought-watch system using NOAA/AVHRR data. *Advances in Space Research* 13 (5): 219–222.

Kogan, F.N., J.T. Sullivan, and Pu Bu Ciren. 1996. Testing post-launch calibration for the AVHRR sensor on world desert targets during 1985-1993. *Advances in Space Research* 17 (1): 47–50.

Kogan, F., A. Gitelson, E. Zakarin, L. Spivak, and L. Lebed. 2003. AVHRR-based spectral vegetation index for quantitative assessment of vegetation state and productivity: Calibration and validation. *Natural Resources* 278 http://digitalcommons.unl.edu/natrespapers/.

Kogan, F., N. Kussul, T. Adamenko, S. Skakun, O. Kravchenko, O. Kryvobok, A. Shelestov, A. Kolotii, O. Kussul, and A. Lavrenyuk. 2013a. Based on earth observation, meteorological data and biophysical models. *International Journal of Applied Earth Observation and Geoinformation* 23: 192–203. https://doi.org/10.1016/j.jag.2013.01.002.

Kogan, F., T. Adamenko, and W. Guo. 2013b. Global and regional drought dynamics in the climate warming era. *Remote Sensing Letters* 4: 364–372. https://doi.org/10.1080/2150704X.2012.736033.

Kogan, F., M. Goldberg, T. Schott, and W. Guo. 2015a. SUOMI NPP/VIIRS: Improve drought watch, crop losses prediction and food security. *International Journal of Remote Sensing*. https://doi.org/10.1080/01431161.2015.1095370.

Kogan, F., W. Guo, A. Strashnaia, A. Kleshenko, O. Chub, and O. Virchenko. 2015b. Modelling and prediction of crop losses from NOAA polar-orbiting operational satellites. *Geomatics, Natural Hazards and Risk*. https://doi.org/10.1080/19475705.2015.1009178.

Kogan, F., Z. Popova, and P. Alexandrov. 2016. Early forecasting corn yield using field experiment dataset and vegetation health indices in Pleven region, North Bulgaria. *Ecologia i Industria (Ecology and Industry)* 9 (1): 76–80.

Kogan, F., W. Guo, and W. Yang. 2017. SNPP/VIIRS vegetation health to assess 500 California drought. *Geomatics, Natural Hazards and Risk*. https://doi.org/10.1080/19475705.2017.1337654.

Kogan, F., W. Guo, W. Yang, and H. Shannon. 2018. Space-based vegetation health for wheat yield modeling and prediction in Australia. *Journal of Applied Remote Sensing* 12 (2). https://doi.org/10.1117/1.JRS.12.026002.

Kuciauskas, A., J. Solbris, T. Lee, J. Hawkins, S. Miller, M. Surratt, K. Richardson, R. Bankert, and J. Kent. 2013. Next-generation satellite meteorology technology unveiled. *Bulletin of the American Meteorological Society* 94: 1824–1825. https://doi.org/10.1175/BAMS-D-13-00007.1.

Kushnir, V., G. Korotaev, F. Kogan, and A. Powell. 2009. Consequences of land and marine ecosystems interaction for the Black Sea coastal zone. *Jeophysical Research Letter* 2: 38–46.

LeComte, D., and F. Kogan. 1988. *Assessment of crop condition in Africa. NOAA Technical Memorandum, NESDIS/AICS 13*, 66. January: Washington D.C.

Myers, V.I. 1970. Soil, water, and plant relations. In *Remote sensing with specific reference to agriculture and forestry*, 253–267. Washington: National Academy of Sciences.

NOAA (Satellite and Information Service). 2017. JPSS has new name: NOAA-20. https://www.nesdis.noaa.gov/JPSS-1.

PotashCorpo. 2013. Agriculture: Crop overview. http://www.potashcorp.com/industry_overview/2011/agriculture/16.

Rao, C.R.N., and J. Chen. 1995. Inter-satellite calibration linkages for the visible and near-infrared channels of the advanced very high resolution radiometer on the NOAA-7, -9 and -11 spacecraft. *International Journal of Remote Sensing* 16: 1931–1942.

———. 1996. Post-launch calibration of the visible and near infrared channels of the advanced very high resolution radiometer on the NOAA-14 spacecraft. *International Journal of Remote Sensing* 17: 2743–2747.

———. 1999. Revised post-launch calibration of the visible and near- infrared channels of the advanced very high resolution radiometer on the NOAA-14 spacecraft. *International Journal of Remote Sensing* 20: 3485–3494.

Salazar, L., F. Kogan, and L. Roytman. 2008. Using vegetation health indices and partial least squares method for estimation of corn yield. *International Journal of Remote Sensing* 29 (1): 175–189. https://doi.org/10.1080/01431160701271974.

Saleous, N.E. 2005. An extended AVHRR 8-km NDVI data set compatible with MODIS and SPOT vegetation NDVI data. *International Journal of Remote Sensing*. 26 (20): 4485–5598.

Solbrig, J.E., and T.E. Lee. 2013. Advance in remote sensing: Imaging the earth by moonlight. *Eos* 94: 349–350. https://doi.org/10.1002/2013EO400001.

Tarpley, J.P., S.R. Schneider, and R.L. Money. 1984. Global vegetation index from NOAA-7 meteorological satellite. *Journal of Applied Meteorology and Climatology* 23: 491–494.

Trishchenko, A.P., J. Cihlar, and Z. Li. 2002. Effects of spectral response function on surface reflectance and NDVI measured with moderate resolution satellite sensors. *Remote Sensing Environment* 81: 1–18. (PII: S 0034-4257(01)00328-5. https://doi.org/10.1016/S0034-4257(01)00328-5.

Tucker, C.J., and P.J. Seiler. 1986. Satellite remote sensing of total dry matter production. *International Journal of Remote Sensing* 7: 1395–1416.

Tucker, C.J., C. VanPraet, E. Borwinkel, and A. Gastron. 1983. Satellite remote sensing of primary production in Senegalese Sahel. *Remote Sensing Environment* 13: 461–474.

Tucker, C.J., J.E. Pinzon, M.B. Brown, D.A. Slayback, E.W. Pak, R. Mahoney, E.F. Vermote, and N. El Salcous. 2004. An extended AVHRR 8-km NDVI data set compatible with MODIS and SPOT vegetation NDVI data. *International Journal of Remote Sensing* 9: 41–55.

Unganai, L.S., and F.N. Kogan. 1998. Drought monitoring and corn yield estimation in southern Africa from AVHRR data. *Remote Sensing Environment* 63: 219–232.

Vargas, M., F. Kogan, and W. Guo. 2009. Empirical normalization for the effect of volcanic stratospheric aerosols on AVHRR NDVI. *Geophysical Research Letters* 36: 35–41.

4 Vegetation Health Method

4.1 Introduction

The Earth's system is an integrated component of ocean, land, biosphere, and atmosphere. The terrestrial biosphere is a very important part of the Earth providing strong contribution to its productivity. In general, the biosphere is the principal source of food, energy, fuel, and fiber. All of these components are an integral part of Earth, providing life for the population. Vegetation is the most important part of the biosphere, contributing to all components of biosphere productivity (Hield et al. 1995). Vegetation provides food and fiber (crops and pastures), fuel and energy (trees, bushes, etc.), which depends strongly on the environmental resources (Monteith 1972; Price 1992; Prince and Goward 1995). Therefore, it is very important not only to understand the dependence of vegetation and crop production on the environment but also to quantify this relationship and model it for advanced prediction of crops and pasture production and food security.

As it is known, crop and pasture production is strongly controlled by climate, weather, hydrology, and ecosystem laws (Running et al. 1999; Orlovsky et al. 2009; Kogan 1987, 1989; Kuciauskas et al. 2013; NCDC 2011). Among these parameters, ecosystem, climate, and hydrology are long-term (multiyear) interactive factors setting up the principals for distribution of global vegetation over space and time, setting up ecosystem productivity. Weather is a short-term component affecting vegetation amount and distribution in a matter of days, weeks, and months (up to 1 year). Both short- and long-term components of vegetation have large spatial and temporal heterogeneity in appearance and production. For example, ecosystems, developed over many years following climate and hydrology, range from deserts (no vegetation) to rainforest [plentiful vegetation (Running et al. 1999; Friedl et al. 2002; Kogan 1987, 1989)]. Detailed analysis indicates 13 large ecosystem classes (Friedl et al. 2002), including evergreen needle leaf forests, evergreen broadleaf forests, deciduous needle-leaf forests, deciduous broadleaf forests, mixed forests, closed shrubs, open shrubs, woody savannas, savannas, grasslands, permanent wetlands, croplands, and urban-built-up lands. Climate is a strong long-term contributor to the distribution and productivity of vegetation. Such climate parameters as multiyear mean precipitation and temperature are among the most influential factors for the development of ecosystems (Runyon et al. 1994). For example, the evergreen broadleaf forest was developed under conditions of plentiful precipitation and warm temperatures, while the savannas were formed under dry and hot climate conditions. Similarly, hydrology contributes to vegetation through surface-water partitioning, storage, and river flows (Prentice et al. 1992).

F. Kogan, *Remote Sensing for Food Security*, Sustainable Development Goals Series,
https://doi.org/10.1007/978-3-319-96256-6_4

Short-term weather components control vegetation productivity within a 1-year period (Monteith 1972). For example, agricultural fields can change from bare soil to complete canopy coverage in a matter of a few months. The quantity and quality of biomass, its density, height, phenology, and production (in case of crops and pastures) created by any vegetative ecosystem, depend strongly on the amount and distribution of solar energy and such environmental parameters as rainfall, temperature, soil moisture, humidity, and others. An important fact is that vegetation influences back on the environment (Sellers et al. 1995; Kogan and Guo 2014; Kogan 1989). More than half of precipitation that falls on land returns to the atmosphere by moving from roots through plant tissue to leaves, from which it evaporates (Monteith 1972). Following this physiological process, the plants cool off, changing the ambient conditions. Another important consideration is that about one-third to one-half of land surface has been disturbed and modified by humans (CGIAR 2015; Friedl et al. 2002; Running et al. 1999). Humans are planting crops, cut trees, irrigate crops and pastures, etc. Following the recently identified slowdown tendency in production of the traditional agricultural and expansion of crops and pasture to the new areas, the fraction of existing natural vegetation is going to decrease and new vegetation will increase in upcoming decades. In Ghana (Northern Region, Upper West and Upper East), for example, 23% of forests and 16% of shrub land were converted into cropland in the last 20 years (CGIAR 2015). Anthropogenic land changes and use influence back on climate, modifying ecosystem and hydrology regime (Gillins 2014).

Finally, it is important to underline that (1) vegetation is the principal component of land and biosphere; (2) vegetation is strongly influenced by a group of environmental factors, especially climate, ecosystem, hydrology, and weather; (3) vegetation changes in response to the environmental variations can be described by such measurements as vegetation amount, density, heights, weight, production, and also by vegetation conditions (greenness, vigor, development, health, and others); (4) from the environmental factors, which affect vegetation amount, climate and hydrology are setting up a long-term (years) level of vegetation amount expressed in ecosystem type; weather provides a short-term contribution to vegetation changes; (5) compared to long-term (climate, ecosystem and hydrology) factors, short-term weather contributions are frequent (days, weeks, etc.) within a 1-year period and must be separated from long-term components in order to monitor weather impacts on vegetation (Kogan 1987, 1989, 1990, 1995, 1997, 2001).

4.2 Theoretical Base of Vegetation Health Method

Vegetation health is the method designed to derive vegetation condition (favorable, unfavorable, normal, etc.) or health (healthy, unhealthy, stressed, etc.) in response to changing weather (precipitation, temperature, and others) in each ecosystem and climate. In general, if vegetation is well developed, green, vigorous, and uniformly covers an area, then vegetation is considered to be healthy (or in favorable condition). Oppositely, if vegetation is not well developed, sparse, less vigorous, and less green (or even yellow), then vegetation is considered unhealthy or in poor condition (Kogan 1987, 1989, 1990, 2001, 2002). Since precipitation and air temperature are the two major weather parameters controlling vegetation health, the VH method was designed for space-based monitoring moisture and thermal conditions. A very simple principle was established: if vegetation is very green and vigorous its satellite-derived numerical conditions in the form of moisture–thermal indices set up as very healthy. If vegetation is much less green, not vigorous, or even wilting, the vegetation state is set up as unhealthy or stressed. Such assessment principals are applied to any stage of the vegetation development for any ecosystem, climate, and hydrology. Between these two extreme conditions (healthy and stresses), vegetation health is estimated continuously from extreme stress to very healthy, providing different levels of vegetation health, such as, for stressed level: light, moderate, severe, extreme, etc. and for healthy level: normal, above normal, healthy, etc. The vegetation health method was introduced in the mid-1980s (Kogan 1987) and was applied globally, since 1981, through present using observations from the NOAA operational afternoon polar-orbiting satellites. This method turned out to be quite popular and in the past 20 years was applied to the records of a number of scientific satellites, such as MODIS, LANDSAT, SPOT, and others (Saleous 2005; Orhan et al. 2014; Rajasekar and Weng 2009; FAO 2017; NIDIS 2017).

4.2.1 Biophysical Considerations

The vegetation health (VH) numerical method stems from properties of green vegetation to reflect sunlight and emit absorbed solar radiation by the Earth's surface, which is measured by sensors on NOAAs operational polar-orbiting satellites. The importance of these NOAA satellites in their operational status. They have been in operational service (serving users' community), using the original sensor for 32-year, intermedia system, testing the new sensor, and will continue with the new system until the mid-twenty-first century. The VH method was developed first from the data measured by the Advance Very High Resolution Radiometer (AVHRR) on the NOAA afternoon polar-orbiting satellite since 1980. From 2013, the VH method uses data from the Visible Infrared Imaging Radiometer Suite (VIIRS) on the new generation of NOAAs Suomi NPP (S-NPP) afternoon operational polar-orbiting satellites. Since 2018, the VH method continues to be produced from the JPSS-1/VIIRS afternoon operational satellite, which was launched in November 2017 and called NOAA-20 (JPSS 2014). This system plans to continue its VH operational services for the next two and a half decades.

The two instruments, VIIRS and AVHRR, on the NOAA operational polar-orbiting afternoon satellites, measure reflected and emitted solar radiation from the entire earth surface, including land vegetation and agricultural crops. The most useful for VH is reflected solar radiation in the visible (VIS) and near-infrared (NIR) spectral bands and emitted radiation in infrared (IR) spectral band. The amount of reflected and emitted radiation depends on chlorophyll and moisture content in vegetation. If vegetation is healthy, it contains much chlorophyll and water and reflects little radiation in the VIS part of solar spectrum (a higher amount of chlorophyll absorbs more radiation and reflects less). Near-infrared light penetrates leaf canopy, interacting with multiple canopy levels before reflecting bask to the atmosphere to be measured by a satellite sensor. The NIR is strongly sensitive to vegetative biomass and vigor and through them to water content (ASP 1975; Tarpley et al. 1984; Cracknell 1997). Healthy vegetation reflects more in the NIR compared to unhealthy vegetation. Vegetation emission in the IR band also depends on the amount of water, since transpiring canopy is cooler and emits less. The biophysical explanation is opposite for unhealthy vegetation. As described in Chap. 3 (Rouse et al. 1973; ASP 1975; Tucker 1979; Tarpley et al. 1984; Kogan 1987, 1989, 2001; Gray and McCrary 1981; Hield et al. 1995), VIS and NIR measurements are used to calculate the normalized difference vegetation index (NDVI): NDVI = (NIR − VIS)/(NIR + VIS). NDVI has showed a strong relationship with chlorophyll concentration on land and in water ecosystems, crops primary production, biomass, vegetation health, and other parameters of land surface (Gray and McCrary 1981; Tarpley et al. 1984; Tucker 1979; Tucker and Sellers 1986; Kogan 1987). The IR emission was converted to brightness temperature [BT (Kogan 1987; Kidwell 1997)].

In principal, NDVI (Rouse et al. 1973) is used to assess moisture-based and brightness temperature (BT) to asses thermal-based vegetation health (Kogan 1989, 1990). If vegetation is healthy, NDVI becomes larger (more chlorophyll and water) and BT smaller (cooler vegetation canopy). Oppositely, for unhealthy vegetation, the NDVI is smaller (reduced chlorophyll and water amount) and the BT is larger since the vegetation surface becomes hotter following reduced water content in vegetation and the resulting reduced transpiration. It is important that both NDVI and BT respond to major biophysical characteristics of vegetation: chlorophyll and water content. It should be also emphasized that NDVI estimates gradual changes in vegetation condition (several weeks), while BT estimates both gradual and fast (daily) changes, since in the last case, extreme temperature could decimate vegetation (sometimes in a presence of some water in the soil) in a matter of days. Such situation occurred in Portugal, Spain, Italy, and France in the fall of 2017, when extreme temperature left rivers nearly dry, sparked deadly wildfires, and devastated crops and pastures with huge economic consequences (Vincent 2017; Euronews 2017; News24 2017). In summary, NDVI quantifies an intensity of photosynthetic activity of vegetation productivity, assessing indirectly vegetation greenness, vigor, cover, height, phenology, and ecosystem type, which are strongly related to the amount of chlorophyll and water in vegetation and also energy absorption (Tucker 1979; Kogan 1987, 1989, 1990). BT quantifies changes in vegetation productivity following vegetation response to thermal conditions.

4.2.2 Basic Laws for Extracting Weather Component from NDVI and BT

As described in Chap. 3, weekly NDVI and BT indices were processed comprehensively to remove high- and low-frequency noise producing two no noise indices: smoothed NDVI (SMN) and smoothed BT (SMT). As was discussed earlier with NDVI and BT, SMN and SMT indices similarly characterize two environmental components: long-term (ecosystem, climate, and hydrology) and short-term weather component. Since the VH is designed for monitoring weather impacts only on vegetation, one of the principal tasks of the VH algorithm is to remove the long-term component from the SMN and SMT indices and single out only short-term weather component. The removal was done for the following principals of three environmental laws: Liebig's Law of Minimum (LOM), Shefield's Law of Tolerance (LOT), and the Principle of Carrying Capacity (PCC). In general, these laws require to derive and set up the environmental limits for long- and short-term components that resources or habitat can support (Shelford 1931; Ehrlich and Holdren 1971; Reining 1974; Orians 1990; Kogan 1995; Sinclair 1999; Hui 2006; LOM 2017). These limits, called climatology of SMN and SMT, were determined using the principals of the three laws.

Liebig's LOM states that vegetation growth is strongly controlled by the scarcest resources or the most limiting factor. From many factors (precipitation, sunlight, temperature nutrients, etc.) affecting vegetation (growth, development, and productivity), the most limiting one determines the rate of vegetation change (Sinclair 1999). The advances of biophysical researches have shown that a low level of one factor can be partially compensated by appropriate levels of other factors (Orians 1990; Sinclair 1999; LOM 2017). This was reflected in Shefield's LOT, stating that a performance of an organism is based on a complex of conditions specified by a certain minimum, maximum, and optimum of environmental factors or their combination (Shelford 1931). Following this law, vegetation health of any ecosystem can be evaluated by the levels of the available resources expressed by their multiyear extreme minimum and maximum. These extremes set up the principal of carrying capacity (PCC) for ecosystems (Ehrlich and Holdren 1971; Reining 1974). The PCC of biological species in an environment determines population size that can be supported indefinitely upon the available environmental resources inside their limits (Reining 1974; Hui 2006; Stigebrandt 2011). In case of vegetation, these limits are multiyear absolute minimum and maximum of vegetation characteristics represented by satellite SMN and SMT indices (Reining 1974; Kogan 1987, 1989, 1995). These limits were established from 37-year (1981–2017) AVHRR and VIIRS data.

4.3 Vegetation Health Algorithm

The vegetation health (VH) algorithm (Kogan 1987, 1989) includes several steps: (1) data collection: visible (VIS) and near-infrared (NIR) reflectance and infrared (IR) emission from land surface; (2) data processing: pre- and post-launch radiance calibration, calculation of reflectance and radiative temperature; (3) calculation of NDVI and brightness temperature (BT); (4) weekly compositing of NDVI and BT; (5) complete elimination of high- and low-frequency noise from NDVI and BT weekly time series and development of smoothed NDVI (SMN) and smoothed BT (SMT); (6) approximation of SMN and SMT annual cycle; (7) calculation of SMN and SMT climatology (SMN_{max}, SMN_{min}, SMT_{max}, and SMT_{min}) based on LOM, LOT, and PCC; and (8) calculation of vegetation health indices. Steps (1)–(5) were described in Chap. 3. This part explains steps (6)–(8).

After smoothing NDVI and BT (step 5), year-to-year difference caused by weather variations in SMN and SMT became more apparent (Fig. 4.1). The largest mid-season SMN (SMN_{max}) was 0.64 during 1986 in Illinois and only 0.47 in North Dakota (Fig. 4.1a) ecosystem, which was healthy vegetation characteristics in wet mid-summer (weeks 27–32). In addition to wetness, the summer of these years was also cool, indicated by low values of SMT_{min} in both states (295°K in Illinois and 300°K in North Dakota (Fig. 4.1b). Opposite to healthy vegetation indicators, the 1988 mid-season SMN was the lowest in a 6-year period (0.52 in Illinois and 0.32 in North Dakota) and SMT was the highest (305°K in Illinois and 309°K in North Dakota) from the displayed years in both locations due to extremely hot (and also dry) summer. Dry/hot conditions in 1988 were confirmed by weather observations (NOAA 1988). Meanwhile, Fig. 4.1 also shows that the same SMN and SMT values should be interpreted differently in various ecosystems. For example, the prairie ecosystem in Illinois has the higher level of mid-summer SMN_{min} around 0.52, compared to North Dakota steppe ecosystem, which SMN_{min} is 1.6 times less (0.32). Concerning the largest values of indices, the highest mid-summer SMN_{max} (Fig. 4.1a) and SMT_{max} (Fig. 4.1b) characterizes the contribution of the indices to vegetation health from both long-term ecosystem and short-term weather in both locations. Summarizing this discussion, it is important to emphasize again that Fig. 4.1 shows that the highest values of SMN and SMT quantifies both long-term ecosystem and short-term weather contribution to vegetation greenness, development, completeness, and productivity. The lowest indices quantify ecosystem contribution to the minimum vegetation health only (without weather). Therefore, the difference between SMN_{max} and SMN_{min} and between SMT_{max} and SMT_{min} quantify only weather (weather component) contribution to vegetation health.

It is clear that the weather components in SMN and SMT are superimposed on the ecosystem component. Maximum vegetation is developed in years with optimal (wet and cool)

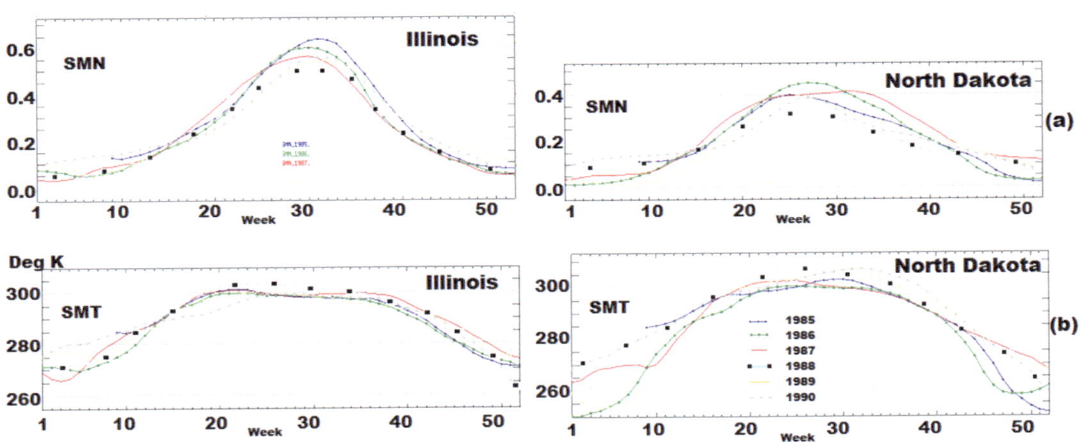

Fig. 4.1 Smoothed annual weekly (**a**) SMN (SMN_{max}, SMN_{min}) and (**b**) SMT (SMT_{max} and SMT_{min}) during 1985–1990 in the central Illinois prairies and North Dakota steppe ecosystems (Kogan 1995)

weather, since such weather stimulates efficient use of ecosystem resources (for example, an increase in the rate of soil nutrition uptake). In contrast, minimum vegetation is developed in years with extremely unfavorable weather (dry and hot), which suppresses vegetation growth both directly and through a reduction in the rate of ecosystem resources use. For example, lack of water and high temperature in drought years reduces solar energy use due to a reduced amount of chlorophyll in leaves resulted in unhealthy vegetation. The absolute maximum and minimum of SMN and SMT calculated from many years of data that contain the extreme weather events can be used as criteria for quantifying these extreme conditions in the indices.

Therefore, having 37 years of blended AVHRR (1981–2012) and adjusted VIIRS (2013–2017) data, the extreme largest and the extreme lowest SMN and SMT values during these years for each of the 52 weeks of the year and for each global 4 km^2 land pixel were calculated. The resulting SMN_{max} and SMN_{min} were used as criteria for estimating the upper (maximum index) level of most healthy vegetation in case of a very favorable wet weather and the lower (minimum index) level for unhealthy or stressed vegetation in case of extremely unfavorable (dry) weather for each pixel and week. For SMT_{max} and SMT_{min}, the lowest and the highest levels were set up oppositely to SMN since SMT_{max} characterizes vegetation condition during the hottest weather, which is unfavorable or stressed for vegetation health. The SMN_{min} (cooler temperature) characterizes a favorable condition or healthy vegetation. These extreme limits for SMN and SMT characterize the "carrying capacity" of an ecosystem in terms of weather impacts on vegetation productivity (Reining 1974). Figure 4.2 shows the SMN- and SMT-based 6-year example (Kogan 1995) of carrying capacity (from SMN_{max} to SMN_{min} and from SMT_{max} to SMN_{min}) for four 16 km^2 land pixels of wet and dry ecosystems in the central Illinois and North Dakota (both USA). It is interesting that the highest SMN_{max} in dry ecosystem (indicates the healthiest vegetation condition during 1985–1990) corresponds to the lowest SMN_{min} (the most stressed vegetation conditions in the wetter ecosystem; red line). Similarly, SMT_{max}, which characterizes stressed vegetation conditions in wetter ecosystem (Illinois), indicate healthy thermal conditions (SMT_{min}) in drier ecosystem (North Dakota).

Following the Fig. 4.2 discussion, the SMN_{max}, SMN_{min}, SMT_{max}, and SMT_{min} were estimated for each 4 km^2 global land. Figure 4.3 shows these 37-year MAX and MIN weekly indices dynamics typical for the main US ecosystems. Each ecosystem has its own SMN and SMT signatures in terms of the indices values, shape of the curves, rate of indices changes during leaf appearance, the highest development and senescence and partitioning of each index into weather and ecosystem (including climate) components (Kogan

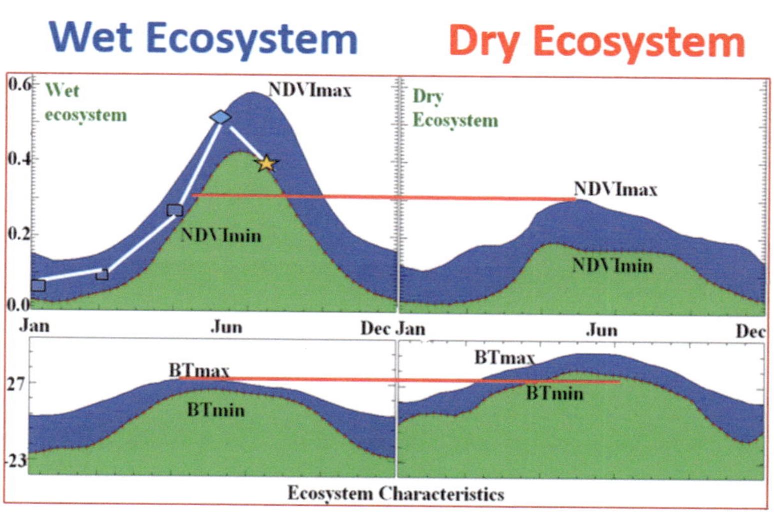

Fig. 4.2 Carrying capacity expressed by SMN_{max}, SMN_{min}, SMT_{max}, and SMT_{min} in dry and wet ecosystems of the central USA during 1985–1990

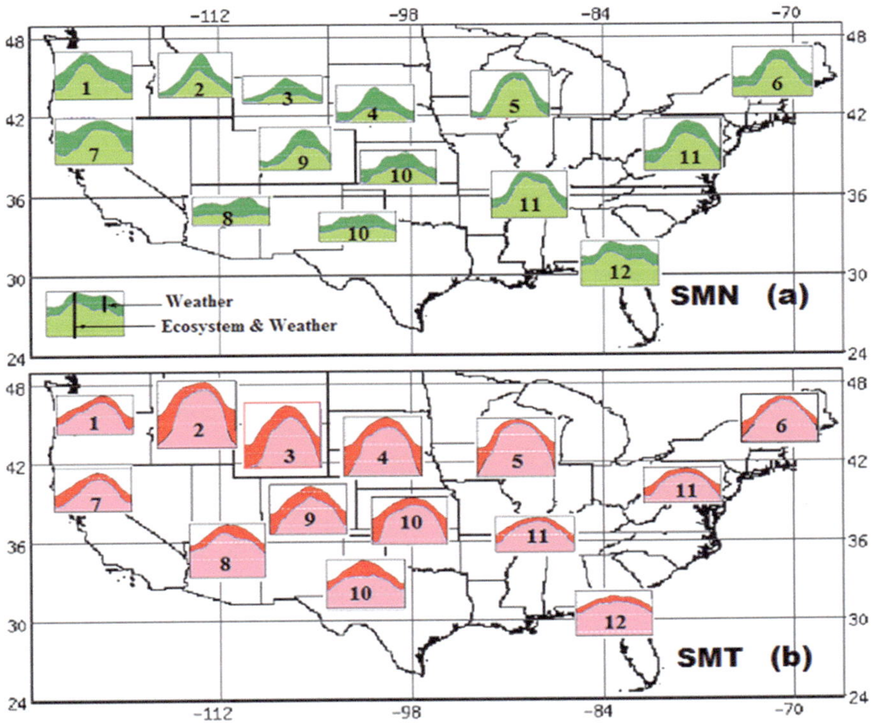

Fig. 4.3 Multiyear (1981–2017) weekly SMN_{max}, SMN_{min} (**a**), SMT_{max} and SMT_{min} (**b**) time series, characterizing main ecological zoned of the USA (Hammond Inc. 1991): (1) coniferous forest, annual precipitation >80 in.; (2) mixed) coniferous forest and steppe, 10–20 in.; (3) steppe, 10–20 in.; (4) prairie, 20–40 in.; (5) mixed coniferous and broad leaf forest, 20–40 in.; (6) mixed coniferous and broad leaf forest in Maine, 40–60 in.; (7) mixed coniferous and broad leaf forest in California, 40–60 in.; (8) desert and desert shrubs, <10 in.; (9) coniferous forest, 20–64 in.; (10) steppe, 10–20 in.; (11) broad-leaf forest, 40–60 in.; (12) mixed coniferous and broad leaf forest in Florida, 40–60 in.

1987, 1995). The eastern and western US states [types 1, 2, 5, 6, 7, 11, and 12, (Fig. 4.3a)] have the highest SMN values with clearly defined seasonal dynamics and smaller total areas for weather (dark green area) versus ecosystem components (light green area). This type of seasonal dynamics is typical for forest ecosystems with some difference between deciduous and coniferous forests. Compared to forest ecosystems, drier climate areas, such as desert, desert-shrubs, steppe, and some prairie (types 3, 4, 8, 9, 10), have the smallest SMN for both ecosystems and weather components (Fig. 4.3a). The central-southern states with the most drier ecosystems and climate have the lowest SMN values without distinctive seasonal dynamics and almost equal contribution of weather and ecosystem components into the integrated area under the maximum curve (types 8, 10). Similar to SMN, there are SMT differences between eastern-western and central ecosystems. Forest ecosystems have the lowest level of both SMT_{max} and SMT_{min} and smaller range of weather component compared to dry ecosystems in the central and southern USA.

Figure 4.4 demonstrates multiyear max-min annual time series for the two extremely different world ecosystems: desert and forest. Following SMN_{max} and SMN_{min}, it is important to emphasize that (1) even the most unfavorable for living Sahara Desert is represented by the two environmental components (ecosystem and weather) in the SMN and SMT indices, with almost equal contribution of each component but without any seasonal cycle; (2) other deserts, providing also almost equal contribution for the two components, have higher SMN values and some seasonal cycle corresponding to the period with the coolest temperature (SMT) conditions (Arizona

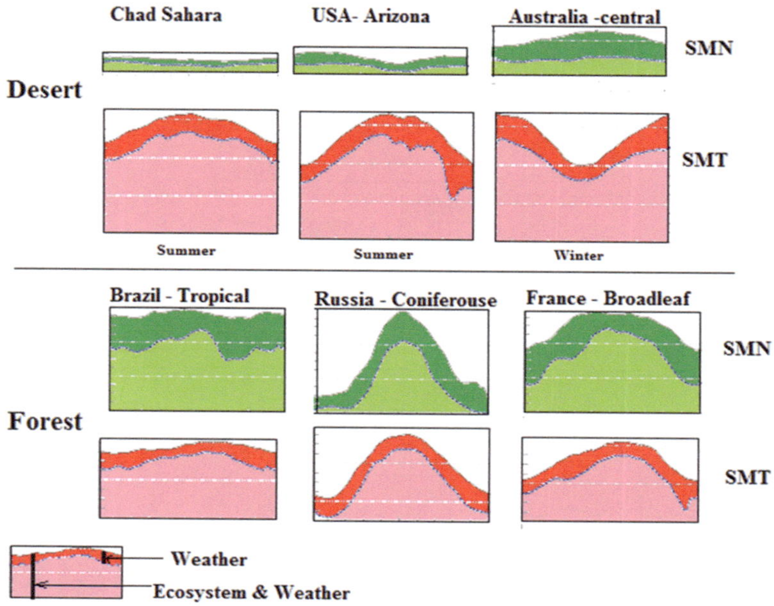

Fig. 4.4 Multiyear (1981–2017) weekly SMN_{max}, SMN_{min} (**a**), SMT_{max} and SMT_{min} (**b**) time series, characterizing desert and forest ecosystems of the world

desert in summer and Australia desert in winter). For forest ecosystems, it is noticeable that, (1) weather component contribution to SMN is much smaller compare to ecosystem; (2) strong seasonal cycle in coniferous and broadleaf forest, and (3) no seasonal cycle in tropics. Similar to SMN, SMT clearly shows two thermal components, ecosystem and weather, with smaller weather contribution, a strong seasonal cycle for broadleaf and coniferous forest and almost no cycle for tropical forest. The most important conclusion from this analysis is that any place on the Earth can be represented by the two environmental components in observations by operational satellites: moisture in SMN and thermal in SMT. However, since these indices characterize two components, ecosystem and weather, with the last one much smaller than the ecosystem, the weather component was separated from the ecosystem by deriving 37-year SMN_{max}, SMN_{min}, SMT_{max}, and SMT_{min} (called climatology), from which vegetation health is estimated. Figure 4.5 shows stratification of the world land based and maximum and minimum climatology of SMN and SMT for mid-July. These images identified a clearly principal world ecosystems. Both parameters show desert areas (Sahara, Australian, southeastern USA, etc.) with low SMN and high SMT, tropical areas with high SMN and low SMT (Brazil, central Africa), crops and grassland areas (eastern half of USA, Europe, southern European Russia, and others), broadleaf and coniferous forest (Canada, northern USA and Russia, Europe, etc.). This spatial distribution of ecosystems and climate zones corresponded well to in situ data (Hammond Inc. 1991; Running et al. 1995; Stigebrandt 2011).

For land surface ecosystems (except deserts) the range between maximum and minimum in the weather component of SMN and SMT indices is much smaller than the ecosystem one. Consequently, the weather impacts on vegetation are not easily detectable if to use the entire indices values. In order to enhance assessment of weather impacts on vegetation health, drought detection and prediction of agricultural production the weather component was (1) separated from the ecosystem component and (2) was linearly scaled from zero to 100 for its entire range. The scaling was done differently for SMN and SMT since they characterize differently moisture and thermal impacts on vegetation health. SMN was scaled from zero for SMN_{min} (unhealthy vegetation) to 100 for SMN_{max} (healthy vegetation); SMT was

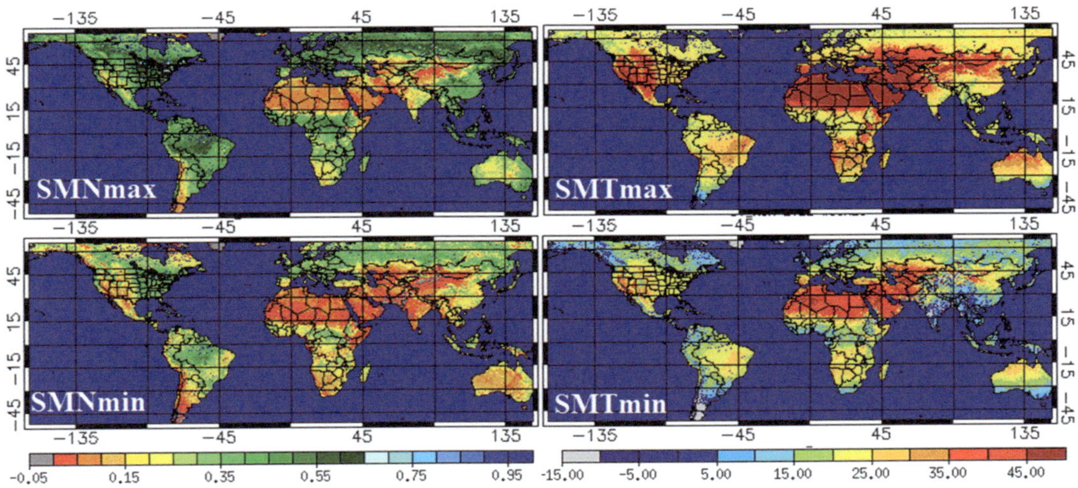

Fig. 4.5 Images of Mid-July (week 28) SMN_{max}, SMN_{min}, SMT_{max}, and SMT_{min} derived from 37-year VH data for each 4 km^2 world pixel

scaled from zero for SMT_{max} (unhealthy vegetation) to 100 for SMT_{min} (healthy vegetation). Such scaling was performed for each grid cell and week. The resulting indices were named the vegetation condition index (VCI), the temperature condition index (TCI), and the vegetation health index (VHI), combining together VCI and TCI. They were defined by Eqs. (4.1), (4.2), and (4.3). VCI and TCI were designed for assessment of moisture and thermal conditions, respectively, from extremely unfavorable to optimal, and assessment of the impact of these conditions on vegetation health (from stressed to healthy). VHI was designed for assessment of total moisture and thermal impacts on vegetation health, stressed-to-healthy (Kogan 1987, 1990).

$$\mathrm{VCI} = 100 \times \left(\mathrm{SMN} - \mathrm{SMN}_{\mathrm{min}}\right) / \left(\mathrm{SMN}_{\mathrm{max}} - \mathrm{SMN}_{\mathrm{min}}\right) \tag{4.1}$$

$$\mathrm{TCI} = 100 \times \left(\mathrm{SMT}_{\mathrm{max}} - \mathrm{SMT}\right) / \left(\mathrm{SMT}_{\mathrm{max}} - \mathrm{SMT}_{\mathrm{min}}\right) \tag{4.2}$$

$$\mathrm{VHI} = a \times \mathrm{VCI} + \left(1 - a\right) \times \mathrm{TCI} \tag{4.3}$$

where SMN, SMN_{max}, and SMN_{min} (SMT, SMT_{max}, and SMT_{tmin}) are no noise weekly SMN and SMT and their 1981–2016 absolute maximum and minimum (climatology), respectively; a is a coefficient quantifying a share of VCI and TCI contribution in the VHI. Since this share is generally not known for specific crop and location, it was assumed that VCI and TCI contributions are equal (a = 0.5). However, a can be calibrated against measured ground vegetation characteristics such as biomass, yield of agricultural crops, and others.

The VCI, TCI, and VHI indices have numerical values between zero and 100, where zero estimates exceptional vegetation stress (extremely unfavorable conditions) and 100 estimates a very healthy vegetation (very favorable conditions). In principal, if vegetation is very green and vigorous, its satellite-derived numerical conditions in the form of the VCI, TCI, and VHI (moisture, thermal, and combined, respectively) indices are set up as 100 (very healthy). If vegetation is much less green, not vigorous, or even wilting, the vegetation state is set up as zero (very unhealthy or extremely stressed). Between these two extreme conditions (very healthy and extremely stresses), vegetation health is estimated continuously providing different levels of vegetation health, such as, for stressed level, light, moderate, severe, extreme exceptional (from 45 to 0), and for healthy levels, normal, moderate, healthy, very healthy, etc. (from 60 to 100). Such assessment principles are applied to any stage of vegetation development and to any ecosystem, climate, and hydrological regime. A vegetation health approach was introduced in mid-1980 and was applied globally from 1981 through the present time, using the NOAA operational afternoon polar-orbiting satellite observations. Satellite-based numerical vegetation health method became quite popular and in the past 20 years was applied to a number of scientific satellites' records, such as MODIS, LANDSAT, SPOT, and others (FAO 2016, 2017; NIDIS 2017; Saleous 2005).

4.4 Vegetation Health at Work

An example of the global assessment of moisture, thermal, and combined weather-related conditions from the VCI, TCI, and VHI is shown in Fig. 4.6. As seen, an intensive moisture (VCI) and thermal (TCI) vegetation stress developed in summer 2012 in the central USA, eastern Brazil, Ukraine and Russia's west and Siberia south. These regions were experiencing a combination of moisture–thermal (VCI–TCI) stress. Such combination of stress is an indicator of drought from a lack of moisture and hot temperature. Such environmental combination deteriorates vegetation health, reduces strongly crop and pasture production, triggers fires, and causes other unfavorable events. The most adverse consequences from mid-summer moisture–thermal stress in those regions would lead to considerable losses in agriculture. Oppositely to stressed regions, a strip of 2° latitude in Sub-Sahara Africa had favorable moisture (VCI) and thermal (TCI) conditions in July 2012. This VCI/TCI values predicted productive crop harvest and strong food security in these countries. Some of the world regions were affected either by moisture or by thermal stress only in 2012. For example, Canada, most of Argentina, western Brazil, Europe, and northern Russia had excellent moisture conditions (VCI in Fig. 4.6) but extreme thermal stress (TCI) in mid-summer 2012. The impact of this thermal stress on moisture conditions and vegetation health would depend on duration of thermal stress. Figure 4.7 demonstrates how the duration of thermal stress (TCI) in Ukraine during 2007 deteriorated moisture conditions (VCI) and the total vegetation health (VHI). The thermal stress in Ukraine (TCI changes between 0 and 10) started at the beginning of the year, while moisture conditions were very favorable (VCI is 90–100). The VCI-based conditions began to deteriorate in April 2007 after thermal stress continued for 4 months. From April to mid-June, during one and a half month only, VCI was reduced from 90 (very healthy moisture conditions) to 10 (extremely unhealthy conditions). The combined moisture–thermal conditions (VHI) deteriorated as well following quick VCI reduction and continuation of severe thermal stress. As the result of such indices' dynamics, the mid-summer 2007 in Ukraine turned out to have the worse moisture–thermal stress (VHI = 5) during the most important period of winter wheat growth (grain formation) resulting in considerable losses of grain production (Adamenko and Prokopenko 2009; Kogan and Guo 2009).

Figure 4.8 compares global, much less severe, moisture–thermal joint vegetation stress in the summer of 2017, with the strongest stress in 2012 (NOAA/NESDIS 2018). Total drought-triggered vegetation stressed area in 2012 (VHI < 40) occupied 28% of the world's vegetated land, including 9% of an extreme vegetation stress (VHI < 15), caused by exceptional drought. Compared to 2012, in 2017 total drought-induced vegetation stress (mostly lights) area occupied 17% of land with less than 1% area under extreme vegetation stress. Following the 2012 VH-based estimation of drought impacts, global grain production was 4% below consumption (USDA/FAS 2017; USDA/NASS 2013; USDA 2017; WB 2017). Among grain-producing countries, the most drought affected (red color) in 2012 was the USA (15.8% contribution to global grain production), Kazakhstan (0.8% contribution), and eastern Brazil. Mid-level drought-based vegetation

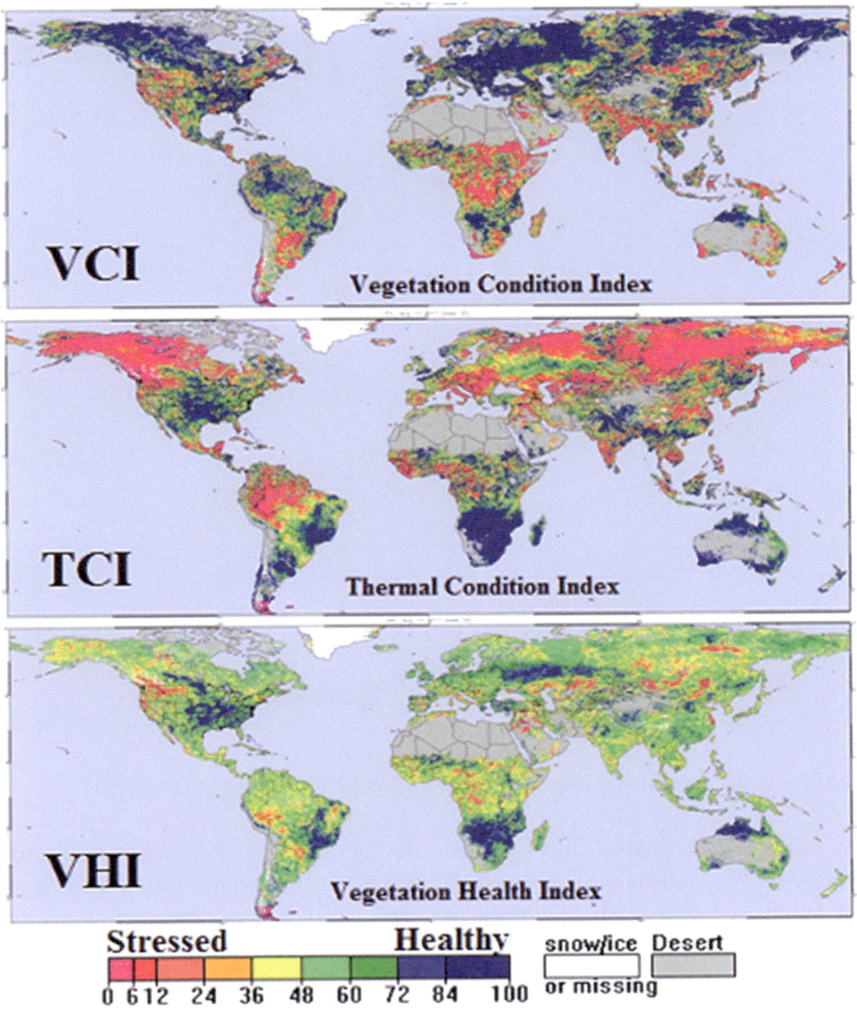

Fig. 4.6 Global 4 km^2 resolution VCI, TCI images for July 16, 2012

Fig. 4.7 Dynamics of weekly vegetation health indices (VCI, TCI, VHI) during 2007 in winter wheat area of Ukraine

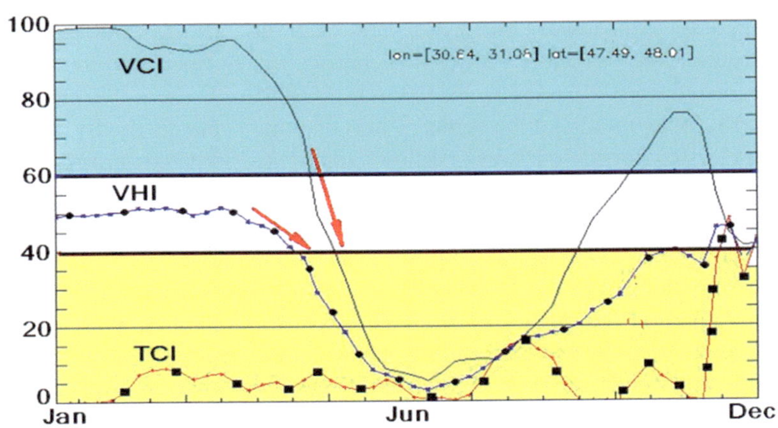

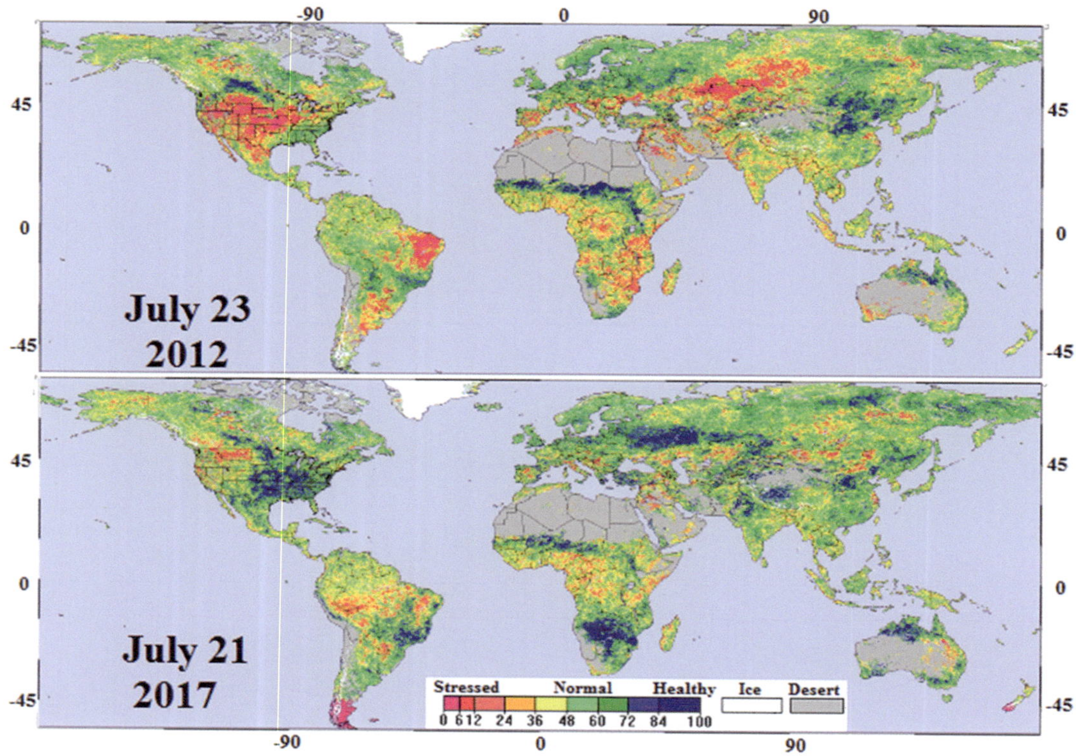

Fig. 4.8 Vegetation health index in mid-July (week 28) 2012 and 2017 (NOAA/NESDIS 2018)

stress (light red and dark yellow color) was in most of Ukraine (2.2% contribution), in southern Russia, and southern Europe. The majority of African countries (except small strip area in Sub-Sahara), experienced moderate-to-strong drought-triggered vegetation stress (VHI < 35), which affected food security (USDA 2017). Grain production was reduced, especially in Tanzania, Kenya, Ethiopia, Somalia, Zaire, Angola, Zimbabwe, Morocco, and Algeria. Required food assistance for these countries was predicted from VH assessments (NOAA/NESDIS 2018) in the middle of the summer. In Asia, mid-level 2012 vegetation stress (yellow color) was experienced by most of India (10.5% contribution to global grain production) and southern China. India had moderate problems with food security in 2012.

Although the summer of 2017 was less affected by drought-related vegetation stress (Fig. 4.8), some periods in 2017 were characterized by extreme vegetation stress due to a devastating drought, which affected the food security situation in Europe. Devastating or exceptional drought in VH nomenclature are generally quite rare, especially if they affect the economy of an entire country. However, such drought occurs from time to time, and if damages are extreme and affect the entire country, drought attracts exceeding attention and requires an advanced estimation of the impact. Such unusual thermal droughts affected western European countries, especially Portugal and Spain in the fall of 2017. In these two countries, the 2017 drought left rivers nearly dry, sparked deadly wildfires, and devastated crops and pastures (Vincent 2017). According to meteorologists (IPMA 2017), the 2017 drought was classified as a very unusual, since 94% of Portugal was estimated as being under "extreme" drought with less than 50% of rainfall. Plus, unusually excessive temperatures in both countries created exceptional drought conditions. Very devastating drought consequences were cited in the world news (Euronews

2017; TheGuardian 2017). A few of the most unusual drought-related environmental damages were cited in the following way: Douro River, which is one of the symbols of the Iberian Peninsula, was 60 percent dry by November 2017. In Portugal, 28 of the country's water reservoirs were at less than 40% of their storage capacity in October. The shortage of water was so severe that one hundred fire trucks began transporting water from one dam in northern Portugal to another that was running dry. Besides, following strong shortage of water, trucks delivered water to Viseu, a city of around 100,000 residents (Vincent 2017). Severe water deficiency in the second half of 2017 were aggravated by excessive fall temperatures in many places of Iberian Peninsula, Italy, and France. The temperatures hit and exceeded 40 °C, which was the most sustained heatwave in Portugal since 2003 (IPMA 2017; TheGuardian 2017).

Vegetation health assessments of thermal stress (TCI) in Europe supports this drought's devastations. Figure 4.9 demonstrates mid-day TCI-derived exceptional thermal stress in four West-European countries. European news was focused on these countries, especially Portugal and Spain. Such critical extreme temperatures and water conditions, in addition to problems for population living, created unusual economic situation to grow enough food (Euronews 2017). Grains, olives, and grapes were suffered from thermal and water stress (USDA/FAS 2017). About 1.38 million hectares of grains, sunflowers, grapes, and olive trees were affected by drought in Spain as of the end of October and especially in early November 2017 (IPMA 2017). The 2017 water and thermal stressed situation worsened drought impacts on the economy since the current year drought consequences were intensified through continuation of the 2016 dry conditions, which started in the second half of the year (USDA 2017; USDA/FAS 2017). Following TCI estimates, the 2017 grain production was predicted to be much below historical average levels in Portugal and Spain. The current countries' estimates indicate that, compared to the last year, grain harvest is expected to plunge up to 70% and olives harvest is expected to drop in half, compared to regular production (Euronews 2017; Vincent 2017). In addition to crops and trees devastation, grass on the pastures was very scares and farmers have to buy much fodder for their cattle to survive. The strong food demands by the domestic livestock industry, along with limited pasture's grass availability will contribute to an increase in the countries' grain import needs (Vincent 2017; TheGuardian 2017). Following a reduction in 2017 production of grapes, Spain will produce much less wine. The situation worsened since drought affected not only Spain but also reduced grape production in Italy and France. These countries produce nearly 50% of world wine, and due to 2017 thermal stress (Fig. 4.9), the entire world will be facing a shortage of wine. Disastrous grapes harvest means wine prices could be going high (Euronews 2017). Besides production losses, severe thermals stress triggered a few fires, which killed 109 people in 2017 in Portugal (Euronews 2017; Vincent 2017).

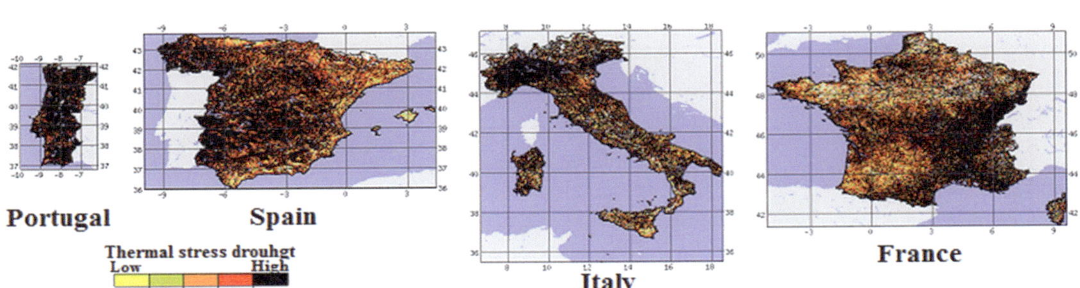

Fig. 4.9 Exceptional VIIRS/TCI-based thermal stress in Western Europe at the end of October and early November 2017 (NOAA/NESDIS 2018)

4.5 Validation

Any developed method must be validated comparing algorithm estimations with corresponding ground data. Many ground measurements, especially such as weather, soil moisture, crop production, biomass data, and other assessments (human health, fire risk, malaria, and others) may be used for these purposes. This subchapter contains these results, received in cooperation with other institution and users. Most of these results have been published.

Figure 4.10a clearly emphasizes that SMN time series in Sudan changed considerably along with weather during 1984–1987. The lowest SMN (0.12) in 1984 was related to below normal precipitation (51% of normal). When precipitation was above or near normal, as in 1987 (115%) and 1985 (95%), respectively, Sudan's maximum mid-summer SMN increased almost three times (to 0.31). In 1986, 27% below normal rainfall (dP is 73% of normal) reduced maximum mid-summer SMN to nearly 0.27. Figure 4.10b supports SMN-based finding in Sudan, showing that VHI derived severe-to-extreme drought (VHI < 35) during the 1988 corn season which coincides with the area of a strongly reduced corn yield (30% below long-term yield trend) and 41% below normal summer precipitation (Kogan 1990, 1995).

The National Agricultural Statistical Services (NASS) of the United States Department of Agriculture (USDA) has developed some methods to model crop and pasture conditions and estimate soil moisture in the USA using weather, weather indices, and other observations. They applied these models to estimate such conditions every year during the growing season. These estimates in May 2013 (USDA/NASS 2013) were compared with VH-based drought estimates from NOAA operational satellites in early May (NOAA/NESDIS 2018). Following Fig. 4.11a, in both cases (1 and 2) the largest percent of the area with P&VP conditions of pasture/rangeland (40 in Nevada to 91% in New Mexico, in Fig. 4.11a1) and winter wheat (40% in Kansas and 74% in Taxes, in Fig. 4.11a2) coincides with the strongest drought (severe-to-exceptional, VHI < 30). The smallest percentage of P&VP conditions in both cases (1 and 2) were estimated for the states with no or light-to-moderated drought (VHI > 35). Identical results of a match between VHI and soil moisture deficit were obtained (Fig. 4.11b). In states with favorable VHI-based conditions (blue color), soil moisture deficit was less (3–37%) compared to drought areas in the west with a deficit of 45–98%.

United States Drought Monitor (USDM) is a low-resolution method used for monitoring drought in the USA (USDM 2017). USDM is

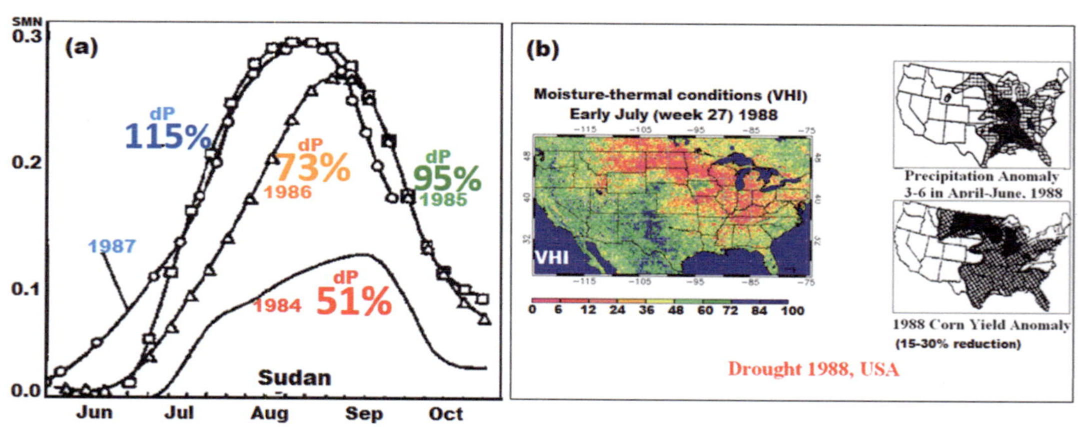

Fig. 4.10 Comparison of satellite data with ground observations: (**a**) smoothed annual weekly no noise NDVI (SMN) in 1984–1987 along with summer precipitation anomaly (deviation from normal, dP) in Sudan and (**b**) area of the USA affected by drought in summer 1988 following VHI-based moisture–thermal drought index, below normal 1988 corn yield and below normal summer precipitation

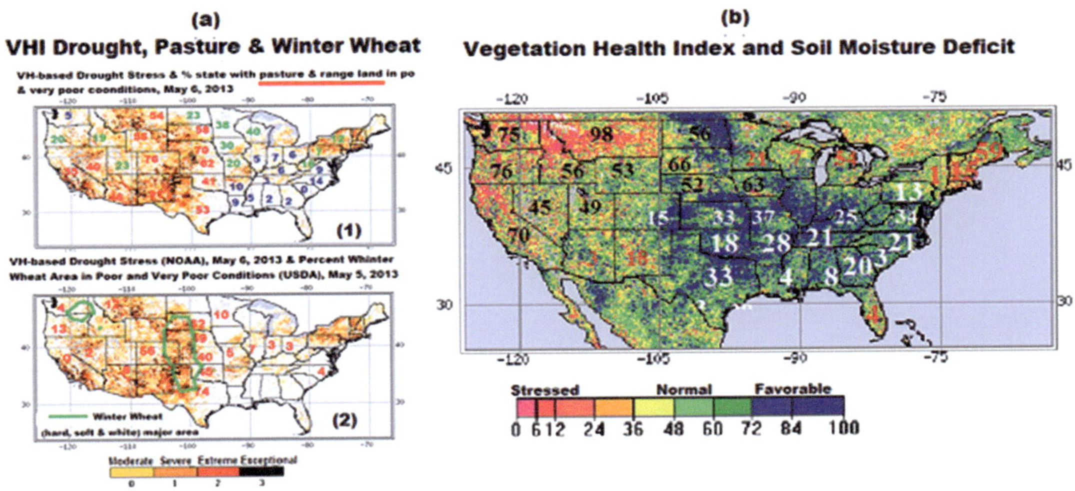

Fig. 4.11 The USA's (**a**) VHI-based drought area and intensity image in early May 2013 and USDA's estimates (1) percent of each state with pasture and rangeland in poor and very poor (P&VP) conditions and (2) percent winter wheat area in P&VP conditions (numbers) and (**b**) vegetation health index and soil moisture deficit (numbers)

based on weather data, indices, and models, which include 3- and 1-month precipitation anomaly, Palmer Z-index, Soil Moisture Model, and Palmer Drought Index. USDM is produced weekly and classifies dryness in five categories from abnormally dry to exceptional drought (Svoboda et al. 2002; USDM 2017). Following the importance of the USDM product, it was compared with the VH-estimated droughts. Figure 4.12 shows comparison of these two products for 2 years, with strong drought in 2012 and light drought in 2016. The result of the comparison clearly indicates that these products are matching well, both in area and intensity. In 2012, central USA was affected by drought in the strongest category, exceptional (D4) from USDM and the same category from VHI (4). In 2016, the same areas in the western one-third of the USA and the eastern part were affected by principally moderated drought (except some areas, such as southern California, where drought was strong). It is interesting that the match was between low-resolution (approximately 1300 km^2) USDM data and high resolution (4 and 1 km^2). One of shortcoming of the USDM is a very low resolution of its data, since the number of weather stations for US territory is limited. Drought impacts assessment with low-resolution USDM data in the central California, where agricultural activity benefits, from irrigation on small areas of 200–300 acres, is not useful. The VH data provides better drought assessments and estimation of irrigation needs with 1 km^2 data.

Dealing with satellite data verification, crop yields are excellent ground indicators to validate the VH assessments. A few years of comparison of vegetation health conditions in Ukraine and Illinois with crops, shown in Fig. 4.13, indicates that yield reduction is associated with stressed conditions of vegetation, which are predicted by satellite-based vegetation health index data. From 2002 to 2004, Ukraine winter wheat data (Fig. 4.13a) showed that the lowest yield (1.8 t/ha) was in 2003 when moderate-to-strong vegetation stress (VHI < 25) was estimated in the principal crop-growing area. In the other 2 years, not affected by vegetation stress (VHI > 50), winter wheat yield was 1.5 time higher (2.7 t/ha in 2002 and 2.8 t/ha in 2004). Similarly, lower corn yield in Illinois (105 b/ac) was in 2012 due to stressed conditions in July indicated by VHI time series (VHI = 20–35, strong drought, Fig. 4.13b). In 2009, VHI-estimated conditions for that period were healthy (VHI = 50–75) and corn yield was 1.6 times higher (175 b/ac). These examples indicated that crop yield changes in a directions corresponding to variations of satellite-derived vegetation health indices.

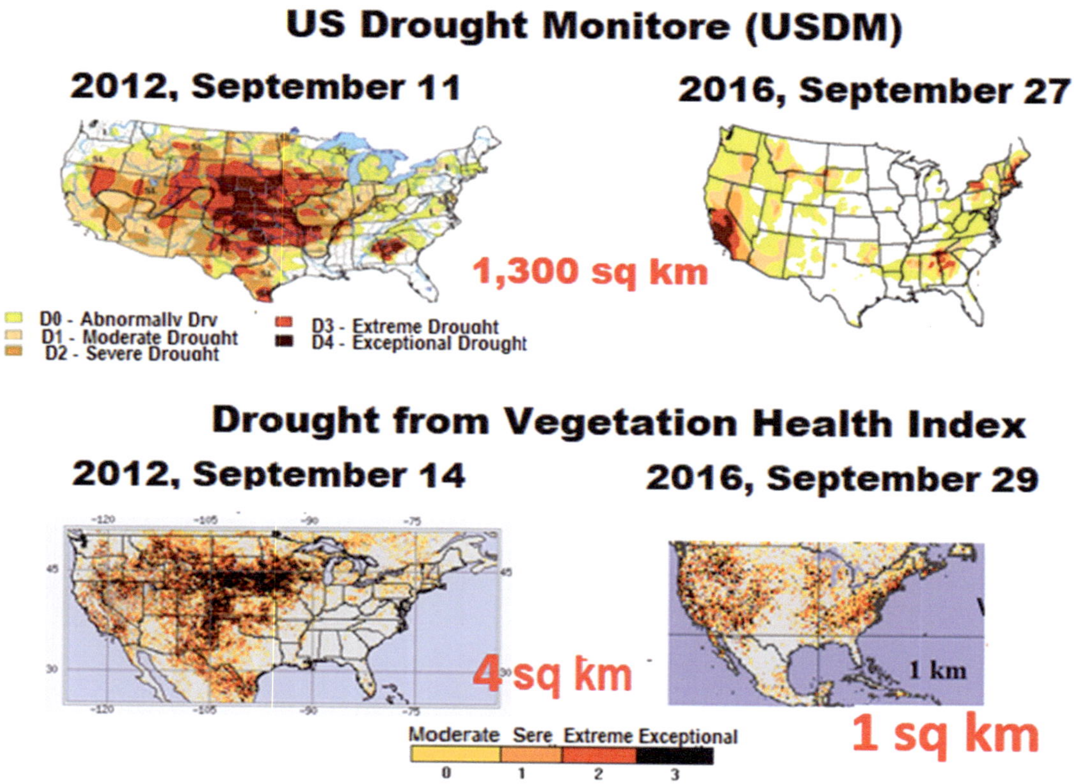

Fig. 4.12 Drought area and intensity from (**a**) US Drought Monitor (USDM) and (**b**) vegetation health index (VHI)

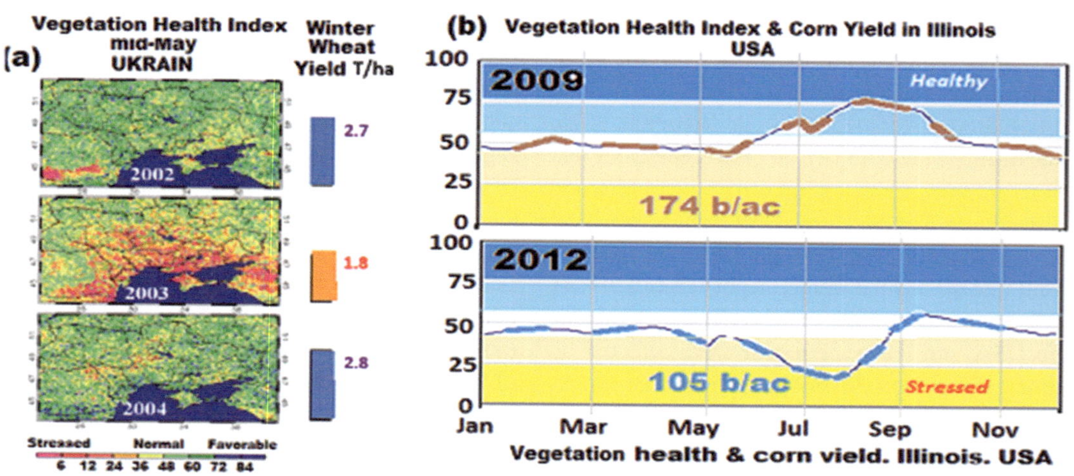

Fig. 4.13 Vegetation health index and crop yield (**a**) winter wheat in Ukraine in 2002–2004 and (**b**) corn in Illinois (USA) in 2009 and 2012

An individual-year match of crop yield with VHI estimations inspired statistically grounded investigation. Figure 4.14 displays multiyear correlation for mean Kansas corn yield, which is strong (correlation coefficient, $R = 0.78$ for VCI and 0.81 for TCI). When moisture (VCI) and thermal (TCI) conditions derived vegetation stress (severe and stronger) corn yield in Kansas dropped below trend and in case of exceptional stress the yield was as low as 40% below trend. When VCI and TCI derived favorable conditions, corn yield in Kansas was up to 35% above the trend level. Table 4.1 presents results of multiyear (at lease 15) correlation analysis between yield anomaly (dY, yield deviation from technological trend line) and mean (inside crop-grown areas) VH indices for different crops in different countries, climate, and ecosystem. In most of Table 4.1 cases, Pearson correlation coefficients (R) are quite high 0.75–0.90, indicating that moisture (VCI), thermal (TCI), and a combination of conditions (VCI/TCI) indices are very good indicators of crop yield. This is true for such different ecosystems/climates as warm tropical (Brazil, India) and dry steppe (central USA, European Russia's south and southern Ukraine). In general, indices below 40 are good predictors of a yield to be below multiyear technological trend value and indices greater than 60 are predictors of an above-trend yield. A very interesting fact is that in some cases, combined (moisture/thermal, VHI) index showed similar (Russia) or even stronger (China, Ukraine, Spain) correlation with dY compared to a single index (moisture or thermal). Time wise, the strongest correlation of dY with VH indices normally occurred during the critical period of crop development (for example, heading-flowering for wheat), which is almost 2 months ahead of harvest. Therefore, VH indices showed to be quite accurate predictors of crops and pasture production and advanced indicators of food security situation. Moreover, VH indices can be used effectively for drought detection, analysis of its intensity and duration, and, what is the most important, they can be used for early prediction of drought-related crop losses, food insecurity in developing countries, and required food assistance.

Following a multiyear correlation between dY and VH indices, an intensity of drought stress was classified based on corn yield reduction (Kogan 1995). The scatter diagram in Fig. 4.15 shows a drought classification scheme. The drought begins when VCI drops below 45 (brown line) since that is the level when relative yield began to drop below 100, which is a yield reduction from regression black line. Yield between 90% and 81% (10–19% deviation from black regression line) corresponds to the VCI between 35 and 26 (between green and yellow lines) indicating moderated drought. Severe drought (between yellow and pink lines) corresponds to a yield between 80% and 75% (reduction from 20% to 25% from regression line), which can be predicted by VCI estimates between 25 and 15. The area between pink and red lines identifies extreme drought, when yield is 66–75% (drops 25–34%) and VCI is between 15 and 6. Finally, exceptional drought is identified when yield is below 65%, which corresponds to the lowest level of VH, 0–5. From Fig. 4.15, the lowest corn yield might be 45% (or drop 55% from regression line).

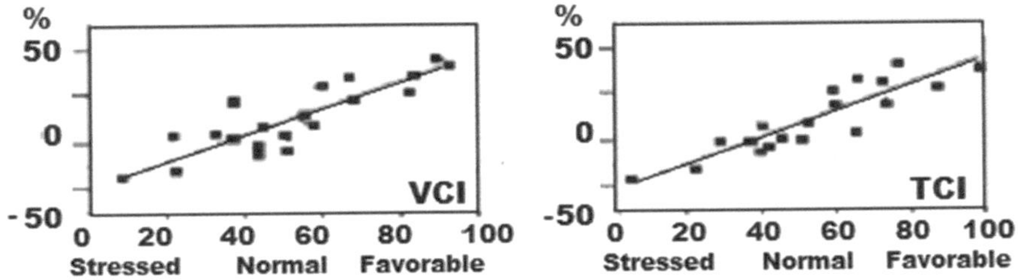

Fig. 4.14 Correlation between Kansas' mean corn yield anomaly (dY, % deviation from 20-year technological trend) with VCI and TCI in mid-July (week 28)

Table 4.1 Relationship between yield anomaly [(dY) deviation from technological trend] and VH (VCI, TCI, VHI) indices (Ecosystem 2016; Kogan et al. 2009, 2013a, b, 2015; Kogan 1987, 1989, 1990, 1995, 1997, 2001, 2002)

General information				Correlation coefficient between dY and VH indices		
Country	Province	Ecosystem	Crop	VCI (moisture)	TCI (thermal)	VHI (moisture–thermal)
USA	Kansas	Steppe	Corn	0.86	0.86	0.84
		Steppe	Winter Wheat	0.87	0.73	0.87
		Steppe	Sorghum	0.86	0.84	0.82
China	Jillian	Temperate drylands	Corn	0.58	0.77	0.81
India	Rajasthan	Warm Tropical Drylands	Wheat			
India	Gujarat	Warm Tropical Drylands	Multiple Crops (wheat, millet, sunflower, pulses, groundnut)			0.86
Russia	Saratov	Temperate Steppe	Spring Wheat	0.76	0.77	0.77
Ukraine	Odessa	Steppe	Winter Wheat	0.49	0.48	0.57
Australia	Entire area	Temperate drylands	Hard Wheat	0.80	0.69	0.77
Bangladesh (1993–2005)	Provinces: Meghna, Jamuna, and Ganges	Warm Tropical Sub-humid	Rice	−0.75	−0.65	
(1993–2011)	Entire area	Warm Tropical Sub-humid	Rice	0.61	−0.62	
	Entire area	Warm Tropical Sub-humid	Potato	−0.39	0.58	
Zimbabwe		Dry steppe	Corn	0.81		
Argentina	Cordoba, Dep 8	Subtropical and Temperate drylands,	Wheat	0.70	0.65	0.73
	Cordoba, Dep 16	Subtropical and Temerate drylands	Wheat	0.57	0.61	0.68
	Cordoba, Dep 19	Subtropical and Temperate drylands	Wheat	0.61	0.81	0.77
Poland		Temperate drylands				
Bulgaria	Pleven region - Early Cultivar		Corn	0.51	0.70	0.65
	Pleven region - Late Cultivar		Corn	0.42	0.60	0.51
Kazakhstan	Central (6 stations)	Dry Steppe	Spring Wheat Biomass	0.87		
Brazil	Capinopolis	Warm Tropical Sub-humid	Soybeans (model)		0.89	
Spain	Cluster 1 (Zamora, Valladolid, Palencia, Segovia, Burgos)	Steppe	Winter Wheat	0.82	0.75	0.91

(continued)

Table 4.1 (continued)

General information				Correlation coefficient between dY and VH indices		
Country	Province	Ecosystem	Crop	VCI (moisture)	TCI (thermal)	VHI (moisture–thermal)
	Cluster 1	Steppe	Winter Barley	0.81	0.77	0.90
	Cluster 2 (Toledo, Cuenca, Ciudad, Real, Albacete)		Winter Wheat	0.74	0.62	0.75
	Cluster 2		Winter Barley	0.78	0.68	0.79
Morocco		Dry steppe				
Hungary		Temperate drylands				
South Africa		Temperate drylands				
Mongolia	Bulgan	Dry steppe	Pastures Biomass			0.91
	Tumenspogt	Dry steppe	Pastures Biomass			0.83
	Bayanunjuul	Dry steppe	Pastures Biomass			0.89
	Partisan	Dry steppe	Pastures Biomass			0.88

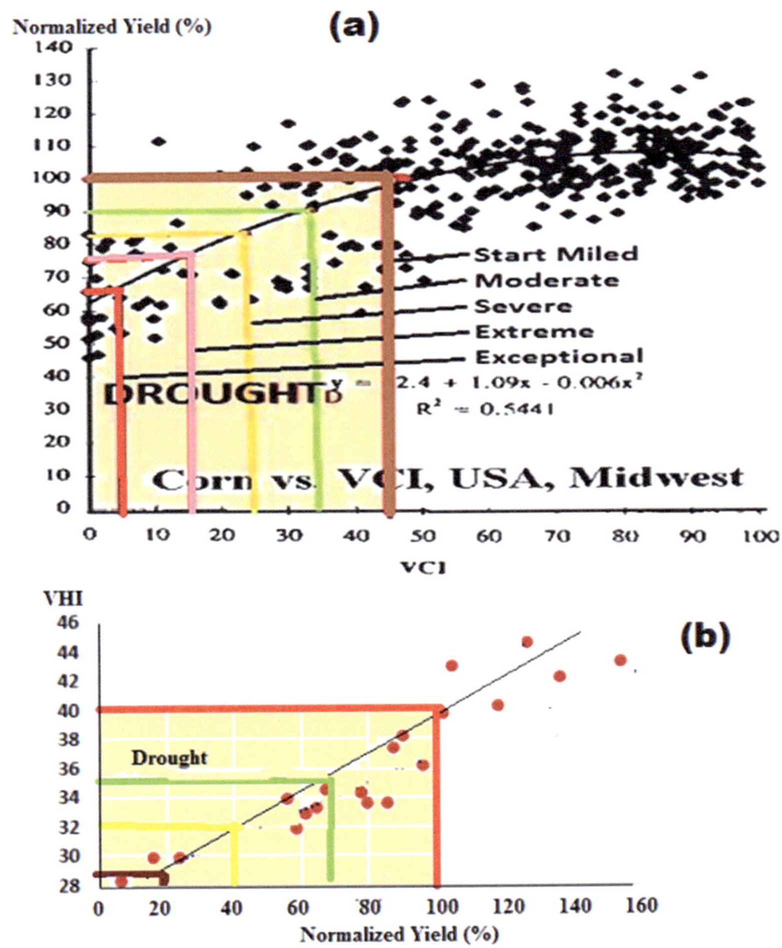

Fig. 4.15 Correlation of Crop Reporting Districts (CRD) mean corn yield anomaly (departure from 1982–1990 mean) in the USA's midwest (Illinois, Kansas, and Nebraska) with CRD-mean VCI (USDA 1991, NOAA/NESDIS 2018)

4.6 Conclusion

Nearly a quarter of the world population does not have enough food for normal living and nearly one billion people become hungry every year. One of the reasons for undernourishment and hunger is drought, which reduces agricultural production. Drought is one of the most adverse environmental disasters affecting the largest number of world population with a deteriorating global food security (FS). During the twenty-first century, eight out of the first 17 years, large-scale droughts were responsible for global FS problems, due to huge losses of grain (major food supply for the world) resulting in global production to be below steadily grown consumption (WB 2017; USDA 2017; PotashCorpo 2013). The new satellite-based vegetation health (VH) method can be successfully used for prediction of grain and pasture production very early to avoid unfavorable global FS situation. The most important is that the VH method will also provide advanced warnings on drought start, area, intensity, and duration, important for making the best decisions on early assessment of drought impact on losses in agriculture and developing measures to mitigate their negative consequences for global FS.

The VH method is unique because (1) operational satellites proved to be a very good single source of information (without weather) for environmental monitoring and prediction of such natural disaster as drought; (2) the vegetation health (VH) method is a new powerful tool, which comprehensively validated on quite a few crops, in different world climates and ecosystems; (3) many developed VH products were successfully used for prediction of crop and pasture production, including losses, helping to improve food security and sustainability; (4) VH data and products have been extremely popular since 2015, 2016, and 2017, with nearly 64,000, 66,000, and 69,000 world users, respectively, accessing the WEB to apply this information in their work; (5) users represent a wide spectrum of interests, including agriculturalists, traders, commodity dealers, bankers, stock Co., farmers, researchers, and others; (6) the first time all of the VH-based application dealing with global FS are presented in one publication.

In November 2017, a new generation of NOAA operational satellite, JPSS-1 (NOAA-20), with VIIRS instrument on board was launched. Among many tasks, NOAA-20/VIIRS was designed to advance drought detection and improve prediction of grain losses using updated vegetation health (VH) method. The VIIRS-based VH will detect drought early, monitor accurately (with 0.5 km^2 resolution) area, intensity, duration and predict agricultural losses 2 months ahead of crop harvests, which will be used for advanced food security assessments. The VH method has been applied to VIIRS data from May 2012, showing considerable improvement in monitoring vegetation condition and drought. It has been found that the new system provides more precise drought spatial resolution, leading to a more accurate assessment of crop losses. These features are very important since large-scale drought reduces global agricultural production every 3–5 years, causing an imbalance between crop production supply and demand, which leads to shortages of food, human health deterioration, poverty, regional disturbances, population migration, and death (Hoerling et al. 2014; PotashCorpo 2013; FDD 2013). The new generation of operational satellites permit users to enhance long-term environmental data records, thereby improving the ability to estimate global warming, land cover changes, and better monitoring of environmental resources.

The new operational satellite system has already proved that data will be used for the development of new products, which will considerably improve climate, weather, water, and socioeconomic services. Among them, will be an enhanced drought watch, including up to 2 months of advanced drought detection on every 500 m^2 of the world land, with a shortest temporal interval. Moreover, data will be calibrated to provide modeling agricultural production and assessment of crop losses up to 3 months ahead of harvest. Forestry will benefit from the prediction of fire danger or risk and the crop ecosystems, in addition to already developed products, benefiting soil science in the estimation of areas saturated with water from excessive rainfall and expected crop losses. An enhanced quality of

VIIRS data will be used for monitoring climate, forcing (e.g., ENSO) impacts on land cover and productivity and also for regional and global assessments of climate and land cover changes, which is important for the prediction of degraded environmental resource. Improvements are expected in the area of mosquitoes-borne diseases, specifically their early detection, assessment of risk, intensity, area under threat, and the number of people affected. Such important parameters in plant process modeling as the fraction of photosynthetic active radiation (FPAR) and leaf area index (LAI) will be measured with greater accuracy. The 37-year global vegetation health data and products of moderate (4 km^2) and high (1 km^2) resolution are currently available from the WEB: https://www.star.nesdis.noaa.gov/smcd/emb/vci/VH/vh_browse.php.

References

Adamenko, T., and A. Prokopenko. 2009. Monitoring drought and impacts on crop yield in Ukraine from weather and satellite data. In *Use of satellite and in-situ data to improve sustainability*, ed. F. Kogan, A. Powell, and O. Fedorov, 3–9. New York: Springer.

ASP (American Society of Photogrammetry). 1975. *Manual of Remote Sensing*, 47. Falls Church, VA: ASP (American Society of Photogrammetry).

CGIAR. 2015. Water, Land and Ecosystems in Africa. https://waterlandandecosystems.wikispaces.com/file/view/Component2_CIAT.pdf.

Cracknell, A.P. 1997. *The Advanced Very High Resolution Radiometer*, 534. New York: Taylor & Francis.

Ecosystems. 2016. Global Ecosystems. https://www.google.com/search?q=global+ecosystems&tbm=isch&source=iu&ictx=1&fir=Gidp3yXe7hK5NM%253A%252CmveIbLDsA0igPM%252C_&usg=__2xzFKa2DvyEyAy3bBne31wR8-8U%3D&sa=X&ved=0ahUKEwi39YGexdrYAhVoct8KHbeDBAcQ9QEIODAD#imgrc=Tk0LW1QZkAqABM.

Ehrlich, P.R., and J.P. Holdren. 1971. Impact of Population Growth. *Science* 171: 1212–1217. https://doi.org/10.1126/science.171.3977.1212.

Euronews. 2017. Devastating drought. http://www.euronews.com/2017/11/09/drought-across-spain-and-portugal-raises-alarm.

FAO. 2016. Earth observations. http://www.fao.org/giews/earthobservation/asis/index_2.jsp?lang=en.

———. 2017. How close we are to zero hunger. http://www.fao.org/state-of-food-security-nutrition/en/.

FDD (FarmdocDAILY). 2013. http://farmdocdaily.illinois.edu/2013/02/locating-the-2012-drought.html.

Friedl, M.A., D.K. McIver, J.C.F. Hodges, X.Y. Zhang, D. Muchoney, A.H. Strahler, C.E. Woodcock, S. Gopal, A. Schneider, A. Cooper, A. Baccini, F. Gao, and C. Schaaf. 2002. Global land cover mapping from MODIS: algorithms and early results. *Remote Sensing of Environment* 83: 287–302.

Gillins, J. 2014. U.S.A. climate has already changed, study find, citing heat and floods. *The NewYork Times*, May 6.

Gray T.T., and D.G. McCrary. 1981. The environmental vegetative index: the tool potentially useful for arid land management. *Proc. 5th Conf. on Biometeorology*, Anaheim CA, 205–209.

Hammond Inc. 1991. *World Atlas, Gemini Edition*, 189–192. Hammond: Hammond Inc..

Hield, C.B., J.T. Randerson, and C.M. Malmstrom. 1995. Global net primary production: Combining ecology and remote sensing. *Remote Sensing Environment* 51: 74–88.

Hoerling, M., J. Eischeid, A. Kumar, R. Leung, A. Mariotti, K. Mo, S. Schubert, and R. Seager. 2014. Causes and predictability of the 2012 great plains drought. *Bulletin of the American Meteorological Society* 95: 269–282. https://doi.org/10.1175/BAMS-D-13-00055.1.

Hui, C. 2006. Carrying capacity, population equilibrium, and environment's maximal load. *Ecological Modelling* 192: 317–320. https://doi.org/10.1016/j.ecolmodel.2005.07.001.

IPMA (Instituto Portugues Meteo A). 2017. MeteoGlobal. https://www.ipma.pt/en/ and http://meteoglobal.ipma.pt/relatos/nevoeiro-13.

JPSS. 2014. Joint polar satellite system. http://www.jpss.noaa.gov.

Kidwell, K.B. 1997. *Global vegetation index user's guide*, 65. Washington, D.C.: NOAA Tech. Rep., Department of Commerce.

Kogan, F.N. 1987. Vegetation health index for areal analysis of NDVI in monitoring crop conditions. *Preprint 18th Conference on Agricultural and Forest Meteorology* AMS, Boston, 103–114.

———. 1989. Remote sensing of weather impacts on vegetation in non-homogeneous areas. *International Journal Remote Sensing* 11 (8): 1405–1419.

———. 1995. Droughts of the late 1980s in the united states as derived from NOAA polar orbiting satellite data. *Bulletin of the American Meteorological Society* 76: 655–668.

Kogan, F. 1997. Global drought watch from space. *Bulletin of the American Meteorological Society* 78: 621–636.

Kogan, F.N. 2001. Operational space technology for global vegetation assessment. *Bulletin of the American Meteorological Society* 82 (9): 1949–1964.

———. 2002. World droughts in the new millennium from AVHRR-based vegetation health indices. *Eos Transaction of American Geophysical Union* 83 (48): 562–563.

Kogan, F., and W. Guo. 2009. Early detection and monitoring droughts from NOAA environmental satellites. In *Use of satellite and in-situ data to improve sus-*

tainability, ed. F. Kogan, A. Powell, and O. Fedorov, 11–18. New York: Springer.

———. 2014. Early twenty-first-century droughts during the warmest climate. *Geomatics, Natural Hazards and Risk* 1–11. doi:https://doi.org/10.1080/19475705.2013.878399.

Kogan, F., T. Adamenko, and M. Kulbida. 2009. Satellite-based crop production monitoring in Ukraine and regional food security. In *Use of satellite and in-situ data to improve sustainability*, ed. F. Kogan, A. Powell, and O. Fedorov, 99–104. New York: Springer.

Kogan, F., T. Adamenko, and W. Guo. 2013a. Global and regional drought dynamics in the climate warming era. *Remote Sensing Letters* 4: 364–372. https://doi.org/10.1080/2150704X.2012.736033.

Kogan, F., N. Kussul, T. Adamenko, S. Skakun, O. Kravchenko, O. Kryvobok, A. Shelestov, A. Kolotii, O. Kussul, and A. Lavrenyuk. 2013b. Based on earth observation, meteorological data and biophysical models. *International Journal of Applied Earth Observation and Geoinformation* 23: 192–203. https://doi.org/10.1016/j.jag.2013.01.002.

Kogan, F., M. Goldberg, T. Schott, and W. Guo. 2015. SUOMI NPP/VIIRS: improve drought watch, crop losses prediction and food security. *International Journal Remote Sensing*. https://doi.org/10.1080/01431161.2015.1095370.

Kuciauskas, A., J. Solbris, T. Lee, J. Hawkins, S. Miller, M. Surratt, K. Richardson, R. Bankert, and J. Kent. 2013. Next-Generation Satellite Meteorology Technology Unveiled. *Bulletin of the American Meteorological Society* 94: 1824–1825. https://doi.org/10.1175/BAMS-D-13-00007.1.

LOM. 2017. Liebig's law of the minimum. https://en.wikipedia.org/wiki/Liebig%27s_law_of_the_minimum.

Monteith, J.L. 1972. Solar radiation and productivity in tropical ecosystems. *Journal Applied Ecology* 9: 747–766.

NCDC (National Climatic Data Center). 2011. Billion dollar U.S. weather disasters. Accessed 2013. http://www.ncdc.noaa.gov/oa/reports/billionz.html.

News24. 2017. https://www.news24.com/Green/News/spain-portugal-struggle-with-extreme-drought-20171121.

NIDIS. 2017. Global drought information system. https://www.drought.gov/gdm/current-conditions.

NOAA. 1988. *Summary of drought conditions and impacts. Drought Advisory 88/12*, 25. Washington, DC: U.S. Department of Commerce, NOAA.

NOAA/NESDIS. 2018. Vegetation health data and products. https://www.star.nesdis.noaa.gov/smcd/emb/vci/VH/vh_browse.php.

Orhan, O., S. Ekercin, and F. Dadaser-Celik. 2014. Use of landsat land surface temperature and vegetation indices for monitoring drought in the salt lake basin area, Turkey. *The Scientific World Journal* 11. https://doi.org/10.1155/2014/142939.

Orians, G.H. 1990. Ecological sustainability. *Environment* 32: 10–15, 34–39. http://www.potashcorp.com/industry_overview/2011/agriculture/16.

Orlovsky, L., F. Kogan, E. Eshed, and C. Dugarjav. 2009. Monitoring drought and oasture productivity in Mongolia using NOAA-AVHRR data. In *Use of satellite and in-situ data to improve sustainability*, ed. F. Kogan, A. Powell, and O. Fedorov, 69–80. New York: Springer.

PotashCorpo. 2013. Agriculture: crop overview. http://www.potashcorp.com/industry_overview/2011/agriculture/16.

Prentice, I.C., W. Cramer, S.P. Harrison, R. Leemans, R.A. Monserud, and A.M. Solomon. 1992. A global biome model based on plant physiology and dominance, soil properties and climate. *Journal of Biogeography* 19: 117–134.

Price, J.C. 1992. Estimating vegetation amount from visible and near infrared reflectance. *Remote Sensing of Environment* 40 (1): 29–34. https://doi.org/10.1016/0034-4257(92)90058-R.

Prince, S.D., and S.N. Goward. 1995. Global primary production: A remote sensing approach. *Journal of Biogeography* 22: 815–835.

Rajasekar, U., and Q. Weng. 2009. Spatio-temporal modelling and analysis of urban heat islands by using Landsat TM and ETM+ imagery. *International Journal of Remote Sensing* 30 (13): 3531–3548.

Reining, P. 1974. The use of ERTS-1 data in carrying capacity estimates for sites in Upper Volta and Niger. *Proc. Annual Meeting of Amer. Anthropological Ass*, Mexico City, Mexico.

Rouse, J.W., R.H. Haas, J.A. Schell, and D.W. Deering. 1973. Monitoring vegetation systems in the Great Plains with ERTS. *Proceedings of the third ERTS symposium*, 309–317.

Running, S.W., G.J. Collatz, J. Washburne, and S. Sorooshian. 1999. Land ecosystems and hydrology. In *EOS science plan*, ed. Michael D. King, 197–260. Washington: NASA.

Runyon, J., R.H. Waring, S.N. Goward, and J.M. Welles. 1994. Environmental limits on above-ground production: observations for the Oregon transect. *Ecological Applications* 4: 226–237.

Saleous, N.E. 2005. An extended AVHRR 8-km NDVI data set compatible with MODIS and SPOT vegetation NDVI data. *Internatonal Journal of Remote Sensing*. 26 (20): 4485–5598.

Sellers, P.J., F. Hall, H. Margolis, B. Kelly, D. Baldocchi, G. den Hartog, J. Cihlar, M.G. Ryan, B. Goodison, P. Grill, K.J. Ranson, D. Lettenmaier, and D. Wickland. 1995. The boreal ecosystem overview and early results. *Atmosphere Study (BOREAS) 15* (6): 114–138.

Shelford, V.E. 1931. Some concepts of bioecology. *Ecology* 12 (3): 455–467. https://doi.org/10.2307/1928991 ISSN 1939-9170.

Sinclair, T.R. 1999. *Limits to crop yield. Plants and population: Is there time?. Colloquium*. Washington, D.C.: National Academy of Sciences.

Stigebrandt, A. 2011. Carrying capacity: general principles of model construction. *Aquaculture Research* 42 (1): 41–50. https://doi.org/10.1111/j.1365-2109.2010.02674.

Svoboda, M., D. LeComte, M. Hayes, R. Heim, K. Gleason, J. Angel, B. Rippey, R. Tinker, M. Palecki, and D. Stooksbury. 2002. The drought monitor. *Bulletin of the American Meteorological Society* 83: 1181–1190.

Tarpley, J.P., S.R. Schneider, and R.L. Money. 1984. Global vegetation index from NOAA-7 meteorological satellite. *Journal of Climate and Applied Meteorology* 23: 491–494.

TheGuardian. 2017. Extreme heat warnings issued in Europe as temperature pass 40C. https://www.theguardian.com/world/2017/aug/04/extreme-heat-warnings-issued-europe-temperatures-pass-40c.

Tucker, C.J. 1979. Red and photographic infrared linear combination for monitoring vegetation. *Remote Sensing Environment* 8: 127–150.

Tucker, C.J., and P.J. Sellers. 1986. Satellite remote sensing of primary production. *International Journal of Remote Sensing* 7: 1395–1416.

USDA. 1991. Crop production: 1990 summary. USDA, NASS, ASB, Cr Pr2-1. January, Washington, D.C., 28.

———. 2017. *Food security*. https://www.usda.gov/topics/food-and-nutrition/food-security.

USDA/FAS 2017. Gain Report. Global Agricultural Information Network. SP1714, June 2.

USDA/NASS. 2013. Crop progress and conditions. https://www.nass.usda.gov/Charts_and_Maps/Crop_Progress_&_Condition/2013/index.php.

USDM (United States Drought Monitor). 2017. http://droughtmonitor.unl.edu/.

Vincent. 2017. Spain, Portugal struggle with extreme drought. Terra Daily News. Madrid. http://www.terradaily.com/reports/Spain_Portugal_struggle_with_extreme_drought_999.html.

WB (World Bank). 2017. Agriculture & rural development. https://data.worldbank.org/topic/agriculture-and-rural-development.

5 Monitoring Drought from Space and Food Security

5.1 Introduction

Drought is a typical phenomenon of the Earth's climate. Losses from droughts, especially in agriculture, are staggering. Even in the USA, a country of the most advanced technology, the average annual cost of drought is around \$6 billion. However, in extreme drought years such as 1988, costs jumped to \$60 billion. During 2001–2017, nearly 20% of global lands were drought-stricken, almost every year and in some years this number jumped up much higher. Developing countries of Africa and Asia have always been the most drought-affected, especially if drought reduced agricultural production and they are facing food security problem. In the twenty-first century, the Horn of Africa experienced several-year droughts, which affected 13 million people in 2011 causing very serious food shortages and hunger. Mongolia's rangelands suffer from very intensive droughts resulting in a lack of feed for livestock every 2–4 years. Unusual summer dryness affected the main grain-producing countries of the Black and Caspian Seas regions in 2007, 2009, 2010, 2012, and 2013 (AMIS 2017; NOAA 2017; Kogan and Guo 2014; NCDC 2011).

Weather data is traditionally used for drought monitoring. However, a weather-based drought-watch system has a few serious shortcomings. First, weather data represents point locations rather than an area. Second, the number of meteorological stations around the world are limited and they are not distributed uniformly, especially in African countries, which have frequent problem with food security. Even in the United States, which has a well-developed weather network, the density of stations is not sufficient for effective drought monitoring. Considering the number of weather stations in the US territory (without Alaska), the US Drought Monitor (USDM), a weather-based US drought product, provides drought information for an area of 1000–1300 mile^2 (640,000–832,000 acres) across the US territory, while California, producing nearly 90% of the USA's vegetables, fruits, berries, and nuts, needs this information for each 200–500 acres occupied by each crop in the irrigated Central Valley. This chapter discusses how the new vegetation health (VH) technology uses high-resolution, 0.5, 1, and 4 km^2 (123.6, 247.1, and 988.4 acres) 38-year NOAA operational polar-orbiting satellite data for early drought detection, monitoring its features (intensity, duration, area, origination, impacts), and prediction of agricultural losses, which are used to assess global and regional food security (Kogan 1989; Kogan et al. 2016, 2017).

5.2 Drought as Natural Disaster

Nearly 85% of all natural disasters on Earth are related and associated with extreme weather events (Obasi 1994). Every day our planet

F. Kogan, *Remote Sensing for Food Security*, Sustainable Development Goals Series,
https://doi.org/10.1007/978-3-319-96256-6_5

experiences droughts, floods, tropical cyclones, tornadoes, heat waves, bushfires, and many other weather-related disasters. In addition to weather, the Earth also experiences geological disasters (earthquakes, volcanoes, etc.). Unfortunately, natural disasters are a part of the environment we live in. They adversely affect the lives of a large number of people and cause considerable damage to the economy, nature, and property worldwide. Moreover, natural disasters are much greater now than 30 years ago and developing countries have much larger losses from disasters compared to developed countries. World leaders, especially in developed countries and international organizations, are very concerned about the disasters, providing economic assistance and developing some measures to reduce or even prevent the adverse impact of disasters. In 1980, the United Nations called for global efforts to minimize the harm from natural disasters, declaring the last 10 years of the twentieth century, the International Decade for Natural Disaster Reduction (IDNDR).

Following IDNDR (1999) in Table 5.1, the World Meteorological Organization (WMO) has estimated that compared to all natural disasters, drought is the most damaging to humanity as an environmental phenomenon. In 25 years since 1967, drought affected 50% of the 2.8 billion people who suffered from all major disasters. Moreover, nearly one billion people were killed by drought between 1967 and 1991, more than from hurricanes, earthquakes, and floods (Kogan et al. 1996). It has been already indicated in Chap. 2 that drought affected food security, reducing global grain production below consumption in almost 50% years of the twenty-first century. Since 2001, drought losses in the 42 highest ranking countries are estimated around $932 billion (Emdat 2011). Between 1980 and 2008, drought globally affected 1.5 billion people (0.3 billion in India only) and killed 0.5 billion (0.3 in Ethiopia). Economic damages were assessed at $13.7 billion in China and $6.0 billion in Australia (Drought 2016). In developing countries, drought impact is much stronger, contributing severely to food shortages, famine, population displacement, and mortality. In Africa, between 1981 and 2010, drought affected 252 million people leading to 0.5 million deaths (Masih et al. 2014; Rojas et al. 2011). In the USA, a country of the high economic and technological development, there have been 203 weather and climate disasters between 1980 and 2016 (NOAA 2017), which were extremely costly. Unfortunately, drought was the principal contributor to costs in the USA and therefore is considered as a "14 billion-dollar" annual event in terms of incurred losses in agriculture, other industries and human activities (NCDC 2011; Wilhite 2000). Table 5.2 shows drought cost and number of people killed during the last 5 years (NOAA 2017).

Table 5.1 World population affected by natural disasters in 1967–1991 (Obasi 1994)

Disaster type	Affected (% of global)	Killed (% of global)
Drought	51	38
Flood	38	9
Hurricanes, etc.	8	27
Earthquakes	2	18
Volcanos	<1	<1

Total number of people affected: 2.8 billion
Total number of people killed: 3.5 million

5.3 What Is Drought?

5.3.1 Drought Definition

Since drought is a very complex and least understood phenomenon with many specific features (Wilhite 2000), there is no precise and universal definition of drought except for the notion that "drought is a period with hot and dry weather" (Wilhite and Glantz 1985). Although this definition is short, it does not provide enough information for characterization of such drought features as beginning/end, intensity, area, duration, origination, and the most important aspect of impact. These problems make it difficult not only to estimate drought impacts but also to predict drought development and effective ways to fight drought's consequences. Besides, a few months of hot and dry weather does not necessarily indicate drought, if the preceding period to dryness was very wet.

Table 5.2 US billion-dollar droughts and fire disasters in 2011–2016 (NOAA 2017)

Year	Disaster	Cost (bill $)	Death (#)	States affected	Damages
2016	Drought and fires Drought	2.0 3.5	21	Southeast & CA North & Southeast	Agriculture Humans & Agriculture
2015	Drought Fires	4.5 3.0	0 12	Ca, NV, OR, WA, ID, MT, UT, AZ, AL	Agriculture Fires
2014	Drought	4.0	0	TX, OK, KS	Agriculture
2013	Drought/heatwave	10.4	53	AZ, CA, CO, IA, ID, IL, KS, MI, MN, MO, ND, NE, NM, NV, OK, OR, SD, TX, UT, WA, WI, WY	Agriculture & humans
2012	Drought/heatwave Fires	30.0 1.7	123 8	NV, ID, MT, WY, UT, CO, AZ, NM, TX, ND, SD, NE, KS, OK, AK, MO, IA, MN, IL, IN, GA, ID, WY, MT, CA, NV, OR, WA	Agriculture humans
2011	Drought/heatwave Fires	12.0 1.4	95 5	TX, OK, NM, KS, LU, AZ	Agriculture humans

Plus, owning to the creeping nature of drought, its effect might take weeks, months, seasons, and even years to appear. The most recent example is the 12-year drought (2006–2017) in California (Kogan et al. 2017), which still continues in 2018 (NOAA/NESDIS 2018). Another drought definition was based on a combination of the essence of drought as hot and dry conditions with biological impacts assessments and was formulated as "shortage of water to meet essential needs" (Wilhite 2000). However, it is not precise either because both "shortage of water" and "essential needs" might be different reflecting specifics of the ecosystems, industry type, applications, and users. For example, in agriculture, "essential needs" means to satisfy water requirements for crop and pastures. However, different crops respond unequally to the same water deficit. Some like sorghum are more drought resistant than corn. For urban communities "needs" might include availability of drinking water or/and water for crop irrigation (Wilhite 2000; Kogan et al. 2017). Another problem is how to measure "shortage of water" for drought detection. In some cases, drought-related vegetation stress due to strong and long heatwaves might heavily damage agricultural crops even in the presence of water in the soil. Therefore, specific definitions have been developed in some countries reflecting areas, industries, and applications (USDM 2017; NIDIS 2017; NOAA/NESDIS 2018). Although there is a problem with drought definition, post-factum drought estimates are working well based on the consequences. Some of these estimates are presented in Tables 5.1 and 5.2.

In the past 50–100 years, droughts have strongly affected global food security. In the first decade of the twenty-first century, severe-to-exceptional droughts covered up to 20% of the globe and continued intermittently for a few years on all continents (Masih et al. 2014; AgriLife 2011). Based on the widely used vegetation health method for food security monitoring (Chap. 3), between 2011 and 2017, drought area had not shrunk. Most recently, 2016 large-area droughts covered northern Brazil, central Africa, northern Russia, and western USA (the western US drought continued from 2006 (Kogan et al. 2017)). Furthermore, following NOAA (2017), droughts affected Sub-Sahara Africa, Mongolia, China, and western Europe in 2015; eastern Brazil, west-central Africa, and Australia in 2014; South Africa and Australia in 2013; and eastern Brazil, South Africa, central Russia, and India in 2012. The best and most recent example of a 12-year drought is in the western USA, specifically California, which between 2012 and 2015 lost a lot of surface and subsurface water, agricultural production, land resources, etc. (USDA/ERS 2017; UWCD 2017; Kogan et al. 2017). Drought often covers a large area over a short time (days, weeks) or a long (several months) period of time, including cases when it lingers for years (Kogan et al. 2017). Finally, in the first 6 months of 2018, intensive drought affected strongly grain-producing areas of the USA, Mexico, Australia, and central India; Mongolia's pasture land was under strong vegetation stress, creating problem for maintaining cattle feeding; strong drought also affected the countries of Sub-Sahara Africa, which are likely to have food security problems.

5.3.2 Drought Features

Unlike other natural disasters, drought has a lot of unique features. In addition to global coverage, excessive cost, and the number of affected people, drought is a part of Earth's climate, occurs every year, and spreads across borders, not recognizing political and economic differences. There are also more specific drought features: (a) it is multidimensional phenomenon, (b) starts unnoticeably, (c) builds up slowly, (d) develops cumulatively, (e) produces cumulative impact, and (f) drought impacts are not immediately observable (NOAA 2017; Kogan 1997). Among these specific features, multidimensionality, cumulative development, and adverse influence features create difficulties in early drought detection, monitoring its appearance, performance, and origination for timely estimation of the impacts. Therefore, weather-based parameters and indices that are often used for drought monitoring are insufficient not only for characterization of spatial and temporal drought features but also such characteristics as drought start, intensity, and duration; all these features are extremely important for timely initiation of mitigating measures. Due to cumulative development, the impact of drought is also cumulative and not immediately observable. By the time damage is visible, the drought is already in full swing and it is too late to mitigate the consequences. This issue is very important in regions and countries with limited and/or variable water supplies and where the economy is highly depended on agriculture.

5.3.3 Measuring Drought

There are four distinctive method of measuring and estimating drought: meteorological, land cover (vegetation), soil moisture, and remote sensing. In order to detect and monitor drought, measures should include such characteristics as drought start/end, intensity, duration, magnitude, area, season, origination, and impact. Drought start and end is identified when some measurements cross certain environmental thresholds, which are experimentally determined from in situ observations; this could be deviation of some weather parameter from climatic norm (Example: precipitation (P) < 80% of multiyear mean or yield (Y) reduction below some level, etc.). Drought intensity commonly refers to the severity measured by a reduction of selected parameters (Example: $P < 60\%$ indicates severe drought or more than 30% Y losses specifies extreme drought). Drought duration is measured in the number of days (weeks, months, etc.) with drought. Magnitude accounts for a combination of drought intensity and duration (example: the number of days with severe drought) or a combination of drought area, intensity, and duration. Drought area can be measured as a percentage of an administrative region important for decision makers to develop drought mitigation measures (example: imposing some limitations for water use). Drought timing is also an important component since it determines if urgent measures are needed to mitigate drought impacts (for example: an early detection of summer drought is needed in determining the amount of agricultural purchase to avoid food security problems). Drought origination specifies if drought is moisture-based, thermal-based, or a combination for planning crop irrigation. Finally, drought impacts should be specified based on a type of affected economy (for example: crop yield reduction, water use limitation, etc.). Each drought is unique, but common features of the strongest droughts that have far-reaching human and environmental impacts include long duration, large moisture deficits, severe thermal stress, large areal extent, and strong impact.

5.3.4 Drought Type

Drought type can be classified by the affected area, ecosystem, economy, climate, weather, produced impacts, and other considerations. However, several more common droughts are referenced in scientific literature: meteorological, hydrological, agricultural, and socioeconomic (Wilhite and Glantz 1985). Meteorological drought is defined by a shortage of precipitation (below climatic norm), and elevated temperature (above climatic norm) for a location. Hydrological drought is defined by shortages of water based on the impacts of precipitation shortfall on surface and subsurface water deficit, expressed in a reduction of river discharge, lack of ground water, decrease in reservoir level, and others (Kushnir et al. 2009). Other two drought type classes emphasize social aspects related to drought impacts. Agricultural drought links availability of water to meet crop and pasture requirements for development of good production. The available water might be measured by the amount of precipitation and/or soil moisture deficit and temperature anomaly, although more often it is measured by crop/pasture losses and other biological parameters (Verigo and Razumova 1963). Finally, socioeconomic drought aspects are associated with supply and demands for some commodity, economic goods and services, including meteorological, hydrological, and agricultural elements (NDMC 2017).

5.4 Drought Detection and Monitoring Methods

There are several methods used for drought detection and monitoring. This book is focused on the most popular methods in the past 50 years. They are classified as meteorological, vegetation, soil moisture, and remote sensing.

5.4.1 Meteorological Methods

Meteorological methods of drought detection and monitoring are widely presented by two approaches: based on observed meteorological parameters and meteorological indices. *Meteorological parameters.* Weather data has been used since the mid-1930s in the former USSR as a tool for drought monitoring (Selyaninov 1937; Tsuberbiller 1963). Among meteorological parameters, temperature (T) and precipitation (P) were the most widely used for drought detection and monitoring. Very typical periods used for P and T aggregation are 7-day, 10-day, monthly, seasonal, and annual. The principal drought criterion is deviation of P and T from climatic mean or normal (PCM and TCM correspondingly). Climatic mean is normally calculated for the period of at least 30 years. Due to climate change, these means should be currently calculated for the most recent 30-year period. The principle formulas calculating P and T anomaly (dP and dT, expressed in percentage) are: $dP = 100*P/PCM$ and $dT = 100*T/TCM$, where P and T are weekly, 10-day, monthly, etc. precipitation and temperature; climatic mean is taken for the same time criteria (weekly, etc.) The dP-based drought is defined selectively (if dP is below 80%, or $P > 20\%$ below the climatic norm). The T-based drought is defined if T exceeds TCM more than 1 °C for a mild and 2–3 °C for a stronger drought. In addition, there were some attempts to use effective precipitation expressed as a sum of P over a certain period of time or P exceeding certain thresholds (for example, >1 or > 5 mm per day) to characterize less- or non-drought conditions, or accumulated P deficit. Additional P-based drought analysis might be done calculating the number of days with P below some thresholds, such as <1 mm or < 5 mm and others depending on the area's climate. If T exceeds climatic norm by more than 2 ° C for 1–4 weeks, then thermal vegetation stress can be expected, especially if it is accompanied by a deficit of rainfall. In semidry areas of Russia, Ukraine, and Kazakhstan, relative air humidity of less than 30% (or water vapor deficit larger than 20 mm) for 10–20 days indicates atmospheric drought conditions. The same conditions might be pronounced if soil moisture in the 0–20 top layer is less than 10–20 mm of productive water (Tsuberbiller 1963; Verigo and Razumova 1963).

Meteorological indices. P and T parameters were widely used to develop the indices (Selyaninov 1937; Palmer 1965). Many indices were designed to capture drought location, time, and economic sector. The most popular indices are shown below. *Hydrothermal coefficient (GTK)* is the oldest known index, which was developed in the former USSR (Selyaninov 1937). GTK is expressed as $\sum P/(\sum T/10)$, where T and P are summarized for the period with average daily $T > 10$ °C. GTK is a very simple index reflecting the essence of crop-growing processes and was correlated with crop yield. The index estimate both dry (GTK < 1.0) and wet (GTK > 1.0) conditions. *Palmer drought severity index (PDSI)* is the oldest index developed and used in the USA. It is based on a simple water balance model used as a basis for developing regional drought severity indicator following the methodology of weighting antecedent versus current conditions (Palmer 1965). Palmer used monthly precipitation as a water input and a simplified two-layer bucket model to quantify monthly water demand as water output based on P and T. PDSI was developed for the semiarid climates of the US Great Plains and was correlated with soil moisture in Illinois. Although the original index was applicable for a limited, mostly semidry area, it was later applied over the entire USA and promoted in a number of countries around the world, for which some parameters may not work well (Heim Jr 2002). Also, some hydrological processes have been simplified (bucket model) in the PDSI what resulted in upward climatic trends in PDSI, which disappeared when more physically rigorous principles and equations were applied (Sheffield et al. 2012). Besides, the index did not match well with corn yield reduction in Iowa during the 1988 US severe drought (Kogan 1997). Ped index (*Index-Ped,* Ped 1975) is one of the interesting meteorological indices combining P and T normalized by their 30-year standard deviation (SD). Ped = ((T-Tmean)/TSD − (P-Pmean)/PSD)), where T and P temperature and precipitation for the selected period (7-, 10-, 30-day) and TSD and PSD is SD for the corresponding P and T. Ped index detects mostly long-term (monthly) meteorological drought. *Standardized precipitation index—SPI* is based on a probability distribution function applied to the year (from at least 30-year period) total monthly precipitation (P). The cumulative distribution is transformed to a normal distribution (with a mean and standard deviation equal to zero and one, respectively) to standardize the output (McKee et al. 1993). The shortcomings of SPI are well described in Hayes et al. (1999). It is important to emphasize other limitations for drought monitoring: the index is monthly and even multi-monthly (3-month and 6-month are the most popular) although drought monitoring requires at least weekly data. Moreover, SPI has never been validated against in situ data, especially for agricultural drought, nor has it ever been published in a review journal. Other concerns are that it is hard to determine what SPI period to use: 1-, 2-, 3-, 12-month, or more for drought detection and monitoring (NIDIS 2017) and why in some instances 3- and 6-month SPI are selected to characterize a drought (USDM 2013) since sometimes one- or two-week drought (specifically thermal) can decimate vegetation, especially crops. *Standardized precipitation (P)-evapotranspiration index (SPEI)* is based on an estimated monthly climatic water balance, which is P minus potential evapotranspiration (PET). SPEI is expressed as a standardized Gaussian variation with a mean and standard deviation zero and one, respectively (Begueria et al. 2010). It was based on 0.5° gridded area of monthly total P and mean T. PET was calculated based on the Thornthwait method with Kolmogorov test applied to determine the fit between SPEI and the log-logistic distribution at all timescales across the world. The 3-month and 12-month SPEI were used as a short-term drought indicator and a water deficit, respectively. Again, this index has similar to SPI problems. The new *forest drought-stress severity index— FDSI* (Redorbit 2012), which combines winter precipitation amounts, and late summer and fall temperatures and precipitation into single statistics called Red-FDSI matched the records of changing forest conditions in the Southwest remarkably well (Redorbit 2012). The *US Drought Monitor (USDM)* was developed to estimate and display the magnitude and spatial extend of drought in the United States. The USDM is not an index, but is a "composite product developed from climate indices, numerical models and the e-mails' input of regional drought watchers around the country" (Svoboda et al. 2002; USDM 2017). USDM is produced weekly and classifies dryness in five categories (assigned by climate and water expert opinions in the USA) from abnormally dry to exceptional drought. Following http://drought.unl.edu, the short-term drought-blended indicators include Palmer Z-index (35% percentile contribution), 3-month precipitation anomaly ((*P*)—25%), 1-month *P* anomaly (20%), NOAA's Climate Prediction Center (CPC) Soil Moisture Model (13%), and Palmer drought index (7%); the long-term drought-blended indicators include Palmer hydrological index (25%), 24-month *P* anomaly (20%), 12-month *P* anomaly (20%), 6-month *P* anomaly (15%), 60-month *P* anomaly (10%), and CPC modeled soil moisture (10%). USDM is currently considered as the main drought product in the USA (USDM 2017), although its resolution is very low.

5.4.2 Soil Moisture and Vegetation Methods

Vegetation (or land cover) and soil moisture (LCSM) method of drought detection and monitoring are based on field observations. These methods are the most reliable since they characterize drought through direct measurements of vegetation performance and soil moisture availability, which are cumulative indicators (Verigo and Razumova 1963). However, one problem with this method is a lack of observations over space and time because these measurements are labor and time intensive. Meanwhile, in some countries (USSR (currently Russia), Mongolia, and China) these measurements (soil moisture, corps' phenology (phases), height, density (number of plants per 1 m^2), and biomass (cutting and weighting on an area of 1 m by 1 m at a limited number of weather stations) have become standard via a 10-day procedure. These parameters normally estimate drought conditions if their values in a particular year are below a multiyear mean (normal). *Normalized biomass (NorBio)* was designed as: $NorBio=(Bio_i - Bio_{min}).)/(Bio_{max} - Bio_{min})$, where Bio_i is measured (cutting and weighting) biomass and Bio_{max} and Bio_{min} are multiyear biomass maximum and minimum, respectively, for a location. NorBio indicates if a current year biomass is closer to multiyear absolute minimum of biomass, indicating unfavorable vegetation conditions, or if it is closer to absolute maximum biomass, indicating favorable conditions. *Soil moisture anomaly (SMA)* was measured in mm of productive (used by crop) moisture (Verigo and Razumova 1963) for every 10 cm to the depth of 1 m. SMA used for monitoring drought is derived as a deviation of total soil moisture in 20 cm soil layer from multiyear mean values ($SMA=SM_i/SM_{mean}$), where SM_i is soil moisture (mm) for the selected depth and a period of crop development.

Although meteorological, soil moisture, and vegetation methods were widely used, it is important to indicate some of the most common shortcomings: (a) sparseness of weather station networks, especially in climate, ecosystem, and population marginal areas; (b) selection of the right period of data aggregation to reflect cumulative drought feature; (c) the identification of drought start; (d) the estimation of contribution of pre-drought conditions into the current drought; and (e) the absence of drought validation (most of the developed methods, including those with the strong theoretical background, were not validated against in situ data which diminishes their reliability). Therefore, in the recent 50 years, satellite methods have been developed.

5.4.3 Satellite-Based Methods

Operational polar-orbiting environmental satellites, launched in early 1960s, were designed for high-resolution daily weather monitoring around the world. In early years, they were mostly applied for cloud monitoring, measuring precipitation and air temperature, and also for developing and advancing skills in satellite data development, understanding, and applications. New era opened with a series of operational TIROS-N launched in 1978 which are continued up to now. Among other sensors, these satellites have the Advanced Very High Resolution Radiometer (AVHRR), which has shown to be extremely useful for observing and monitoring the earth surface, weather, and climate. Since the 1980s, specific advances were achieved in understanding ecosystems (vegetation distribution) and weather's impact on vegetation. In the past four decades, considerable experience has been gained in measuring, modeling, and interpreting vegetation conditions from satellite numerical data and images. These have helped develop many new applications, including operational, high-resolution drought detection, monitoring and measuring impacts on agriculture. These achievements have helped to understand how to use operational satellites for estimation of the impact of climate and weather on food security.

In the last four decades, satellite technology has successfully filled in drought monitoring and impacts assessment gap. The current operational satellite data and indices provide numerical approximation of global and regional drought with very high (0.5 and 1 km) and moderate (4 and 16 km^2) spatial resolution (Kogan et al. 2017). These are important advantages of satellite technology in drought monitoring compared to low-resolution weather station data (on the average one weather station in the USA covers 1000–1300 km^2 in Africa—10,000–30,000 km^2). In the past 10–15 years, satellite remote-sensing Vegetation Health method has proven (through validation and applications) to be perfect for operational drought management as a separate tool and as well as in combination with weather data (Kogan 1989, 1995a, b, 1997, 2002). Finally, it is important to emphasize that the remote-sensing drought detection and monitoring method has considerably improved with the development of a new generation of operational satellite technology that started at the end of 2011. The new Visible Infrared Imager Radiometer Suite (VIIRS) on the National Polar-orbiting Operational Satellite System (NPOESS) Suomi Preparatory Platform (SNPP) has provided observations for each 375 m^2 of the globe, with four times more spectral bands (compared to its predecessor, AVHRR operational sensor). Besides, NOAA operational afternoon polar-orbiting satellites has had and will have a long service history since they began in 1981 and have continued for 38 years, including intermediate generation of SNPP/VIIRS and the new J-1 generation, which was launched in 2017, was renamed as NOAA-20 and will continue (NOAA-21 to NOAA-24) until the mid-2040s. Besides NOAA operational polar-orbiting afternoon satellites, some land cover investigations were produced by NASA/MODIS (Terra—morning and Aqua—afternoon) research satellites.

The most popular product used for land cover monitoring since the early 1980s was the *normalized difference vegetation index (NDVI)* applied widely (Deering 1978; Tucker 1978; Kogan 1989, 1995a, 1997; Cracknell 1997; Kidwell 1995). The NDVI, expressed as NDVI = (NIR-VIS)/(NIR + VIS), is the most popular index because it was produced operationally for the entire globe, has a reasonable match with vegetation distribution, is simple, and explains well land ecosystem and environmental conditions (Kogan 1989; Kidwell 1995; Cracknell 1997). NDVI is a dimensionless measure of a combination of reflectance in the visible (VIS) and near-infrared (NIR) solar radiation spectral bands (Cracknell 1997). The most popular operational AVHRR sensor on NOAA polar-orbiting satellites has wide spectral range or spectral response function, for these channels: 0.58–0.68 μm for VIS and 0.72–1.1 μm for NIR, compared to the other similar sensors (MODIS), including the new generation of operational sensor, VIIRS NOAA/SNPP and J-1. Most satellite indices developed for land surface and drought monitoring were based on NDVI and a few on a combination of NDVI and thermal channels. A few of the most popular indices are presented below. *Normalized multiband drought index—NMDI* (Wang and Qu 2007) is based on empirical parameterization of relationship between NDVI and Normalized Difference Water Index (Gao et al. 2000). The NMDI = $(R_{800} - (R_{640} - R_{2150}))/(R_{800} + (R_{640} - R_{2150}))$, where R is reflectance and number indicate nanometers. This index is using very narrow bands from VIS (640), NIR (800) and mid-IR (2150). *Normalized difference drought index—NDDI* is based on NDVI and NDWI (Gao et al. 2000). The NDDI is expressed as NDDI = (NDVI-NDWI)/(NDVI+NDWI). Other drought indices were developed using NDVI and one of the thermal channels deriving land cover temperature. *Vegetation supply water index—VSWI* (Carlson et al. 1990) is based on NDVI and land surface temperature (LST). The VSWI = LST/NDVI. The LST/NDVI ratio normally increases during drought because LST becomes hot, while NDVI loses greenness, following shortage of water. *Vegetation supply drought index—VSDI* is based on a combination of blue, red, and shortwave IR optical spectral bands (Zhang et al. 2013). The VSDI = $(1 - [(R_{swir} - R_{blue}) + (R_{red} - R_{blue})])$. Following these authors, the index is suitable for drought monitoring throughout the growing season in America. *Wide dynamic range vegetation index—WDRVI* (Gitelson and Kufman 1998), developed at the University of Nebraska, Lincoln. Unlike most other methods, this one was comprehensively based on field hyperspectral measurements with portable spectral radiometer and also measurements of biophysical parameters (plant density, number of green leaves and dead leaves, density of reproductive organs, and leave area index) during several-year on wheat, maize (both irrigated and dry land), and soybeans crops. The WDRVI = $(a$*NIR-VIS$)/(a$*NIR + VIS), where a is ranging between 0.05 and 0.2 depending on crop's vegetation fraction for each crop. The WDRVI performed much better than NDVI in estimation of lush vegetation. *Enhanced vegetation index—EVI* (Huete et al. 2002). The EVI = G*(NIR− VIS_{red})/[(NIR + C1*VIS_{red}) − C 2*VIS_{blue} + L], where VIS_{red} and VIS_{blue} and NIR are atmospherically corrected (or partially corrected for Rayleigh and ozone absorption) surface reflectance, L is the canopy background adjustment that addresses nonlinear, differential NIR and VIS_{red} radiant transfer through a canopy, and $C1$, $C2$ are the coefficients of the aerosol resistance term, which uses the blue band to correct for aerosol influences in the red band. The coefficients adopted in the EVI algorithm are, $L = 1$, $C1 = 6$, $C2 = 7.5$, and G (gain) = 2.5. The EVI method was applied to the MODIS (narrow bands) data to develop surface reflectance and EVI products with resolution: 16-day composite over time and 250, 500, and 1000 m over space. EVI were compared with limited biophysical data (reflectance, NDVI, EVI, and their seasonal profiles) collected over four validation test sites to investigate temporal and spatial variability in vegetation properties and condition. The EVI values are lower than NDVI. *Vegetation drought response index—VegDRI* (Brown et al. 2008). This index integrates climate- and satellite-based drought indicators with some biophysical parameters on 1 km^2 map with a weekly frequency. The input data include climate indices: 52- and 40-week SPI for spring-summer and fall, respectively, PSDI (self-calibrated and improved); NDVI-based parameters: percent (relative to 16-year mean) average seasonal greenness (PASG—integrated over the growing season with biweekly steps); start of seasonal anomaly expressed as a deviation from seasonal 16-year mean; land cover (1 km resolution from USGS); soil available water capacity; irrigated agricultural land (from the USDA's 1997 Census); and ecological regions (from the Omernik Level III ecoregion data). VegDRI was developed for Montana, North and South Dakota, Wyoming, Nebraska, Colorado, and Kansas, where climate data was collected from 776 weather stations between 1989 and 2005 and NDVI data was collected at 1 km^2 resolution. The VegDRI models were composed based on "if then" rules assigned by the values of the described parameters. VegDRI was applied to the entire USA and validated comparing the index spatial distribution with USDM in the study area (Brown et al. 2008). *Evaporative drought index (EDI):* Yan et al. (2014) proposed an evaporative drought index (EDI) to monitor droughts over the conterminous United States. They used MODIS and NCEP reanalysis data, and statistical methods to estimate ET and PET at 4 km^2 spatial resolutions and a monthly time step, and used the deviation of the ET/PET ratio from unity to define the EDI. The integrated remote-sensing data in the EDI is sensitive to the vegetation drought response and enhances EDI capabilities for drought monitoring and detection. In addition to the described indices, there are others such as crop moisture index (CMI), surface water supply index (SWSI), reclamation drought index (RDI), simple difference index, simple ration index, temperature vegetation drought index (TVDI), perpendicular drought index (PDI), anomaly vegetation index (AVI), standard vegetation index (SVI), and scale drought condition index (Palmer 1965).

Unlike meteorological, vegetation, and soil moisture methods, satellite indices have high spatial resolution, were published in review journals, and some were validated with in situ measurements, including weather data, pasture biomass, soil moisture, and radiometric measurements in the fields (Gitelson and Kaufman 1998). On a negative side, it is important to indicate that most of these indices (a) were not developed and validated with data obtained from operational satellites and were not projected to be in service for a foreseeable future; (b) had only regional applications; (c) did not include cumulative component for drought monitoring; (d) did not reflect pre-drought condition; (e) were not calibrated to assess drought start/end, intensity, duration, and impacts; (f) did not provide the entire spectrum of vegetation condition in order to assess accurately drought start based on transition from favorable/normal to stressed conditions. Therefore, the new method called vegetation health (VH) was developed in the early1980s (Kogan 1987, 1989).

5.4.4 Operational Satellite-Based Vegetation Health (VH) Method and Drought Monitoring

The new VH numerical method (described in detail in Chap. 3 and 4) was developed from operational NOAA/AVHRR satellite data. Briefly, it is based on a combination of the VIS (0.58–0.68 μm) and NIR (0.72–1.1 μm) radiances (channels), converted to NDVI, and infrared (10.3–11.3 μm) channel measurements, converted to brightness temperature ((BT), Kidwell 1995; Kogan 1989, 2001). These data were pre- and post-launch calibrated, daily and weekly-composite maps were created, and high-frequency noise was completely removed from weekly NDVI and BT time series, converting them to Smoothed NDVI (SMN) and Smoothed BT (SMT). The VH method is built on the three basic environmental laws: Law of Minimum (LOM), Law of Tolerance (LOT), and the Principle of Carrying Capacity (CC), which defined the last 37-year (since 1981) environmental limits that resources of habitats can support (Orians 1990; Kogan 1995a). Following these laws, climatology (the lowest and the highest ecosystem levels that characterize CC of the environmental resources or extreme fluctuations of weather-related vegetation conditions) in each pixel was developed. The ecosystems were stratified and high-to-low frequency vegetation fluctuations associated with weekly weather variations were identified (Kogan 1989, 1997). Finally, the three indices characterizing moisture (VCI = 100*((SMN–SMNmin)/(SMNmax–SMNmin)), thermal (TCI = 100*((SMTmax–SMT)/(SMTmax–SMTmin)), and vegetation health (VTI = a*VCI + $(1-a)$*TCI) conditions were approximated, where SMN, SMNmax, and SMNmin are no noise weekly NDVI, 36-year absolute maximum and minimum, respectively; SMT, SMTmax, and SMTmin are no noise data characterizing brightness temperature; a is a coefficient that quantifies a share of VCI and TCI contribution to the total Vegetation Health.

The new method was first applied to the 16 km^2 NOAA AVHRR-based global vegetation index (GVI) data set issued routinely since 1985 (Kogan 1989; Kidwell 1995). Currently, the method, called global vegetation health (VH) is applied to NOAA/AVHRR 4 km^2 resolution global area coverage (GAC), SNPP/VIIRS, and NOAA-20/VIIRS data sets. The AVHRR global VH is produced from 1981 through present for each 4 and 16 km^2 land pixels between 75° N and 55° S and between 180° W and 180° E by sampling the AVHRR-based 1.1 km^2 daily radiances and compositing them over 7-day period for each 4 and 16 km^2 global land pixels. Since 2013, the

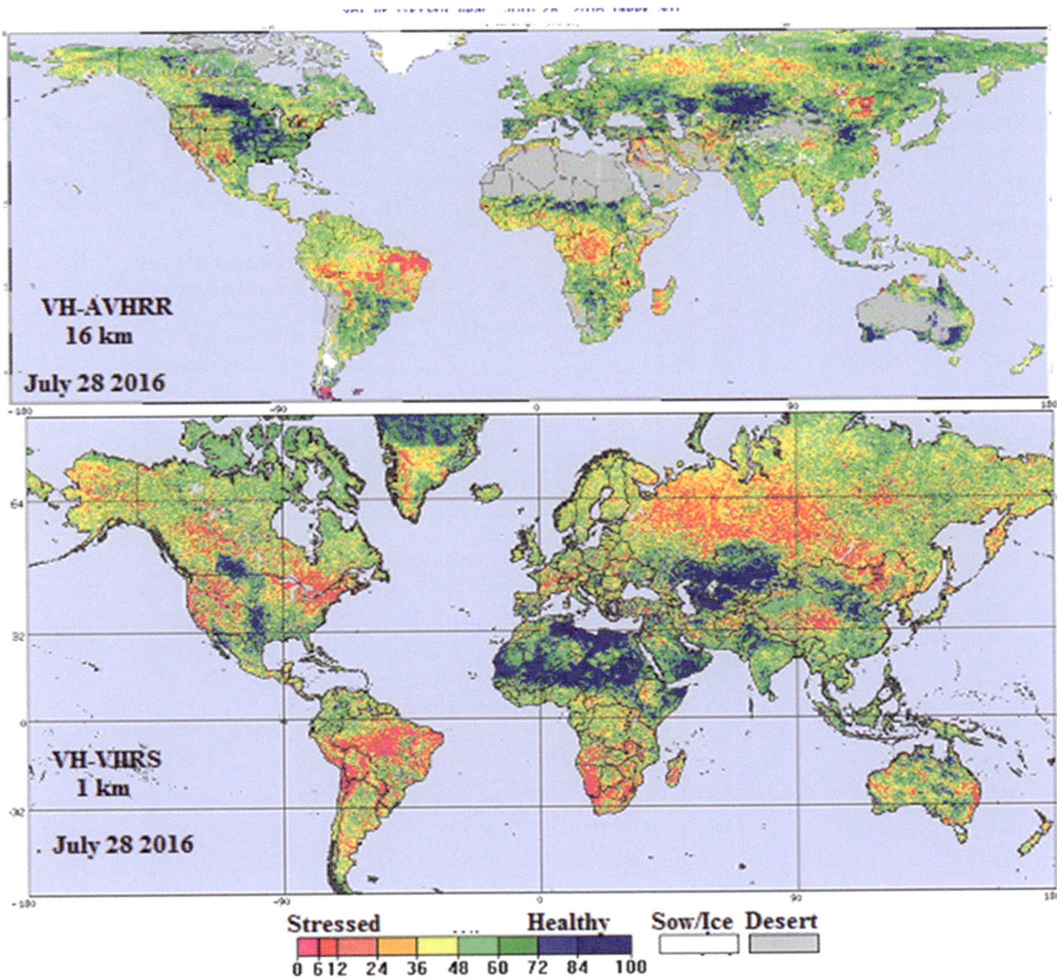

Fig. 5.1 Global moisture–thermal index (VHI) for the end of July 2016 (the hottest year on earth in the past 100 years) from AVHRR's 16 km^2 and VIIRS's 1 km^2 data

global weekly VH is sampling the VIIRS-based 375 m² daily radiances and compositing them over 7-day period globally for every 1 and 4 km^2 pixels and from 2018 the same technique is producing NOAA-20/VIIRS data. Figure 5.1 shows global vegetation health index (VHI) for the end of July 2016 (the hottest year on earth in the past 100 years) from NOAA/AVHRR 16 km^2 and SNPP/VIIRS 1 km^2 data. These data were used to (1) distinguish drought from moderate and healthy conditions and (2) classify drought from non-drought conditions and determine drought intensity based on the impacts.

Drought classification: The advantages of VH (VHI, VCI, and TCI) against another satellite-based method is in providing numerical classification of vegetation conditions, which distinguishes drought from non-drought (based on moisture, thermal, and combine condition) and assesses drought start/end, intensity, area, duration, origination, and impacts. VH indices and products used for timely assessments of vegetation stress (both moisture and thermal), crop and pasture production losses and through them food security situation were formalized based on the VH-assessed crop yield reduction. Following

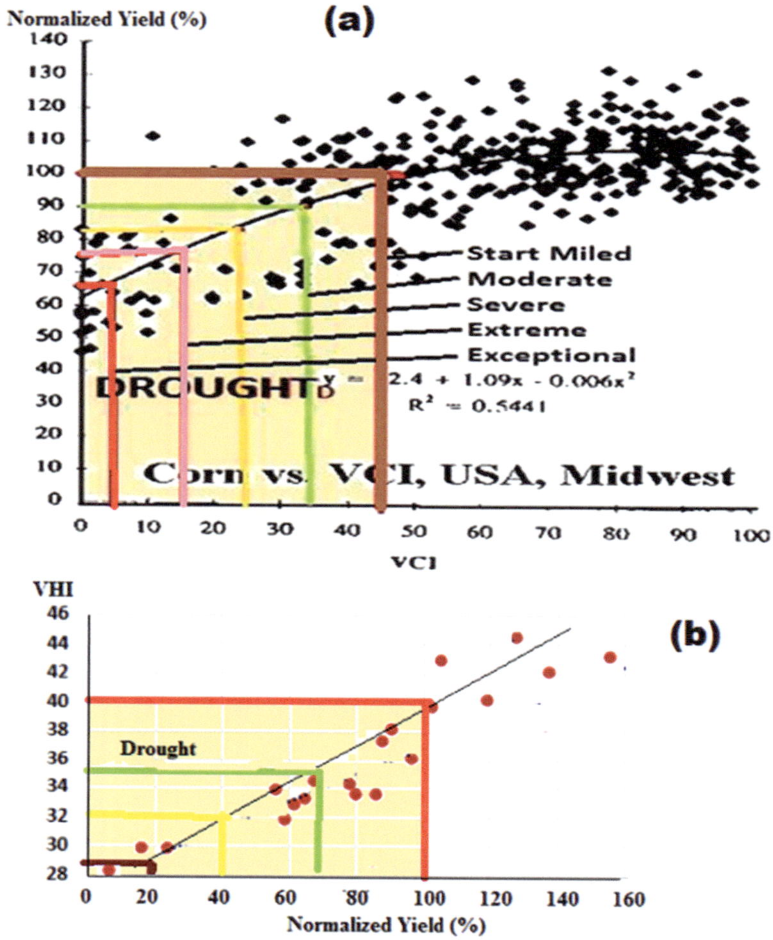

Fig. 5.2 VH drought classification based on correlation: (**a**) between normalized mean US counties (Illinois, Kansas, and Nebraska) corn yield with CRD-mean NOAA/AVHRR vegetation condition index (VCI, (NOAA 2018)) during 1981–1993 and (**b**) between normalized mean spring wheat yield in the European Russia and vegetation health index (VHI) during 1982–2001

the referenced correlation (Hayes and Decker 1996, Kogan 1997, 2006) between crop yields and VH indices (Fig. 5.2), droughts are derived when yield is reduced below a long-term trend level. As seen, the corn yield reduction begins when VCI (Fig. 5.2a) drops below 45 (red vertical line). Drought intensity also corresponds to a certain level of yield reduction. For VCI between 45 and 35 (green vertical line) yield is reduced up to 10%, indicating moderate drought. If yield is reduced by an additional 8% (from 90 to 82), VCI is reduced from 35 to 25 indicating severe drought (yellow line). For the next 8% yield reduction (82–74), VCI drops from 25 to 15, indicating extreme drought (orange line). And finally, reduction of corn yields from 84 to 66 and below is caused by VCI reduction below 15 (brown line) indicating exceptional drought. Similar criteria were established for VHI correlation with wheat yield in Russia (Fig. 5.2b).

Drought products and validation. In addition to assessment of drought intensity, which is one of the most important classification parameters, VH indices are used to develop such drought products as start/end, area, duration, dynamics, causes, impacts and even a 1–2 week prediction of drought initiation, tendencies and yield losses. Figure 5.3 shows that some products provided real time every week at NOAA/WEB (NOAA/NESDIS 2018). Drought causes might be from moisture stress (VCI), thermal stress (TCI), or combine (VHI). These estimates are shown in Fig. 5.3a for southern USA in 2006 and 2011. The 2011 drought produced much stronger

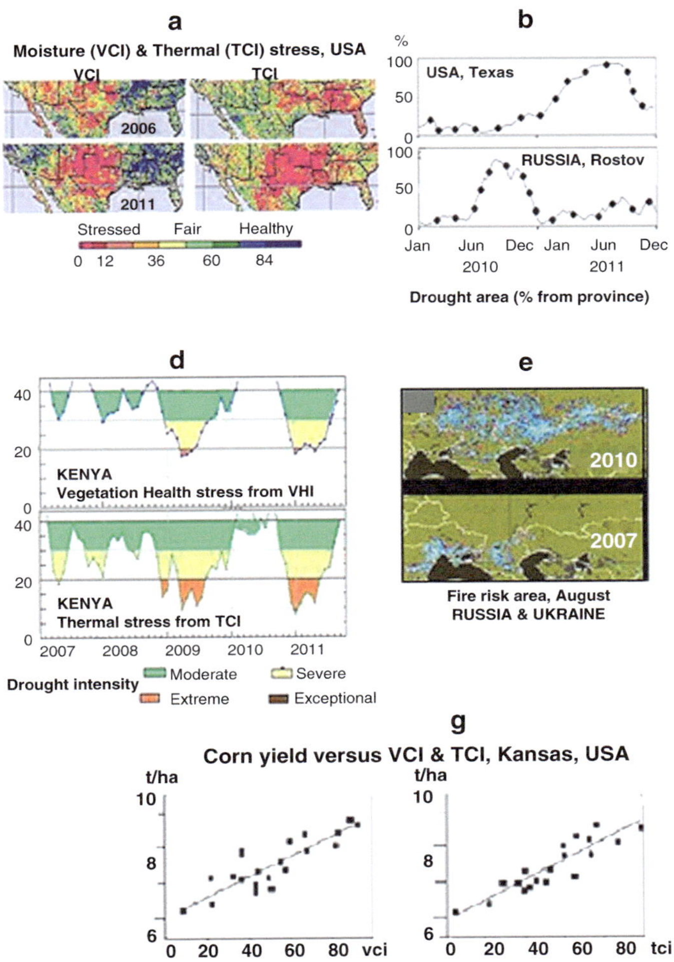

Fig. 5.3 Vegetation Health indices and products for drought monitoring: (**a**) 2006 and 2011 moisture (VCI) and thermal (TCI) conditions in the USA south, (**b**) 2010 and 2011 drought area in the USA's Texas and Russia's Rostov, (**c**) Drought duration and intensity during March-July 2009 and 2011 in Kenya, (**d**) Moisture–thermal (VHI) and thermal (TCI) stress weekly time series from 2007 through 2011 in Kenya, (**e**) VH-based mid-summer 2007 and 2011 fire risk area in Russia and Ukraine, (**f**) Drought start from moisture–thermal (VHI) and thermal stress (TCI) in wheat belt of Australia during 2006 and 2007, (**g**) correlation of Kansas (USA) corn yield with moisture (VCI) and thermal (TCI) indices

impacts due to both moisture and thermal stress, while in 2006, southeastern USA had thermal stress (red area) but no moisture stress (favorable moisture conditions shown by blue color). Drought area is a very important parameter of estimate. As seen in Fig. 5.3b, droughts of 2010 in Russia's Rostov province and during 2011 in Texas (USA) covered almost 100% of these regions, leading to considerable losses in agriculture, especially grain production. In African countries, estimated drought advanced start, intensity, duration, and area provide normally an early indication of food insecurity. VH-estimated thermals stress (TCI) and total vegetation health stress (VHI) with different duration and intensity in Kenya are shown in Fig. 5.3c, d. During a 5-month (March-July) period of 2009 and 2011, the minor (March–April) and major (May–August) growing seasons were affected by moderate and severe droughts. In 2009, southern and in 2011, central Kenya were drought-affected. However, extreme drought continued for almost 2 months and the exceptional, for 1 month. These combined assessments indicated considerable losses of agricultural production for less intensive but long drought, creating more problems with food shortages. This information is very important for agriculture and for assessment of food security. Summarizing, it is possible to conclude that short and intensive drought could be detrimental, reducing crops and pastures production in grassland ecosystems. Midterm drought in forest ecosystem can cause enhanced fire risk. Long-term drought leads to a reduction of ground water and water in rivers and lakes (example, the recent VH-based 12-year California drought presented at the end of this chapter). VH drought products also include assessments of fire risk (Fig. 5.3e), which provide an advanced warning on forested area and also crop and pasture damages (for example, Russia 2010 or Ukraine 2007). Since there is no reliable drought prediction method, the VH weekly moisture and thermal stress dynamics can be used as a prediction tool. As seen in the Fig. 5.3f, the first signs of the approaching 2006–2007 drought in southwestern Australia wheat belt appeared in March-April 2006, 2-month prior to the TCI and VHI time series crossed down the 45s threshold (the drought start. indicator). The wheat-growing season (November 2006-May 2007) had intensive thermal and moisture stress (VHI = 15–30) leading to huge losses in Australia's wheat production, reported by FAO (2017).

Over the years of VH use for drought detection/monitoring and verification, it has been revealed that this method can be successfully applied for prediction of crop yield and risk of malaria. Testing VH indices in 29 countries of North America, South America, Africa, Europe, Australia, and Asia revealed that yield of such crops as wheat (both winter and spring), corn, sorghum, rice, and soybeans has strong correlation with the VH indices during the critical period of crop development and reproduction. Figure 5.3g demonstrates these results for corn in Kansas. As seen, there is a strong correlation of mean state corn yield with VCI and TCI during the critical period of corn growth in July 1985–2011. During drought years (VHI <40), 25–40% corn production could be lost (depending on drought severity) compared to normally wet years. For example, in 2011, drought in Texas slashed corn yield by 28% compared to an average for the last 10 years and 36% relative to the highest yield received in 2010 (USDA/FAS 2017). The results of application and validation of the VH method have been published in the past 20 years (Kogan et al. 2017; Bhuiyan et al. 2017; Kogan and Guo 2014, 2015, 2016; Kogan 1989, 1995a, 1997, 2001, 2002; Seiler et al. 2000; Unganai and Kogan 1998).

5.5 Vegetation Health-Based Droughts from the Past to Present

Droughts are a typical climate phenomenon occurring every 4–5 years and affecting very important agricultural regions around the globe. The twenty-first century began with a series of widespread, long, and intensive droughts around the world, continuing the twentieth century trend. In eight out of the first 17 years of the current century, global grain production, the main source of food and feed for world, fell below a steadily grown consumption (PotashCorpo 2013; USDA/ERS 2017; WB 2017; FAO 2017). Drought was the major problem of grain shortages because severe-to-exceptional (SE) droughts covered 7–16% and extreme-to-exceptional droughts affected 2–6% of the world productive land (Kogan et al. 2013a, b). A wide range of economic and societal activities have been affected, especially agriculture, forestry, water, energy, ecosystems, transportation, and recreation in multifaceted ways. These droughts had adverse consequences for societal sustainability since they reduced agricultural production, caused a misbalance between grain supply and demand, shortages of food, malnutrition, human health deterioration, poverty, regional disturbances, population migration, and death. This section provides a VH-based description and verification of global and regional droughts during the warmest global temperatures in the past, with specific focus on the twenty-first century.

Global droughts and food security. Following Chap. 2, food security/insecurity is strongly affected by the amount of food produced annually, determining food supply and the amount of food consumed annually, which determines food demand. Since grain (wheat, rice, rye, corn, and coarse grain) is the principal product used globally for food and feed, the amount of global grain produced annually can be considered as a measure of food supply (production). Global food demand (consumption) can be measured by the total amount of grain consumed annually. Therefore, global food security/insecurity can be assessed by the difference between food production and consumption. Food insecurity might appear when grain production falls below consumption (considering the depth and frequency of the gap).

In the past 45 years, at the background of steady grown in both grain production and consumption (Fig. 2.1), in quite a few years, their difference become negative (grain production becomes below the consumption) indicating some potential problem with food security. Moreover, as seen in Fig. 2.2, this gap has been increasing since 1988 (shown by a declining trend), especially in the twenty-first century. This situation creates a huge concern about an enlarging misbalance, which creates the current situation and will lead to future problems with global food security (GFS). Additional global FS concern arises from the anticipated drought intensification, expansion, and penetration to new areas following the current and anticipated future climate change (Solomon et al. 2007), which might increase the negative gap and enhance its frequency. Regarding the opposite side of the gap scale (positive), when grain production is larger than consumption, there are no environmental concerns about food security. However, as seen in Fig. 2.6, global positive differences have gradually decreased in the twenty-first century, indicating that the future situation will likely not improve since the Earth's population growth rate is increasing, while grain production rate is decreasing (leveling off).

In years when droughts affect major grain areas, the global grain production declines below grain consumption, creating potential for development of global FS, which is normally not an issue if the production-consumption grain balance is positive, what normally occurs in no-drought, or minor-drought, or no-major area drought years. Therefore, VH-based drought conditions were investigated in years with a large positive and negative balance between grain production and consumption. The years with the absolute value of the balance (production minus consumption) greater than 2% (relative to grain production) were selected for the analysis. Most of these years were in the twenty-first century. The years with large negative balances (grain production below the consumption) included 1988, 2002, 2003, 2006, 2012, and 2015 and the years with a large positive balance were 2004, 2005, 2009, and 2016. The assumption was that the negative balance was typical for the years with strong droughts in a few major grain-producing countries and the positive balance was in years when the principal grain countries were not (or slightly) affected by droughts. The major grain-producing countries can be identified from Table 5.3. They include the countries contributing more than 10% of annual world grain production (China, the USA, and India). In years when these countries are not affected or only one of them is affected by drought, additional global grain reduction might come from countries with 2–4% contribution to global grain production.

Figure 5.4 displays the area and intensity of global droughts in the twenty-first century, specifically the years with negative and positive balance between grain production and consumption. In general, comparing two opposite situations (columns *a*—negative and *b*—positive) it is evident that in the negative balance years, the area of droughts is bigger and they are more intensive than in the positive balance years (column *b*). It should be emphasized, that the most affected countries are those with more than 10% contribution to annual grain production. One of the worst droughts, which reduced global grain production nearly 7.5% (compared to consumption, (WB 2017, USDA/FAS 2017)) affected the globe in

Table 5.3 Cereal production around the world and in specific countries, produced more than 30,000 million tons in 2014 (WB 2017) and percent country's production from the world

Region	Production (mil. metric tons)	% Production from the world
World	2,818,512	100.0
China	557,407	19.8
USA	442,843	15.8
India	295,360	10.5
Russia	103,136	3.7
Brazil	101,402	3.6
Indonesia	89,855	3.2
France	73,331	2.6
Ukraine	63,378	2.2
Bangladesh	55,759	2.0
Canada	51,301	1.8
Argentina	38,423	1.8
Germany	52,010	1.8
Vietnam	50,178	1.8
Australia	38,423	1.3
Thailand	37,804	1.3
Pakistan	38,394	1.3
Mexico	36,527	1.3
Turkey	32,708	1.2
Poland	31,946	1.1

2012 (Fig. 5.4a). The three main grain-producing countries, the USA (contribute nearly 16% to annual global grain), Brazil (nearly3%), and southern grain areas of European Russia (nearly 4%) were stricken by extreme and exceptional drought. The impact of 2012 drought on the central US environment and economy was enormous, since it caused grain (corn, soybeans), and pasture losses, wildfires, water level reduction in rivers and reservoirs, and many economic disturbances. Following yield losses, the price for food and farmland increased (USDA/FAS 2017). Analysis of 37-year VH records indicate that the US drought of such intensity reappears every 9–20 years (2002 in the western USA, 1988 in the Great Plains) plaguing 10–15% of the contiguous United States. The 2012 drought of slightly milder intensity also affected Argentina and Spain. From smaller grain contributors to global production, it is important to indicate severe-to-exceptional drought in Kazakhstan and also droughts in the countries of southern Africa, creating food insecurity situation. Similar to 2012, a global food security problem developed during the 1988 drought (not shown), when grain production was 7% below consumption (WB 2017). Again, the USA, Brazil, and Argentina (21% total contribution to the global grain production) were countries with grain shortages. The 1988 US drought costs around $60 billion in damages to the US economy, human health, environment, and wildlife (Wilhite 2000). VH-drought estimates (Kogan 1995a, b) indicated that by the end of June 1988, US vegetation experienced stress in the most productive areas of the Great Plains, the US grain basket. In 1988, US grain production felt below domestic consumption probably for the first time in the second half of the last century. It is interesting that in all years of the negative balance, US droughts always contribute to global food shortages. That occurred in 2012, 2006, 2003, 2002 (Fig. 5.4a), and 1988. In addition, Australia also had droughts during those years in wheat-growing areas, some regions of China, India, and Mexico suffered from light-to-moderate droughts in 2006, India from intensive drought in 2002, and Europe was drought-stricken in 2006 and 2003. Following Fig. 5.5b, droughts, in the years when global grain production was up to 4.0% above grain consumption, are mostly moderate-to-mild intensity compare to extreme droughts in a negative balance years, and has smaller areas in the major grain-producing countries (China, the USA, India, Russia, and others). All years, shown in this category in Fig. 5.4b, had a mild drought in the western USA (with smaller areas of principal grain crops), in some parts of China, India, Brazil, and Argentina. Even African countries with frequent regional problems of food security experienced relatively mild droughts in such years (the dessert area is not counted in this analysis).

Figure 5.4 demonstrates global droughts based on an assessment of vegetation health (VH) from a combination of moisture (VCI) and thermal (TCI) conditions. As indicated, during some years of less intensive droughts from a combination of moisture/thermal conditions (VHI), crops might experience either moisture or thermal stress, which might lead to some reduction of

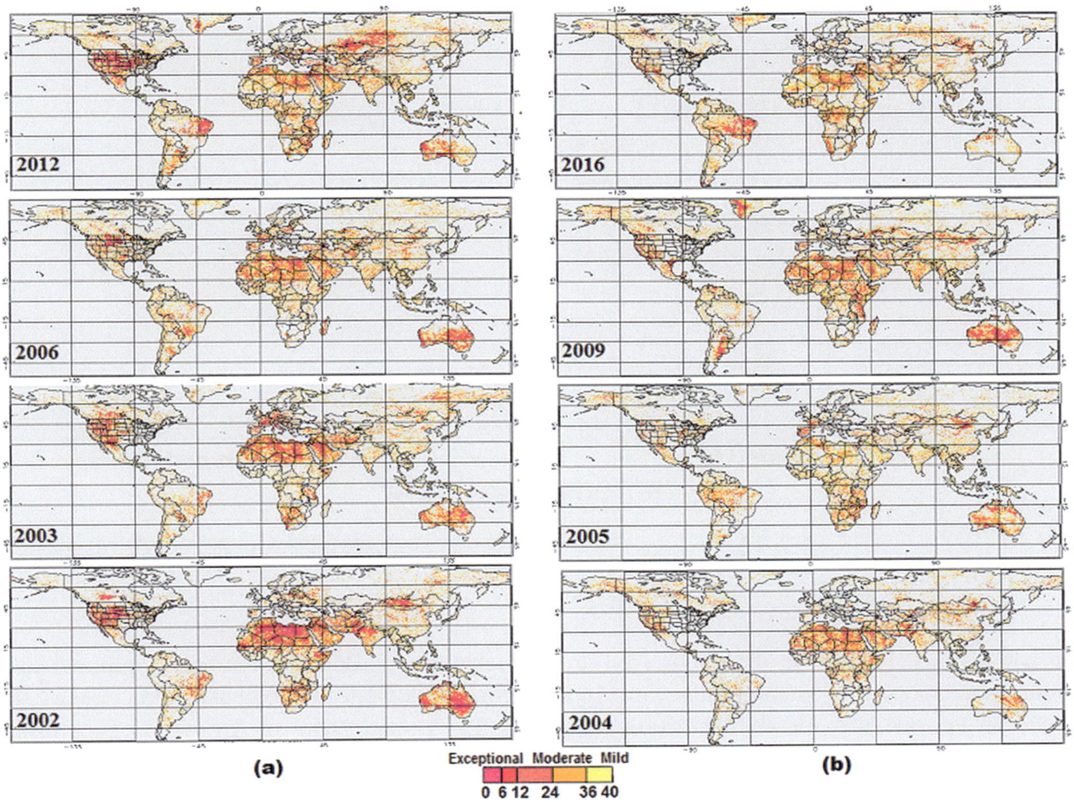

Fig. 5.4 Global droughts (area and intensity) in the twenty-first century for the years of different balance between production and consumption (P–C): (**a**) negative and (**b**) positive (P–C)

grain production depending on stress intensity, duration, affected area, and the period of crop development. Figure 5.5 demonstrates such situation for 2 years, 2005 (shown as less intensive drought year in Fig. 5.5b) and 2015. In both years, many principal grain-growing world countries experienced severe thermal stress in the middle of July, but no moisture stress in 2005 and a small area of moisture stress in 2015. Such situations should be considered for final review of food security assessment. For example, southern Africa, Brazil, and eastern USA in 2005, and India, south-eastern Africa, Brazil, and Europe in 2015 might have reduced grain production depending on the duration of thermal stress, which assessments are available in real time on NOAA WEB: https://www.star.nesdis.noaa.gov/smcd/emb/vci/VH/vh_browse.php.

Unusual regional droughts in the main grain-producing countries, contributing to food insecurity. This section discusses regional drought, which have not had global impacts on food security. However, these droughts were presented because they were unusual, had specific features, reduced a country's agricultural production, and in most cases had some regional economic disturbances. First, the discussion is focused on the major grain-producing countries, China, the USA, India, and Russia, with a total contribution of around 50% of the global grain production (Table 5.3).

Among these countries, China is a number one supplier of global grain, contributing nearly 20% (Table 5.3). Cereal production in China is increasing stably since 1961, reaching nearly 560 million metric tons in 2014 (WB 2017). However,

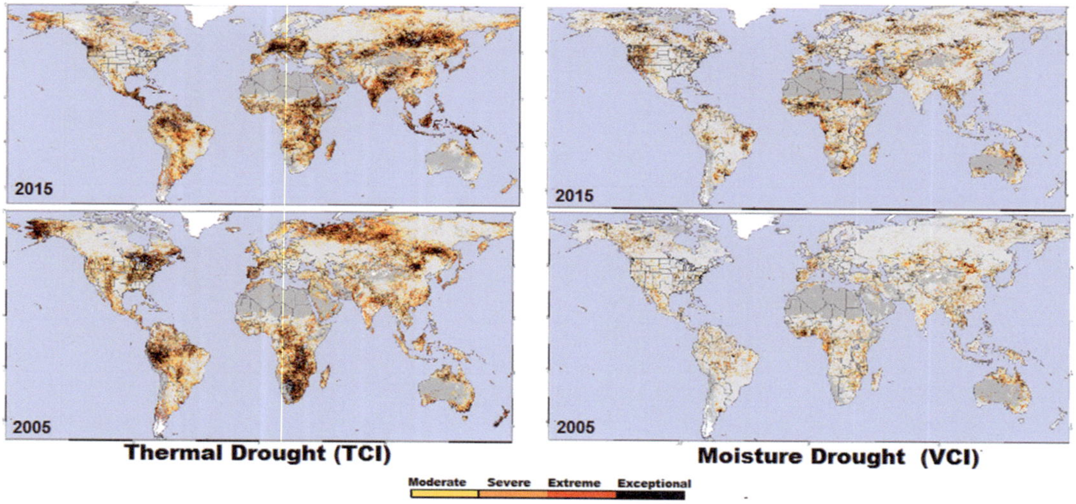

Fig. 5.5 Global droughts (area and intensity) thermal- and moisture-based (TCI and VCI, respectively) in 2005 and 2015

in some years, drought in the major crop area reduced the amount of grain produced in China. Such an episode was recorded in 2003, when total cereal production dropped 15% relative to the 54-year trend of this production growth. Figure 5.6a shows China's vegetation health (VHI) in 2003 compared to 2016 when the production was 12% above the trend. Although the 2003 drought in the major grain area was only mild-to-moderate (yellow-to-orange color) it covered around 70% of the area and continued for 3–5 months. If such a drought would had been an extreme intensity grain production would had dropped 40–60%. Compared to 2003, the 2016 vegetation health was in the highest category (very healthy conditions, light and dark blue color) for 80% of the grain area. This example indicates that relatively moderate drought contributed to losses of grain production due to such drought features as large area and long duration. Since China is the major global grain producer, drought-related reduction of China's grain production in 2003 likely affected global food security. In addition to considering a country's global grain production for evaluation of drought impacts, it is important to also take into consideration a country's grain exports. China is a very good example: being the largest contributor of global grain production but only a minimal contributor (0.2–1.9%) to global exports (Table 5.4).

The United States of America is the world's largest producer and leading exporter of agricultural products, including grains, contributing nearly 16% to annual global grain production in 2014 (Table 5.3) and between 16 and 24% to global grain export during 2007–2016 (Table 5.4). Unfortunately, drought in the USA occurs every 2–3 years and in the past three decades, strong droughts covered principal agricultural areas for 6 years (1988, 1995, 2002, 2003, 2006 and 2012), reducing both US and global grain production (Kogan and Guo 2014) and the contribution to global grain trade (up to 16% compare to 24% in the most favorable year (Table 5.4)). In years of either non-intensive and/or intensive but small drought area, their affect is normally local. Such drought struck Texas, Oklahoma, New Mexico, and some neighboring states in 2011 (Fig. 5.7b). This drought was so localized that only 8% of the contiguous US was affected by exceptional (the strongest) and 15% by extreme-to-exceptional droughts. State level VH estimates indicated that drought of severe-to-exceptional intensity covered 60% of Texas and 70% of Oklahoma; exceptional intensity drought was recorded on nearly 40% of Texas and Oklahoma and 30% of New

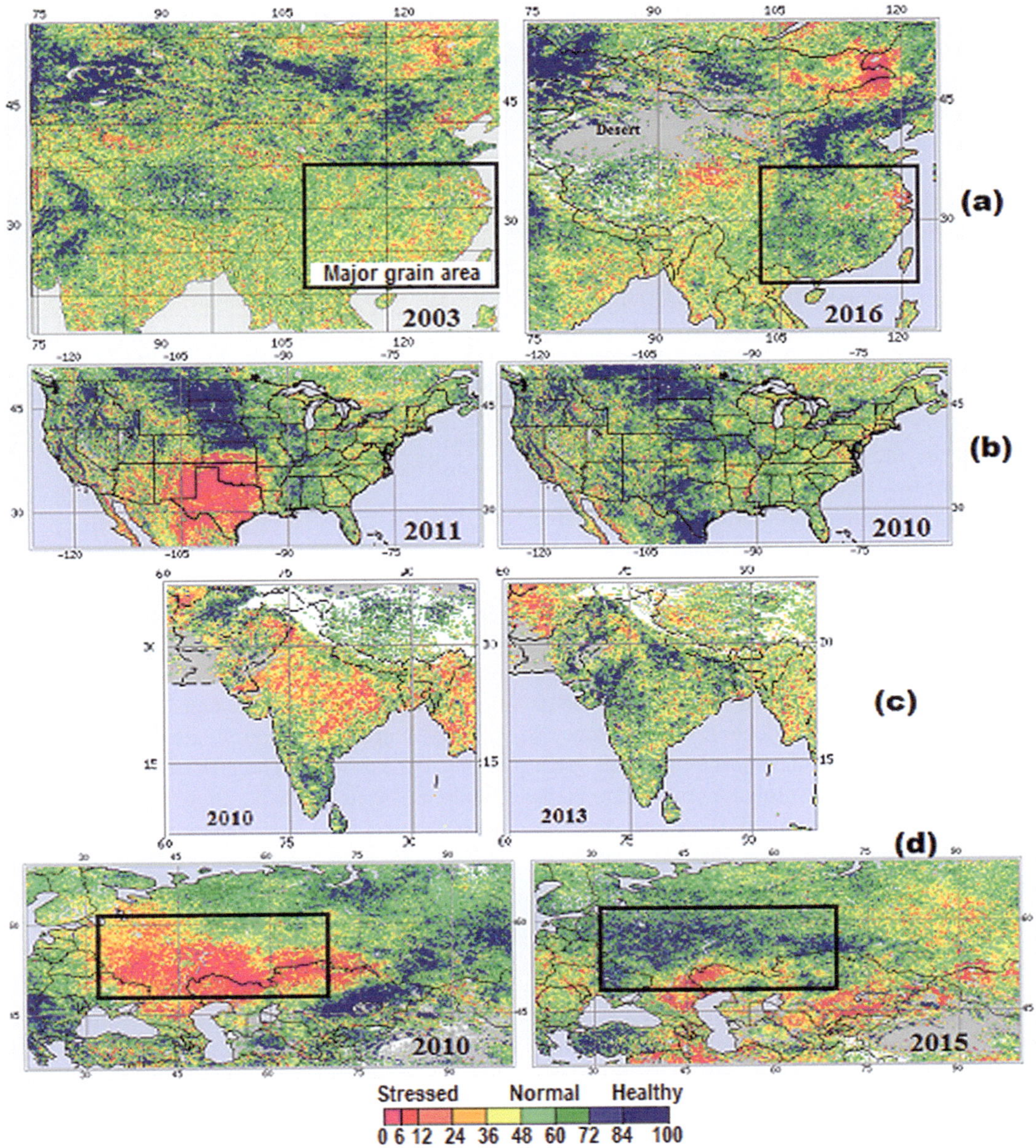

Fig. 5.6 Mid-summer vegetation health (moisture–thermal from VHI) in China, the USA, India, and Russia, the three largest world grain producers

Mexico. Summer in the two hardest hit states was the hottest on record and it was estimated that the Texas drought cost \$5.2 billion in agricultural losses (AgriLife 2011). Additional losses also came from wild fires, depletion of water in lakes and reservoirs, tourism decline, deterioration of human health, and the ways of living.

India is the number three contributor to total global grain collected annually (Table 5.3) although it is only a moderate contributor to grain exports. Cereal production in India increased more than three times (from nearly 90 to 290 million metric tons) between 1961 and 2014 (WB 2017). In spite of such strong and steady grain production increase, in the twenty-first century, India was affected twice by moderate-to-strong droughts in 2003 and 2010, which reduced cereal production 7–9% (relative to the 54-year trend in production growth). Such drop in grain production led to a reduction of India's contribution to the global trade (Table 5.4). Such drought events were analyzed with satellite-based assessments of vegetation health. The combine moisture and thermal vegetation health index (VHI) in 2010 is shown in Fig. 5.6c compared to VHI in 2013, when vegetation condition in the major grain area were favorable and cereal production exceeded the trend by 6%. The VHI characterizes the 2010 drought in the grain area as light-to-extreme (yellow-to-red color). However, only 35% of grain area (eastern India) was affected by extreme drought. Therefore, grain reduction was relatively small. In 2013, the same area in India has a light drought, which prevented a large increase in the country's grain production.

Table 5.4 Grain exports (% from the world) during 2007–2016 (AMIS 2017). The year of Max and Min is in parenthesis

Country	Maximum	Minimum
USA	23.9 (2007)	16.2 (2012)
European Union	12.9 (2014)	6.2 (2007)
Ukraine	10.2 (2016)	4.2 (2010)
Argentina	9.8 (2016)	7.1 (2013)
Brazil	9.2 (2015)	3.0 (2009)
Russia	9.1 (2016)	1.5 (2010)
Canada	8.4 (2013)	7.7 (2010)
Australia	8.4 (2016)	4.1 (2007)
India	7.5 (2012)	1.6 (2009)
Kazakhstan	3.5 (2011)	2.0 (2010)
Turkey	3.5 (2013)	0.6 (2007)
China	1.9 (2007)	0.2 (2015)
South Africa	0.9 (2011)	0.3 (2007)
Mexico	0.7 (2015)	0.3 (2010)
Nigeria	0.4 (2010)	0.2 (2007)

A very unusual, extremely large-area and intensive drought affected Russia in 2010 (Fig. 5.6d), leading to a decline in the country's grain production and the most important, grain export to 1.5% (of total global) compared to 9.1% in a favorable year (Table 5.4). A very specific drought featured that year included intensive thermal stress, which started in April, covering the entire crop area of European Russia and south-western part of Russian Asia. This heat stress quickly dried up soil moisture that severely deteriorated vegetation health. Russia's drought was the worst since 1972, when drought of similar intensity, duration, and impact covered the same spring and winter grain areas (Kogan 1986). In 2010, Russian grain production dropped 23% to 75 million metric tons (compared to 97% of mean grain production when there was no drought in 2009 (FAO 2017)). This drought, which continued until November affected not only the 2010 spring grain harvest but the 2011 grain harvest since the drought depleted soil moisture for the September planting of winter wheat (occupying 65% of total winter and spring wheat area). Following this drought's impacts, the Russian government imposed a grain embargo resulting in global wheat prices to jump up sharply (the biggest increase since 1973) when the former Soviet Union (FSU) purchased unexpectedly huge amounts of grain from the international market following a considerable drop in country's grain production after an extremely strong 1972 drought. The 2010 drought triggered hundreds of fires (in addition to grain losses) accompanied by a very heavy and long-lasting smoke and heat waves, which affected human health and considerably contributed to the death rate in Russia. Droughts of such scale and impacts in Russia as 2010 occur every 25–40 years (1972, 1946, 1921 and earlier).

Strong droughts in the moderate grain-producing and trading countries. Russia's 2010 drought also affected neighboring grain producers Ukraine and Kazakhstan. Although these countries are not major contributors to global grain production, Ukraine is still very important globally as it is the 10th largest wheat producer in the world and is one of the largest contributors (10% in a good crop year) to the global grain export (Table 5.4)). However, in the years of such long, intensive and large-area drought like 2010, Ukraine reduced grain exports 2.5 times (to 4% of the total global annual grain export), which could create some problems with food security in some countries. During the twenty-first century, large-area droughts of moderate-to-severe

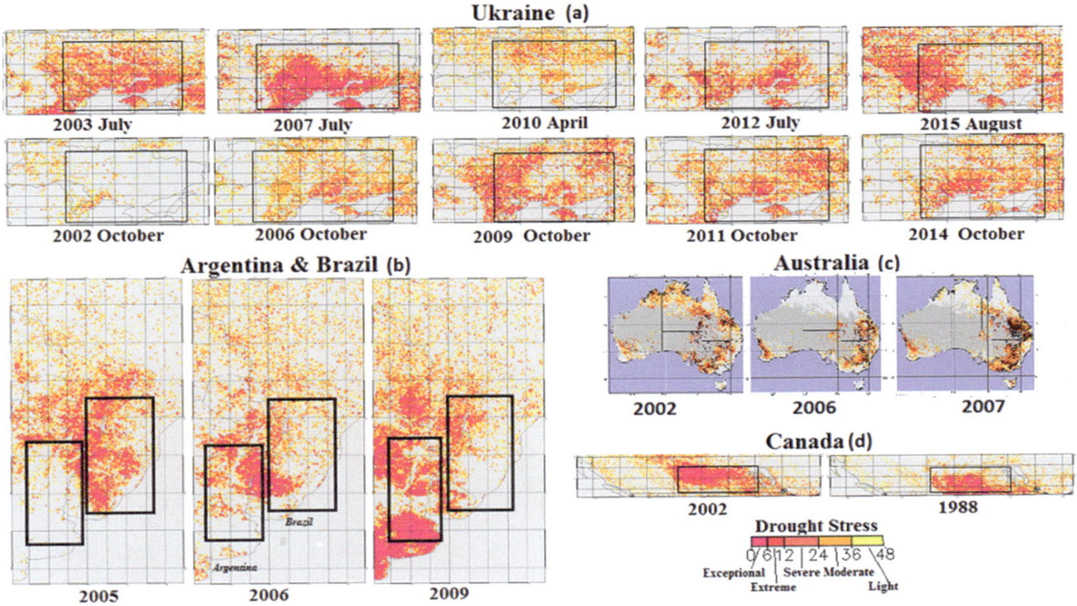

Fig. 5.7 Mid-growing grain season moisture–thermal (from VHI) drought (area and intensity) in (**a**) Ukraine, (**b**) Brazil and Argentina, (**c**) Australia, and (**d**) Canada

intensity occurred in Ukraine every 2–3 years covering more than 20% of country's territory. And based on VH assessments, droughts such as 2003, 2007, 2010, 2012, and 2015 covered between 40% and 60% of the country and up to 80% of the major grain crop area. Ukraine has two grain crop seasons, summer and winter. Summer grain crops occupy 35% of the total grain crop area, are planted in April-May, and harvested in September while winter crops occupy 65% of the total grain area, are planted in September-October, and harvested in June. As seen in Fig. 5.7a, five of the strongest droughts affected Ukraine in the first 16 years of the twenty-first century. Four of them affected both winter and summer crops. A very low crop production occurs when large-area, long, and intensive droughts affect both spring and winter crops. As seen in Fig. 5.7a, very extreme droughts damage winter crops during the early stages of their development in fall of the year preceding harvest and during spring. Summer crops are normally affected in June-August. The most intensive drought of this type was in 2007 (Fig. 5.7a). Summarizing, it should be emphasized that the extreme droughts during 2001–2016 in Ukraine severely affected either spring-planted grain crops during summer (2007 and 2012) or fall-planted winter grain crops during spring (April-May) such as 2003 and/or late fall (drought of 2009 for the 2010 winter grain crops harvest). It is an environmental paradox since Ukraine has the most productive (chernozem) soil excellent for growing grain, but crops suffer from frequent droughts, leading to considerable grain losses (Kogan 1986). Besides, winter grain crops might be damaged or even killed during low winter temperatures (below −20 °C) when snow cover is less than 10 cm.

In South America, the major grain-producing countries are Argentina and Brazil. Cereal crops in Argentina grow in the humid pampas, one of the world's best arable land. They include corn (68% of the total grain production), wheat (24%), barley (7%), and rice (1%). Argentina is a very important country for the world market, being the second largest exporter (after the USA) of corn and coarse grains, and the third largest of wheat (FAO 2017). Major cereal crops in Brazil are corn (66% of total production, based on 2015

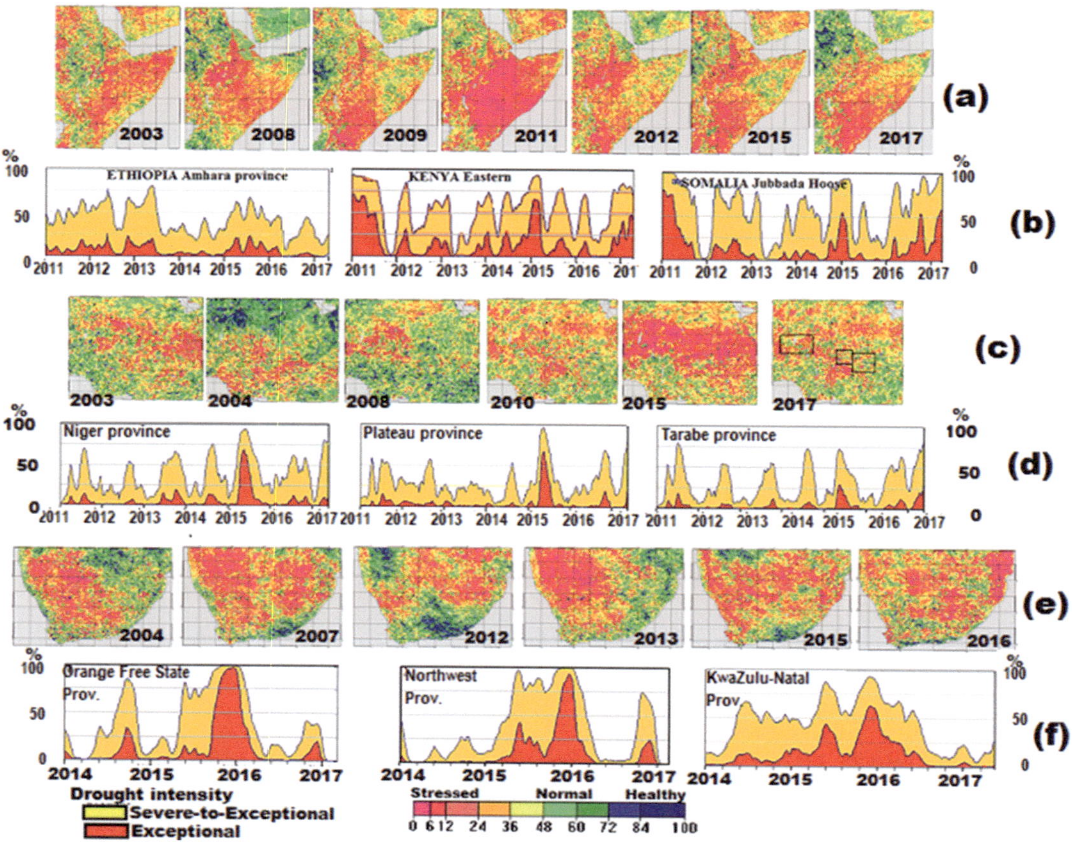

Fig. 5.8 Mid-growing grain season moisture–thermal (VHI) conditions (image) and percent area with droughts in African countries: (**a**, **b**) Horn (Ethiopia, Kenya, and Somalia), (**c**, **d**) Nigeria and (**e**, **f**) South Africa

harvest year), rice (22%), and wheat (12%). They are grown in southern Brazil, which has fertile soil and semi-temperate climate. Corn is collected twice a year during the main rainy season in August and the second time, during dry season in October-November. Droughts and dry spells hit these two countries every 2–4 years reducing their grain production. Following WB (2017), in the last 17 years, Argentina experienced extremely strong reduction of cereal production in 2009, when yield dropped 33% (form the long-term yield time-series trend). The VH (NOAA/NESDIS 2018) estimated that a very large-area, intensive, and long-lasting drought was the main cause of such a strong reduction of 2009 grain production. Following estimates shown in Fig. 5.8b, 70% of the grain-producing area (in the north and especially south) of the country was affected by extreme-to-exceptional drought (VH < 15) during 2–3 month of the growing season (December-February). Another case of intensive but a much smaller-area (22% in north-eastern part) drought occurred in 2006, reducing grain production by 9%. Compared to 2009, the drought of 2006 affected only the northeastern part of the country. The rest of grain area (inside black box) was either free of drought or affected by a very light drought (VHI is 36–48). Therefore, grain reduction in 2006 was 3.5 times less than in 2009. At the end of the twentieth century, two neighboring-year strong droughts occurred during 1988/89 and 1989/90 crop seasons when the country lost between 15 and 20% of the total country's grain production (Seilers et al. 2000). The twenty-first century in Brazil is characterized by three intensive droughts, 2005, 2006, and

2009. Although the last two occurred in the same years of Argentinian droughts, their area was smaller (8 and 13% of the black box area) and grain crop reduction was only 12 and 9%, respectively. The 2005 drought covered much larger area of Brazil (63%) and led to 18% grain reduction. Since the 2005 and 2006 droughts were next to each other, their total effect was stronger in Brazil.

Other unusual cases of the twenty-first century were a few droughts in Australia. Wheat, the major cereal crop, is principally planted in the most southern and eastern parts of the continent. Since Australia contributes up to 8% to the global grain export, in strong drought years, the country's export might drop by half (Table 5.4). The most dramatic drought event during the past 17 years occurred in 2002 and between 2005 and 2007 when drought continued for 15 months without interruption and was very severe. Based on VH assessments (Fig. 5.7), this long drought began in December 2005 and ended in April 2007, affecting the principal wheat and rangeland areas during the two harvest seasons, 2006 and 2007. Between the seasons, the 2006 drought was the worst in recent history because it slashed water supplies (depleting not only soil moisture but also reducing considerably reservoir level) and crop production across large areas; it also effected energy and economy. Total VH-estimated drought area reached 70% (including a nearly 40% area reeling from extreme to exceptional drought intensity) in 2006 and nearly 60% in 2007. Total Australia wheat yield (1.06 q/ha in 2006 and 1.18 q/ha (WB 2017)) dropped 46% (below 1960–2010 yield trend level) in 2006 and 37% in 2007 (FAO 2017). It is interesting that 4 years earlier, in 2002, another drought of similar to 2006 intensity but on a larger area affected Australian agriculture reducing wheat yield (1.08 q/ha) below long-term trend by 45% (WB 2017). From the most recent droughts, the 2013/2014 season began with a very intensive drought in November 2013 (panting time). In January 2014, the expectation was considerable losses in the total country's cereal production, while in February 2014, rains returned saving the harvest. Following VH (NOAA/NESDIS 2018) estimates the 2015/2016 crop season in Australia was affected by regional droughts, and during the 2016/2017 season, Australia is expected to have a very plentiful cereal production.

The major staple grain crop in Canada is wheat. The country also grows rye, corn, barley, and oats. Wheat has two seasons: winter (planting in fall and harvesting in spring) and summer (early spring planting and the end of summer harvesting), the rest of grain crops are summer type. Most crops are planted in the southern part of Canada (provinces Alberta, Saskatchewan and Manitoba). Soils in Canada are deep and have plenty of humus; however, annual precipitation (300–500 mm) are limited to have good harvest every year (Geoforcxc 2017). Therefore, drought reduces grain production every 2–4 years (WB 2017). Since 1970, Canada experienced two extreme droughts, in 1988 and 2002, when the average country cereal yield was reduced by 31 and 28% (relative to the long-term trend (WB 2017)), respectively. Figure 5.7d shows that in both years VH-estimated drought was the major problem, being extreme-to-exceptional (the two highest categories) and covering 70 and 90%, respectively, of the major grain-growing area (black box). Besides, in addition to the area and intensity VH estimated that these droughts continued for 6–8 weeks (Kogan 2002).

Droughts in Africa and food security. Africa is the main continent where regional food insecurity affects its countries following drought-related reduction of crop production and limited imports. In such situations, African countries require the international community's help to prevent hunger and starvation. Grain crops, especially wheat, are the most important food products needed. Following the International Food Policy Research Institute, the wheat economies of Africa are characterized by a growing gap between wheat production and consumption. Following this publication, in 2009–2011, Africa produced 23 and imported 36-million-ton wheat to cover 59-million-ton gap between food supply and demand. Since the 2007–2008 of global food price crises, Sub-Sahara countries have suffered from a limited wheat production and the effects of exports restriction by the major world wheat producers. In 2010, FAO (2017) estimated that 925 million African people suffered from hunger. The 2012 estimates indicated that in Sub-Sahara Africa, 239 million people suffered from hunger and starvation and the future projections to change the situation were not promising (Sasson 2012). Drought was the major problem, reducing grain production. The other consequence of a very severe drought was the spike in food prices, riots and migration of people in the neighboring countries, which occurred when thousands of Somalis had to flee to seek refuge in neighboring Ethiopia and Kenya (Sasson 2012).

In the recent two decades, wheat became the major staple crop in Africa, mostly due to the intensification of population growth, increasing demand for grain and its products. Unfortunately, in most African countries (except Ethiopia, South Africa, and Morocco), domestic grain production is not increasing (Sasson 2012). Considering an intensive population growth, this widens the gap between grain production and consumption. Following this tendency, grain import of the recent years experienced an intensive growth (almost doubling in 15 years) to cover shortages of grain needed for food. The current African population of 1.2 billion is expected to increase nearly 40%, to 1.6 billion by 2030, demanding nearly 70% more food to keep a normal level of living (GRACE 2017). Unfortunately, drought is the main environmental problems affecting African countries each year, reducing grain production and creating food insecurity situation. Therefore, early drought detection and prediction of their development is a very important application of VH tools to predict grain and other crops losses and timely estimation of imports needed and additional help needed to avoid a lack of food, population's malnutrition, and hunger. The new vegetation health method assessing drought start, area, intensity and duration is providing ability to forecast losses of grain in advance of harvests in order to make important decisions about food needs and a shortening the gap between supply and demands.

Droughts affect African countries every year, leading to loss of grain and other crops production. During very unusual droughts and/or multi-year droughts, some countries become insecure in being unable to provide an adequate amount of food for their population. Most African countries grow grain crops, especially wheat, which area has increased in the recent years due to its useful quality to produce some food in order to avoid food crises. The major producers of wheat and other grains in Sub-Sahara Africa region are Ethiopia (nearly 48% of the entire region in 2009–2011) and republic of South Africa (slightly over 30% (FAO 2017)). If they are affected by droughts, the situation with food security deteriorates not only in these two countries but other countries, dependent on grain trade with Ethiopia and South Africa.

During the first 17 years of the twenty-first century, the Horn of Africa experienced seven major droughts as estimated by the VHI for a few countries in Fig. 5.8a. Following these droughts, the countries experienced losses of grain and lack of food. Most of these droughts were severe, but only the 2011 drought covered 100% grain area in Ethiopia, Kenya, and Somalia and 75% grain crop areas experienced extreme-to-exceptional intensity droughts for 3–6-month, which is a very unusual event (Fig. 5.8a, b). In the three Horn of Africa countries producing wheat, there are two seasons: minor, in winter with dry weather when wheat is normally irrigated and therefore occupies smaller area and major (large-area), in summer when crops get moisture from rains. The minor-season harvest provides normally grain and food until the major-season harvest is collected in summer and is ready for use (Kogan 1997). The worse food security problem might appear if both seasons are drought-affected. The VHI estimates in Fig. 5.8b showed that such situation occurred in 2011, when strongest drought was long (especially in Ethiopia), affected both seasons in the Horn, leading to humanitarian crises. Following the crisis, 13 million people suffered from a lack of food, including 750,000 people being at risk for starvation and death (IFRC 2011).

From other strong droughts in the Horn, Ethiopia, the second major cereal producer in Africa after Morocco, was drought-affected in 2003, 2008, 2009 (2 years in a row), 2012, and

2015. These droughts caused 10–15% reduction of grain yield (below long-term trend (WB 2017)). In Ethiopia's Amhara province with large grain crops area droughts severe-to-exceptional and extreme-to-exceptional (extremely severe) intensity covered 50–75% and up to 25% of panted grain crop area, respectively, every year since 2011 (except for 2014 when the area was below 30% and 4%, respectively (Fig. 5.8b)). Following Fig. 5.8b, in two principal grain-growing and pasture provinces of Kenya (Eastern) and Somalia (Jubhada Hooye) droughts of moderate-to-exceptional intensity covered 75–100% of the province area since 2011; extremely strong droughts covered more than 50% provinces in 2011 and 2015. However, the 2015 drought was shorter. From prior years, in addition to the 2011 drought and food security problem, Kenyan agriculture was severely affected by the 2009 drought, when according to FAO (2017) estimates, wheat yield dropped 45% compared to 2010 good crop season. The droughts of 2008 and 2002 were long but affected mostly pastoral areas. Droughts during the minor wheat season were recorded in 2001, 2005, and 2006, although their impact on agriculture was minor since summer conditions were relatively favorable for crops' good harvest.

From the central part of Sub-Sahara, Nigeria was selected for the vegetation health analysis (Fig. 5.8c, d), since the country produces very much food crops such as corn, rice, wheat, sorghum (30–40% of the entire African and 70% regional production), cassava (19% of the world production) and also exports some of the grain each year (FewsNet 2017). Northern Nigeria is the largest producers of cereals and a significant contributor to the African economy. Most of food crops in Nigeria are rain fed except for 17%, which are irrigated. Similar to the Horn, droughts are affecting Nigeria almost every year. But in the first 15 years of the twenty-first century, eight strong and moderate droughts affected Nigeria agriculture in 2003, 2004 2008, 2010, 2011, 2012, 2013 (4-year drought), and 2015 (Fig. 5.8c). From these droughts, the worth damage to agriculture and food security brought an intensive drought-related moisture and thermal vegetation stress (Fig. 5.8c) in spring and summer 2015 (FewsNet 2017). In the main grain-producing provinces (Niger, Plateau, and Taraba) areas affected by moderate-to-exceptional drought reached 75–90%, and in the first two provinces, the strongest drought (sever-to-exceptional) exceeded 50% (Fig. 5.8d). Other strong drought reduced Nigerian grain production between 9 and 16% and turned pastures in a bare-soil area. Some droughts, such as the 2012 drought affected more than 20 million people in 34 countries, stimulating the international community to help. In addition to drought, northern Nigeria, the largest crop-grower area used for food is also affected by desertification (FewsNet 2017; FAO 2017a).

Southern Africa is third larger producers of wheat among all African countries. In the 14 years since the beginning of the twenty-first century, cereal yield in South Africa experienced a technology-related increase from 2 to 4.5 t/ha (WB 2017), which is considerable agricultural progress. However, this progress is often overcome by crop losses from severe droughts. Such situation occurred in 2015–2016, when Southern Africa was devastated by a severe drought triggered by the strongest El Nino (ENSO) event. Since the 1950s such ENSO had the most adverse weather with precipitation much below normal and temperature much above normal (Kogan and Guo 2016, FAO 2017a). In such years, vegetation was under stress although, the worst situation occurred in December-February 2015–2016, when an area of severe-to-extreme drought reached 90–100% in the two major grain-producing provinces, Orange Free State and Northwest, and 50–60% in KwaZulu-Natal province (Fig. 5.8f (NOAA 2017)). Following the United Nations warning, 14 million people were at risk of starvation due to drought-related widespread water shortages, which devastated harvest and livestock production (FAO 2017a). Unfortunately, these agricultural devastations will be felt for the next couple of years as lost grain crop and livestock production return to normal levels. ENSO-triggered unusual drought and crop losses in the republic of South Africa had a devastating impact on many other countries, where nearly 39 million people were at risk of starvation in the south and also in

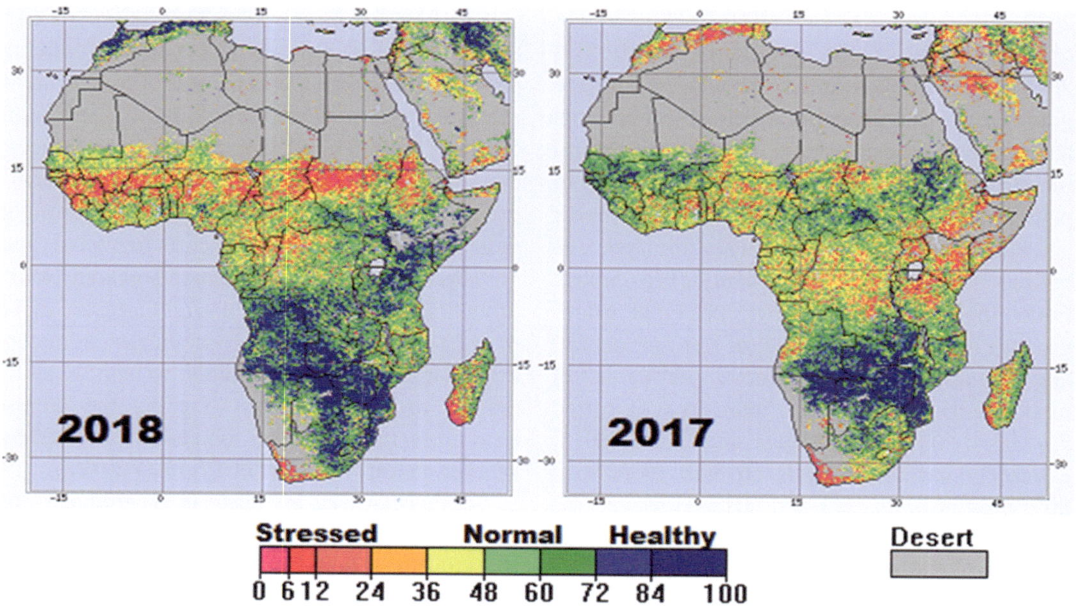

Fig. 5.9 Africa's moisture–thermal conditions (VHI) in early May 2017 and 2018

eastern Africa (especially, Ethiopia—5.7 million people, Kenya—2.6, Somalia—6.2, Sudan - 4.6 (FAO 2018)).

The United Nations recently warned that in 2016–2017, 22 million people in South Sudan, Somalia, Yemen, and Nigeria are at risk of famine and some families are facing the threat of starvation (FAO 2017a, 2018). Therefore, we first analyzed the satellite-based vegetation health situation in the spring of 2017 for the entire Africa continent to see in which countries' vegetation would be under drought-related stress, evaluating drought area, intensity, duration and prospects with food insecurity. Figure 5.9 illustrates vegetation health (VH) in the entire Africa (except deserts) in early May, 2017 and 2018. As seen, in 2017, the entire central Africa (south of Sahara Desert, between 15°N and 6°S) is under vegetation stress (both moisture and thermal). The stress is very severe in the Horn of Africa, neighboring Sudan and some countries (Guinea and Nigeria) of the western Africa. The Horn drought affects all three countries Kenya, Ethiopia, Somalia and also southern Sudan. The drought is continuing since March, covers nearly 90% crop-growing areas of Kenya and Somalia, 63% of Ethiopia and 45% of Sudan. Since at the end of June the drought still continued (NOAA/NESDIS 2018) it reduced grain and other crops production in all these countries leading to some problem with food security. Since Ethiopia is the second largest African grain producer (after Morocco), expected grain losses in 2017 will likely affect other countries food security, which rely on imports from Ethiopia. Opposite to the Sub-Sahara Desert region, south African countries have excellent vegetation health in 2017 (NOAA/NESDIS 2018) and an excellent prospect for a good grain harvest in 2017. Opposite to 2017, May 2018, vegetation continued experiencing healthy conditions in the south (Fig. 5.9). However, countries south of Sahara Desert experienced a very severe vegetation stress (VHI < 25). Following this stress, the prospects for 2018 food security in these countries are not encouraging. This vegetation stress is continuing since the end of 2017 with the strongest impacts on crops in March through June (NOAA/NESDIS 2018).

5.6 Droughts at 0.5 and 1 km² Resolution from VIIRS, the New Operational Sensor

As discussed in Chap. 2, the Visible Infrared Imaging Radiometer Suite (VIIRS), the most advanced, new operational satellite sensor, was launched in 2011 (JPSS 2014). One of the most advanced utilizations of its potential in the food security issues are 1 and 0.5 km resolution drought detection, identifying drought area, intensity, and duration, monitoring irrigation schedule, and also high-resolution prediction of weather-related agricultural losses (Kogan et al. 2015a, 2017; Kogan and Guo 2016). Slightly over 6 years (2012–2018) VIIRS activities have already proven to be useful for many food security tasks, such as identifying countries, provinces, and regions where agricultural production is under the threat of destruction and planning of measures to overcome shortages of food. Fortunately, the new generation system with VIIRS operational activities, which started from 2018 with a launch of NOAA-20 satellite, will provide VH data and products (drought, vegetation stress, yield losses, fire risk, etc.) for the next 25–30 years (NOAA/NESDIS 2018).

The high-resolution VIIRS-based Vegetation Health method was applied to the recent multiyear drought in the United States. As discussed in Kogan et al. (2017), beginning in 2006, the western United States was affected by the 12-year strong mega-drought (Kogan et al. 2017, Kogan and Guo 2015; Howitt et al. 2015). In 2006–2009, the drought affected 45–60% of the USA from 95° W to the west coast. In 2010–2011, the drought area reduced to 40%. But in 2012–2016, the drought intensified considerably and the affected area increased to 70%. This mega-drought (US-WMD) was heavily felt in 17 states' environment, economy, and social life. Consequences included wildfires, strong reduction of water in rivers, lakes, and reservoirs, a severe decline in groundwater, irrigation constraints, agricultural production losses, ecosystem deterioration, intensive job loss, price increase for food and land, and many disturbances due to a lack of water and food (CDWR 2017, Howitt et al. 2015; Mathews 2014). Compared to other catastrophic droughts since 1950, the US-WMD, during its worst 2012 year, affected 71.3% of the western USA, which is comparable to the area under the catastrophic Dust Bowl drought in 1934 (71.6%) and was much larger than the strongest droughts in 1956 (49%) and 1988 (31%). The current US-WMD is longer than the other multiyear droughts (1985–1986, 1988–1992, 1995–1996, and 2001–2003) in the western USA (Kogan and Guo 2015).

Among western states, California was the most severely drought-affected, in 2012–2016, especially in 2013 and 2014, when an area of the strongest vegetation stress reached 70% (Kogan et al. 2017; Kogan and Guo 2015). The drought had considerable impacts on California's environment, economy, and society. Following the Howitt et al. (2015) and CDWR (2017) reports, the total economic drought impact on California is estimated at $2.74 billion. Considerable increase in surface and ground water overdraft is expected to have long-term impacts for the next 2–3 years. Since California's agriculture is widely irrigated, using 52% total water supply in dry year (Hanson 2015), several-year drought greatly affected agriculture, leading to nearly 3% losses in crop revenue and increasing the cost of extra water pumping by 75% (Hewitt et al. 2015). In addition, water shortages increased the amount of drought-related idle land by 45% (currently 540,000 acres) and 21,000 total job losses. Besides agricultural losses and water depletion, the California drought was one of the causes for intense wildfires with larger than before area and property damages.

US Drought Monitor (USDM 2017, Svoboda et al. 2002) is one of the weather-based drought product distributed in the USA. The method is based on weather data (precipitation, temperature) and weather-based indices (standardized precipitation index (SPI) and Palmer drought severity index (PSDI)) and advises based on local drought-watchers. The USDM weather data and indices are calculated from nearly 10,000 US weather station data. Considering the area of the USA and assuming uniform stations' distribution, USDM monitors drought area and intensity with areal resolution of approximately 1000-1300 km^2 of USA territory, including California. However, considering the importance of California's agriculture, which provides more than 90% of US fruits, vegetables, berries, and nuts (MAMC 2015), the area of each crop planted mostly in the Central Valley is much smaller. Besides, many crops are irrigated. Therefore, drought monitoring, characterization of its damage to California's agriculture and calculation of irrigation requirements should be assessed on the area comparable to the size of the planted crops, which is 200–500 acres, or 0.5–2 km^2. The VIIRS-based vegetation health method is capable to detect and monitor drought at each 500 by 500 m (0.25 km^2) land surface (Kogan et al. 2017; NOAA/NESDIS 2018; JPSS 2014). This subchapter demonstrates the 500 and 1000 m resolution SNPP/VIIRS-VH data in monitoring the 2012–2016 strongest California mega-drought, distinguishing drought affected area with and without irrigation and estimating drought intensity and impact, comparing VH drought estimates with the available economic indicators (CNRFC 2016).

Why high-resolution satellite data? High-resolution drought monitoring is especially important in the areas of irrigated crops to estimate irrigation time and the amount of water needed, to evaluate how to redistribute water among the areas in case of water shortages and to assess crop and grasses losses due to a shortage of water. California is the primary state belonging to this category, since nearly 20 commercial crops grown in the Central Valley require irrigation each year (Hanson 2015; Howitt et al. 2015). In drought years, especially such as California's strongest mega-drought, the amount of water used for irrigation increased considerably, requiring drought monitoring at the highest spatial resolution level to determine which of the irrigated areas had good vegetation condition, which suffered from drought stress and how much water was required. Although the example below is focused on California, it is important to emphasize that in some African countries, irrigation is used to intensify crop performance, reduce losses in agriculture, and enhancing food security. High-resolution data (1 and 0.5 km^2) from the advanced operational sensors, such as VIIRS, are very important, especially in Africa for advanced assessment of weather-related losses in agriculture, addressing food security issues, especially those leading to shortages of food, hunger, and people death in very extreme cases.

In the California drought, it is important to compare first SNPP/VIIRS-based 0.5 km^2 resolution vegetation health (NOAA/NESDIS 2018) with 1300 km^2 USDM's drought assessments (USDM 2017) in mid-July (the most intensive drought month) 2012 through 2017 (Fig. 5.10). There are some general similarities in the assessments since both methods show an intensification of drought from 2012 to 2013–2014 and a reduction of drought area and intensity in 2016. However, focusing on spatial details provided by VH-0.5, specifically in the California's Central Valley (CV, the area stretching from north to south in the middle of the state) where all commercial crops are grown, there are huge differences between the two methods. While USDM estimates California as being abnormally dry (light yellow) or having moderate drought (mid yellow color) all over the state in 2012 and 2013, the VH method finds a lot of areas with severely stressed vegetation (dark red). Moreover, most impressive is that in the background of the most intensive drought, a lot of small areas in the CV are having healthy (green and light blue) and even very healthy (dark blue) vegetation. Also important are the blue areas that are consistently stable (the same summer time and in the same locations) from year to year, signaling that crops are irrigated. There are many other differences between the two methods, such as an increase in drought area and intensity in 2015 from USDM and their reduction from VIIRS-VH; more intensive drought in 2016 from USDM than from VH. All of these differences indicate that data resolution plays an extremely important role in estimation of drought dynamics in the areas, where intensive and advanced technologies (irrigation, crop orientation, crops substitutes, etc.) are used to enhance agricultural productivity.

VIIRS/VH-500 drought and impacts in the California's Central Valley. The economic

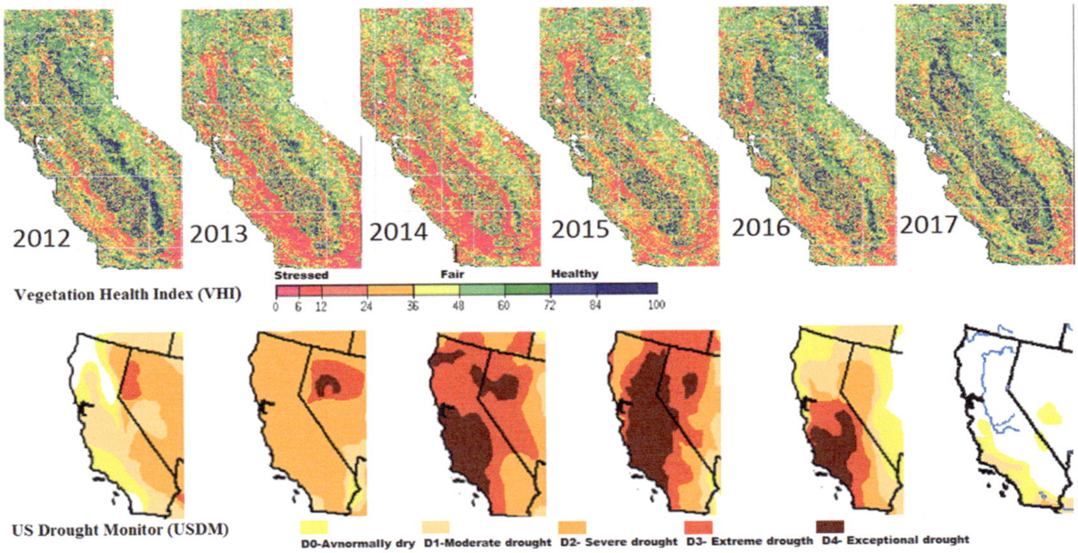

Fig. 5.10 California 0.5 km^2 resolution moisture–thermal conditions (VHI) and US Drought Monitor (USDM) during 2012–2017

impact of the recent California drought in the CV was presented in Howitt's et al. (2015) comprehensive report. The report's economic data estimates in Fig. 5.11c show that drought-related economic losses were not equally distributed in the CV. The least severe drought impact is estimated in the central part of the Valley. The southern part was impacted strongly in crop revenue reduction and land fallow increase (smaller crop area), and the northern part suffered severely from employment loss. Total areal impact from all losses in the CV can be summarized as follows: the strongest was in the southern part (red box in Fig. 5.11c, b), the weakest—in the central part (blue box, the same figures), and moderate—in the northern part (yellow box). These results were compared with the VIIRS/VH-0.5 assessments of the drought area, intensity and change between June 4 and August 1, 2015 (Fig. 5.11a, b). Following VIIRS/VH-500 time-series analysis, drought-related vegetation stress dominated the entire state of California, including both nonirrigated (red color) and some of the irrigated areas (yellow and light red color) of the CV. There is some deterioration of vegetation stress in August (compare to June) in southern and eastern California. However, in some areas, 500 m data shows slight improvement of vegetation health between June and August. In order to determine deterioration/improvement for each 500 m pixel, a difference of VH-500 was taken between these two maps (August minus June). The difference map in Fig. 5.11b shows general similarities with the maps of economic drought impacts in Fig. 5.11c (Howitt et al. 2015), because the three areas of economic impacts in the CV (Fig. 5.11c) are matching with the areas of VH deterioration/improvement of vegetation stress between the two periods (Fig. 5.11b). Specifically, in the southern and northern parts of the CV, where the economic impact was stronger, most of the VH-0.5 pixels showed deterioration of vegetation stress, while in the central part, where the economic impact was milder, half of the pixels indicate deterioration, while the other half shows some improvement due to irrigation. The same three regions were identified on the images of moisture (VCI) and thermal (TCI) conditions (not shown). In addition, the VIIRS/VH-0.5 changes on difference map (between August and June 2015) in southern California were related to strong thermal stress deterioration, while in the northeastern part of the state due to moisture stress.

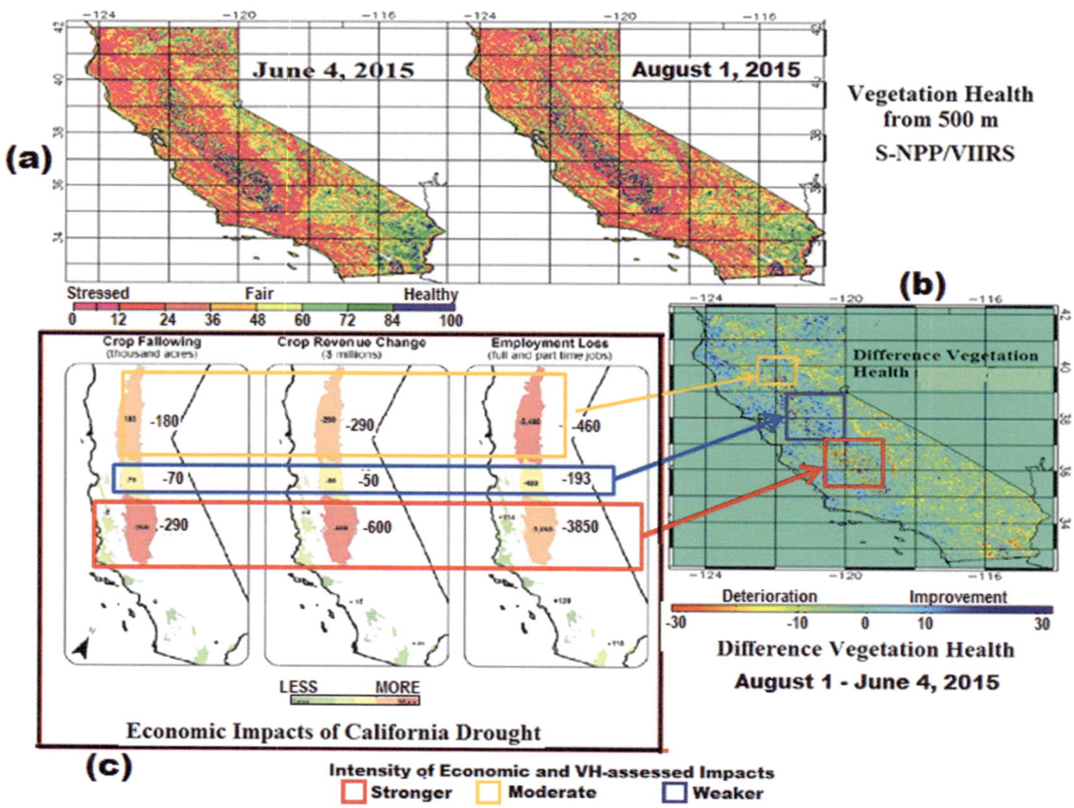

Fig. 5.11 California drought detection and impacts assessments (**a**) 0.5 km^2 moisture–thermal (VHI) conditions in summer 2015, (**b**) difference VH between August 1 and June 4, 2015, and (**c**) economic impacts of California drought (Howitt's et al. 2015)

Vegetation health dynamics in irrigated and nonirrigated areas. It has been already mentioned that, at the background of a very intensive California drought during 2012–2017, VIIRS/VH-0.5 data showed, in addition to regions with stressed vegetation, the areas with healthy vegetation in the CV. The later was not unusual since nearly 7.38% of California lands, especially crops in the CV, are irrigated (USDA/ERS 2017). Unfortunately, it is difficult to find ground data on irrigated individual crops and in situ assessments of crop condition. Therefore, indirect general indicators were used to distinguish irrigated/nonirrigated crop areas. The assumption was that if, during the strongest 6-year (2012–2017) California drought, in some areas vegetation is stably healthy each year, those areas were considered to be irrigated. The VIIRS/VH-0.5 data in Fig. 5.10 has already demonstrated that, at the background of stressed vegetation in the CV, some areas experienced healthy conditions from year to year. Figure 5.12 shows 2012–2016 time-series for the percent of areas (from the entire CV) with unhealthy (red) and healthy (green) vegetation (a) every week and (b) summer-mean (June-August) in two southern regions (% from the total area of each region). The selected southern regions with stable vegetation conditions from year to year were in size 440 km^2 (occupy 0.75% of the entire CV) for nonirrigated and 77 km^2 (0.13%) for irrigated. They had 2172 and 380 of 0.25 km^2 pixels, respectively. As seen in Fig. 5.12a, the areas of vegetation stress in the entire CV were growing for summers (winter is not taken into consideration for this analysis since lower temperatures reduce normally drought impacts) from 2012 to 2014 (drought intensification) with the following reduction by

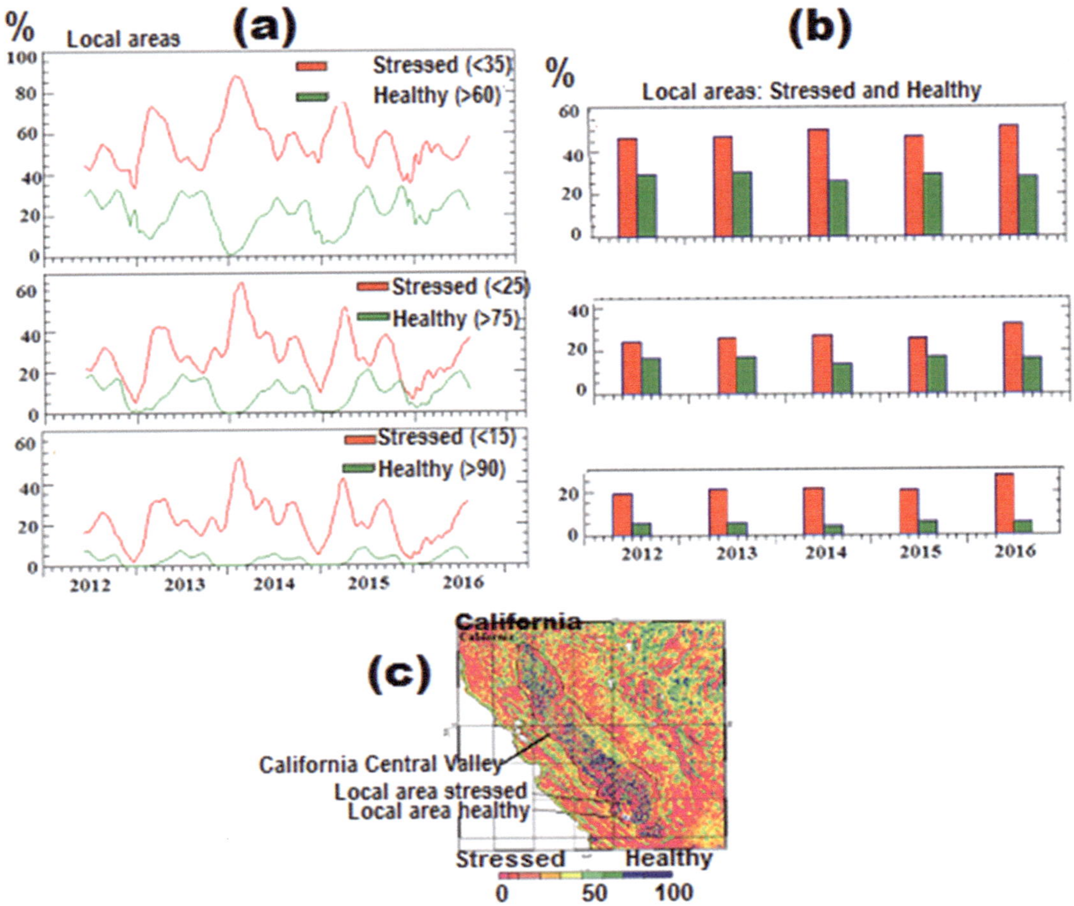

Fig. 5.12 California 2012–2016 stressed and healthy vegetation time series (**a**) weekly and (**b**) summer season in the local stressed and healthy vegetation of the Central Valley, shown on the map (**c**)

2016. This type of dynamics is typical for all levels of stress intensities in California's summer (extreme (VHI < 10) occupying 25–50% of the area, severe-to-extreme (VHI < 15) 30–60% and moderate-to-extreme (VHI < 35) 55–90%). It is important to emphasize that only 25–30% of 0.25 km² pixels in the selected region are irrigated (Fig. 5.12b). However, another fact should be emphasized that the dynamics of healthy vegetation (area and intensity) in the irrigated region (moderate-to-perfect (VHI > 60) 30–35% of the region area, very good-to-perfect (VHI > 75) 18–22% and perfect (VHI > 90) 5–7%) remained practically stable over all 5 years (except a very negligible reduction during the strongest drought in summer 2014). Finally, it is important to emphasize that in the two small southern regions (Fig. 5.12b), stressful versus healthy vegetation dynamics in the areas' size and intensity are generally similar to the entire CV. Specifically, vegetation stress area for all intensities (VHI < 35, 15 and 10) increased slightly in mid-2014 and decreased in 2015. The only difference between CV and the small stressed region is a 5–8% increase in the area of stressed vegetation in summer 2016, which was probably due to local conditions. It is important to emphasize again that the areas with favorable conditions (VHI > 60, 75 and 90) remained stable in size from year to year (except small changes in 2014) for all intensities due to irrigation.

Drought recovery. Summarizing briefly VIIRS/VH-0.5 based long-term drought dynamics in California, it is possible to conclude that a light drought began in 2006 and for the next 4–5 years, transitioned to moderate levels. In 2012, drought intensified considerably causing moderate-to-extreme vegetation stress (VHI < 30), which continued until the end of 2015. During 2015, sea surface temperature (SST) in the central tropical Pacific was warmer than normal indicating a development of El Niño Southern Oscillation (ENSO). Following ENSO southwestern USA, especially California, is normally wet and cool during December-February, inducing development of healthy vegetation. As was expected the 2015–2016 El Niño triggered rains in a very dry California. Following climate data and reports (CDWR 2017; CNRFC 2016), during October 2015-May 2016, northern and central California received the largest amount of rains (85–120% multiyear mean, with up to 175% in some areas) when drought intensity and the affected area started to decrease. As a result, in late 2015 and during 2016, California's reservoirs were recharged up to 186%, the State Water Resources Control Board ended water restriction imposed during severe drought period (Fears 2016) and water withdrawal started to increase again after year reduction (CDWR 2017). However, rains were not distributed uniformly and southern California, which received only 40–60% of normal rainfall, was still experiencing shortages of water. Following 2015–2016 rains, vegetation in many California regions recovered from long stress. Although in some southern areas with a lack of rains, California vegetation still experienced moisture/thermal stress. Following drought reduction, vegetation transitioned to normal conditions in early 2017 (VHI is 50 and larger). The final drought accord occurred because air temperature began to decrease and TCI was transitioned from intensive thermal stress in 2015 (TCI = 12–25) to healthy thermal conditions (TCI = 60–70) in the first several months of 2017. Unfortunately, there is some time-lag between the arrival of rains and vegetation recovery from long-term drought. Therefore, California's CV drought area continued to increase through the mid-2016, reaching 65% by August (Fig. 4.14b). After August 2016, a downhill drought trend was observed for 2–3 months. This downhill trend became very strong, indicating a complete drought recovery in extreme and intensive-to-extreme categories. Only 20% of the drought in a light-to-extreme category remained as of February-March 2017. However, in 2018, an intensive drought in southern California returned (NOAA/NESDIS 2018) with water depletion fires and other events.

5.7 Devastating Droughts in 2017 and Early 2018

Devastating droughts or exceptional, in VH nomenclature, are generally quite rare, especially if they affected the economy of an entire country. However, they occur from time to time and if the damages were extreme and they affected the entire country, the droughts attract exceeding attention since economic damages are enormous and require recovery's estimations. Such unusual droughts affected southern European countries, especially Portugal and Spain in the second half of 2017. In these two countries, 2017 drought has left rivers nearly dry, sparked deadly wildfires, and devastated crops and pastures (Vicente 2017). According to meteorologists (IPMA 2017), the 2017 was classified as a very unusual drought year, since 94% of Portugal has endured "extreme" drought and less than 50% of rainfall in Spain has created severe drought conditions. Very devastated drought consequences were cited in the world news (Euronews 2017; Newsgrid 2017; News24 2017; TheGuardian 2017). A few of the most unusual drought-related environmental damages were the following: Douro River, which is one of the symbols of the Iberian Peninsula is 60% dry by November 2017. In Portugal, 28 of the country's water reservoirs in October were at less than 40% of their storage capacity. Shortage of water is so severe that hundred fire trucks began transporting water from one dam in northern Portugal to another that is running dry. Besides, trucks delivered water to Viseu, a city of around 100,000 residents. Severe water deficiency was aggravated by excessive temperatures in many places of Peninsula, which hit and exceeded 40° C, which are the most sustained heatwave since 2003 (News24 2017; TheGuardian 2017). Such critical water and temperature conditions, in addition to creating problems for population living, created huge economic problems for both countries to grow crops. (Euronews 2017). Grains, olives and grapes are suffered from water and thermal stress (USDA/FAS 2017). About 1.38 million hectares of grains, sunflowers, grapes and olive trees have been affected by drought in Spain as of the end of October and especially in November 2017. The 2017 water and thermal stressed situation worsened the drought impacts on economy since the current year drought consequences were intensified by the 2016 drought, which started in the second half of the year (USDA/FAS 2017). Overall grain production is projected to be much below historical average levels in the discussed countries. Grain harvest is expected to plunge up to 70% from the last year and olives harvest is expected to drop in half, compared to regular production (Vicente 2017; Euronews 2017). In addition to crops and trees devastation, grass on the pastures is very scares and farmers have to buy much fodder for their cattle to survive. The strong food demands by the domestic livestock industry, along with limited pasture's grass availability will contribute to an increase in the countries' grain import needs. Following a reduction in 2017 production of grape, Spain will produce much less wine. The situation is worsened since drought affected not only Spain but also Italy and France grape productions. These three countries produce nearly 50% of world wine, and due to 2017 drought, the entire world is facing a shortage of wine. Disastrous grapes harvest means wine prices could be going high (Euronews 2017). Drought-related severe fires killed 109 people this year in Portugal (News24 2017).

Following such devastating drought damages, it is important to assess (1) if VH method detected this exceptional drought on time, monitored its source, intensity, duration and (2) if it possible to predict this drought in advance of its start, estimate damages, especially agricultural losses and anticipated food security situation. The new generation of NOAA/VIIRS operational satellites, which will be in global services for the next 20–25 years (NOAA/NESDIS 2018), were used for these purposes. Figure 5.13 demonstrates drought stress area in Portugal and Spain from two indices VHI and TCI. Following VHI index (combination of moisture, VCI and thermal TCI conditions), at the end of November 2017 the drought covered the entire area of both countries, although the level 3 intensity (extreme) was observed only in eastern Portugal (covering nearly 40% of the country) and western Spain (around 20%). However, the 2017 drought in these countries was very specific in terms of its source. As seen at the Figure, the two countries were under the strongest thermal vegetation stress (TCI category 4 or exceptional drought). Ninety-six percent of Portugal and 83% of Spain were under the 4-category of thermal (TCI) stress. And such stress might devastate vegetation in a matter of 5–10 days, which is very quick if air temperature exceeds 40° C, what has happened in these countries (News24 2017; TheGuardian 2017).

Although 2018 grain-growing season has just started in the Northern Hemisphere, the first troubling drought indicator has already observed with moisture–thermal (VHI) images. Figure 5.14 shows the end of May image of world drought area of moderate to exceptional intensity. Among major grain-producing countries, the USA has severe-to-extreme drought in the principal grain-growing areas. Since this drought has continued for 2–3 months, the current expectations are some grain losses in the central states. Central India is suffering from severe-to- exceptional drought on quite a big area. The end of wheat-growing season in Australia has some prospects for considerable grain losses and reduction of grain exports. The countries in the northern part of Sub-Sahara Africa are currently suffering from a very intensive drought, which supposed to create some food security problem. Among other world areas, drought is affecting most pasture regions of Mongolia and should be watched for further development.

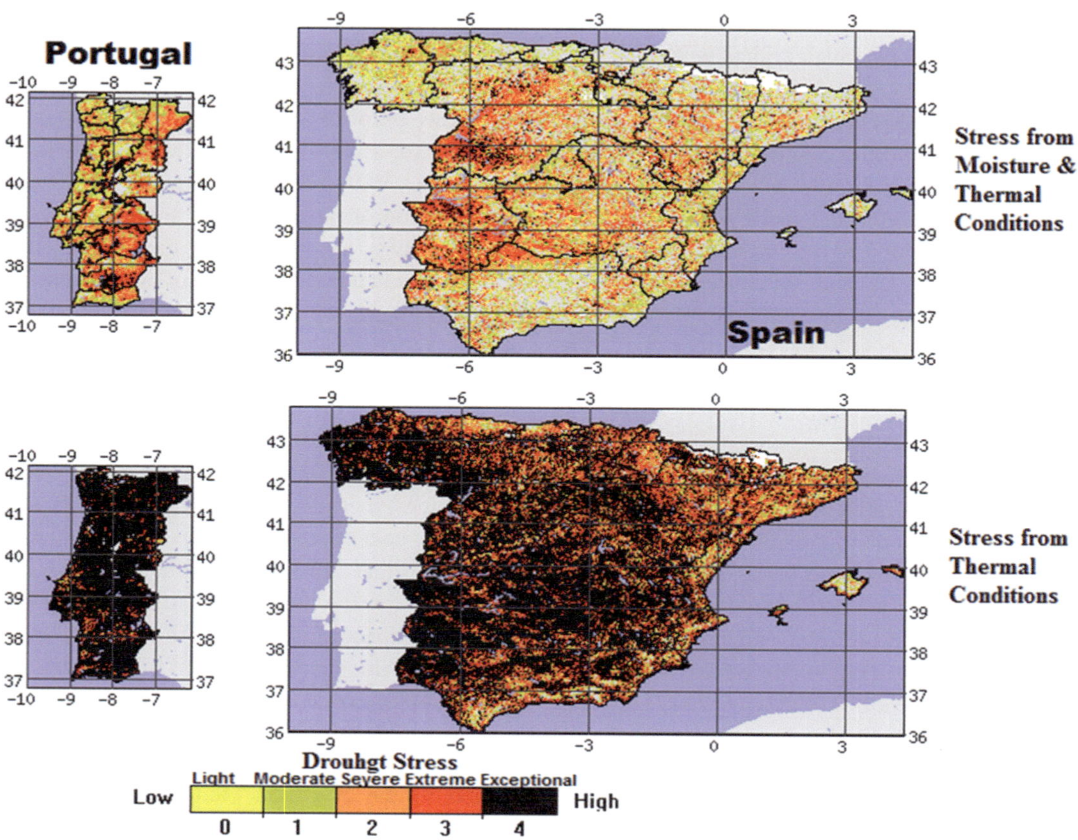

Fig. 5.13 Drought area and intensity from moisture (VCI) and thermal (TCI) vegetation stress in Portugal and Spain on November 25, 2017

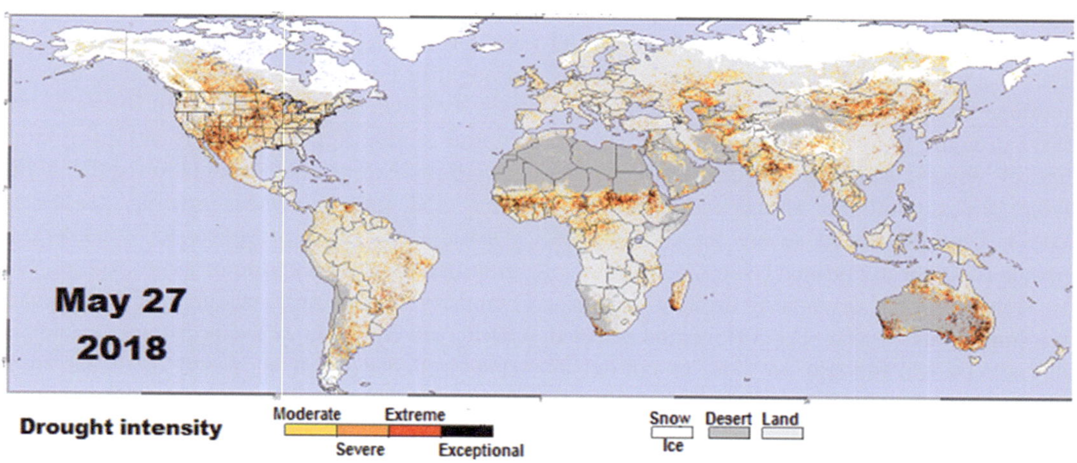

Fig. 5.14 Vegetation health-based global drought area and intensity at the end of May 2018 (NOAA 2018)

5.8 Conclusion

Nearly 85% of all natural disasters on Earth are related and associated with extreme weather events (Obasi 1994). Every day our planet experiences droughts, floods, tropical cyclones, tornadoes, heat waves, bushfires, and many other disasters. In addition to weather, the globe also experiences geological disasters (earthquakes, volcanoes, etc.). Unfortunately, natural disasters are a part of the environment we live in. They adversely affect the lives of a large number of people and cause considerable damage to economy, nature and property worldwide. Drought affects the largest number of people on the Earth (Obasi 1994), reduces agricultural production, and creates the biggest problem with food security. Even in the United States of America, a country of advanced economic development, drought is considered a "14 billion-dollar" annual event in terms of incurred losses from natural disasters (NCDC 2011).

Among all natural disasters, drought is the most damaging environmental phenomenon. In the twenty-first century, nearly 20% of global lands were drought-stricken almost every year, but in some years this number increases considerably. Drought is a typical phenomenon of the Earth's climate. Losses from droughts, especially in agriculture, are staggering. For example, in the USA, the average annual agricultural cost of drought is around $6 billion. However, in extreme drought years such as 1988, drought costs jumped to $60 billion. Developing countries of Africa and Asia have been the most drought-affected, facing food security problem. Chapter 2 showed how drought affects food security, reducing global grain production below the demands for grain in half of the years of the twenty-first century. Drought affects and kills the largest number of people on Earth and in the long run, is a very costly disaster (Wilhite 2000; NOAA 2017). Following IDNDR (1999) and FAO (1999), the World Meteorological Organization estimated that during 25-year since 1967, drought affected 50% of the 2.8 billion people who suffered from all major disasters (Table 5.1).

There are four distinctive methods of measuring and estimating drought: meteorological, land cover (vegetation), soil moisture, and remote sensing. In order to detect and monitor drought, the measures should include such characteristics as drought start/end, intensity, duration, magnitude, area, season, origination and impact. Drought type can be classified by the affected area, ecosystem, economy, climate, weather, produced impacts, and other considerations. Several are very common drought types referenced in scientific literature: meteorological, hydrological, agricultural, and socioeconomic (Wilhite and Glantz 1985).

During the period of operational satellite era (since 1981), the new numerical method called vegetation health (VH) put a firm step into drought detection, monitoring, impacts assessment, trend evaluation during climate warming, and food security prediction. The VH numerical method (described in detail in Chaps. 3 and 4) was developed from the first operational NOAA/AVHRR satellites, their data, and service during 1983–1985 (Kogan 1987, Kogan 1989, 1997, Kidwell 1997). The VH method was improved considerably thereafter and reached the highest stage of improvement and application with the new the SNPP/VIIRS and NOAA-20/VIIRS operational sensors. This new technology provided the highest resolution (0.5 km) real-time data used for an early drought detection, estimating its intensity, area, duration, and impacts, which permitted to predict agricultural production well in advance of harvest and to estimate problems with food security.

Global droughts are a typical climate phenomenon affecting very important agricultural regions. The twenty-first century saw the continuation of the twentieth century drought tendencies (area expansion intensity and duration increase and intensification of the impacts on agriculture) around the world, affecting food security, especially in the underdeveloped countries. In one half of the first 17 years of the twenty-first century, global grain production (the main source of food and feeds for the Earth) fell below a steadily grown consumption

(PotashCorpo 2013; USDA/FAS 2017; WB 2017; FAO 2017). Droughts made considerable contribution to this dis-balance, creating serious concerns about global food security (FS). Additional FS concern arises from the anticipated drought intensification, expansion of their area, and penetration to the new regions due to the current and future climate warming (Solomon et al. 2007). Africa is the main continent where regional food insecurity appears every year in its countries following a drought-related reduction of crop production. In such situation, African countries require the international community's help to prevent hunger and starvation. The NOAA's vegetation health technology is a great help, providing 1-3-month prediction of food security situation.

References

AgriLife. 2011. http://agrilife.org/today/2011/08/17/texas-agricultural-drought-losses-reach-record-5-2-billion/. August 17.

AMIS (Agricultural Market Information System) 2017. *Market database.* http://statistics.amis-outlook.org/data/index.html.

Begueria, S., S.M. Vincente-Ser4rano, M. Angulo-Martinez. 2010. A multiscalar global drought dataset: (SPEIBASE, A new gridded product for the analysis of drought variability and impacts. *Bulletin of the American Meteorological Society*. 1351–1356.

Bhuiyan, C., A.K. Saha, N. Bandiopadhyay, and F.N. Kogan. 2017. Analyzing the impact of thermal stress on vegetation health and agricultural drought – a case study from Gujarat, India. *Advances in Remote Sensing and GIS-based Drought Monitoring* 54 (5): 678–699. https://doi.org/10.1080/15481603.2017.1309737.

Brown, J.F., B.D. Wardlow, T. Tadesse, and M.J. Hayes. 2008. Vegetation Drought Response Index (VegDRI): A new integrated approach for monitoring drought stress in vegetation. *GI Science & Remote Sensing* 45 (1): 16–46.

Carlson, T.N., E.M. Perry, and T.J. Schmugge. 1990. Remote estimation of soil moisture availability and fractional vegetation cover for agricultural fields. *Agricultural and Forest Meteorology* 52: 45–69.

CDWR (California Department of Water Resources). 2017. California Department of Water Resources. http://www.water.ca.gov/drought.

Cracknell, A.P. 1997. *The advanced very high resolution radiometer*, 534. USA: Taylor & Francis.

Deering, D.W. 1978. *Rangeland reflectance characteristics measured by aircraft and spacecraft sensors.* Ph.D. Dissertation, 388. College Stat., TX: Texas A&M Univ..

Drought. 2016. *Drought disasters during 1980-2008.* http://www.preventionweb.net/english/hazards/statistics/?hid=59.

FAO. 2017. *Earth observations.* http://www.fao.org/giews/earthobservation/asis/index_2.jsp?lang=en.

Emdat. 2011. The International disaster data base. http://www.emdat.be/old/Documents/Publications/publication_2004_emdat.pdf.

Euronews. 2017. *Devastating drought.* http://www.euronews.com/2017/11/09/drought-across-spain-and-portugal-raises-alarm.

FAO 2017a. How Close We are to Zero Hunger. http://www.fao.org/state-of-food-security-nutrition/en/.

FAO 2018. FAO Cereal Supply and Demand Brief. World Food Situation. http://www.fao.org/worldfoodsituation/csdb/en/

FAO 1999. Food Insecurity: When people must live with hunger and fear of starvation. *The State of Food Insecurity in the World Report*, Rome, 76 pp.

Fears D. 2016. Drought hasn't eased, but rules have. The Washington Post. 29; p. A3.

FewsNet (Famine Early Warning System Network) 2017. Nigeria Food Security Report. https://fews.net/sites/default/files/documents/reports/Nigeria_OL_10_2016_National%20Final.pdf

Gao, X., A.R. Huete, W. Ni, and T. Miura. 2000. Optical–biophysical relationships of vegetation spectra without background contamination. *Remote Sensing of Environment* 74: 609–620.

Gitelson, A., and Y. Kaufman. 1998. MODIS NDVI optimization to fit the AVHRR data series—spectral consideration. *Remote Sensing of Environment* 66: 343–350.

GRACE 2017. Food security food access. Food Program, http://www.sustainabletable.org/280/food-security-food-access

Hanson B. 2015. *Irrigation of agricultural crops in California 2015.* Available from: http://www.arb.ca.gov/fuels/lcfs/ workgroups/lcfssustain/hanson.pdf. 45 p.

Hayes, M.J., M.D. Svoboda, D.A. Wilhite, and O.V. Vanyarkho. 1999. Monitoring the 1996 drought using the Standardized Precipitation Index. *Bulletin of the American Meteorological Society* 80 (3): 429–438.

Hayes, M.J., and W.L. Decker. 1996. Using NOAA AVHRR data to estimate maize production in the United States Corn Belt. *International Journal of Remote Sensing* 17: 3189–3200.

Heim, R.R., Jr. 2002. A review of twentieth-century drought indices used in the United States. *Bulletin of the American Meteorological Society* 83: 1149–1165.

Howitt, R.E., D. MacWan, J. Medellin-Azuara, J.R. Lund, and D.A. Sumner. 2015. *Economic Analysis of the 2015 Drought for California Agriculture*, 16. Davis, CA: Center for Watershed Sciences, University of California–Davis https://watershed.ucdavis.edu/files/biblio/DroughtReport_23July2014_0.pdf.

Huete, A., K. Didan, T. Miura, E.P. Rodriguez, X. Gao, and L.G. Ferreira. 2002. Overview of the radiometric and biophysical performance of the MODIS vegetation indices. *Remote Sensing of Environment* 83: 195–213.

IDNDR (International Decade for Natural Disaster Reduction) 1999. United Nation Office for Disaster Risk Reduction. https://www.unisdr.org/we/inform/publications/31468

IFRC (International Federation of Red Cross and Red Crescent Society) 2011. Annual Report. http://www.ifrc.org/en/publications-and-reports/ifrc-annual-report---2011/

IPMA (Portuguese Institute of the Sea and the Atmosphere). 2017. https://en.wikipedia.org/wiki/Instituto_Portugu%C3%AAs_do_Mar_e_da_Atmosfera

JPSS. 2014. *Joint polar satellite system*. http://www.jpss.noaa.gov.

Kidwell, K.B., ed. 1997. *Global vegetation index user's guide*. Camp Springs, MD, USA: National Oceanic and Atmospheric Administration, National Environmental Satellite Data and Information Services, National Climatic Data Center.

———., ed. 1995. *NOAA Polar Orbiter Data Users Guide*. Camp Springs, MD, USA: National Oceanic and Atmospheric Administration, National Environmental Satellite Data and Information Services, National Climatic Data Center.

Kogan F. 1986. *The Impact of Climate and Technology on Soviet Grain Production*. Monograph Series on Soviet Union Sponsored by Delphic Associates. DELPHIC 178 pp.

Kogan, F., W. Guo, and W. Yang. 2017. SNPP/VIIRS vegetation health to assess 500 California drought. *Geomatics, Natural Hazards and Risk* 8 (2): 1–13. https://doi.org/10.1080/19475705.2017.1337654.

Kogan, F., and W. Guo. 2016. Strong 2015–2016 El Niño and implication to global ecosystems from space data. *International Journal of Remote Sensing* 38 (1): 161–178. https://doi.org/10.1080/01431161.2016.1259679.

Kogan, F., Z. Popova, and P. Alexandrov. 2016. Early forecasting corn yield using field experiment dataset and Vegetation health indices in Pleven region, north Bulgaria. *Ecologia i Industria (Ecology and Industry)* 9 (1): 76–80.

Kogan F. and W Guo. 2015. 2006–2015 mega-drought in the western USA and its monitoring from space data. *Geomatic, Natural Hazards and Risk*, https://doi.org/10.1080/19475705.2015.1079265.

Kogan, F., M. Goldberg, T. Schott, and W. Guo. 2015a. SUOMI NPP/VIIRS: improve drought watch, crop losses prediction and food security. *International Journal Remote Sensing* 36: 5373–5383. https://doi.org/10.1080/01431161.2015.1095370.

Kogan, F., W. Guo, A. Strashnaia, A. Kleshenko, O. Chub, and O. Virchenko. 2015b. Modelling and prediction of crop losses from NOAA polar-orbiting operational satellites. *Geomatics, Natural Hazards and Risk* 7: 886–900. https://doi.org/10.1080/19475705.2015.1009178.

Kogan, F., and W. Guo. 2014. Early twenty-first-century droughts during the warmest climate. *Geomatics, Natural Hazards and Risk* 7: 1–11. https://doi.org/10.1080/19475705.2013.878399.

Kogan, F., N. Kussul, T. Adamenko, S. Skakun, O. Kravchenko, O. Kryvobok, A. Shelestov, A. Kolotii, O. Kussul, and A. Lavrenyuk. 2013a. Based on Earth Observation, Meteorological Data and Biophysical Models. *International Journal of Applied Earth Observation and Geoinformation* 23: 192–203. https://doi.org/10.1016/j.jag.2013.01.002.

Kogan, F., T. Adamenko, and W. Guo. 2013b. Global and regional drought dynamics in the climate warming Era. *Remote Sensing Letters* 4: 364–372. https://doi.org/10.1080/2150704X.2012.736033.

Kogan, F. 2006. Early drought detection, monitoring and assessment of crop losses from space: global approach. In *Disaster forewarning diagnostic methods and management*, ed. F. Kogan, S. Habib, and M. Matsuoka, 91–95. Bellingham: SPIE.

Kogan, F.N. 2002. World droughts in the new millennium from AVHRR-based Vegetation Health Indices. *Eos Transaction of American Geophysical Union* 83 (48): 562–563.

———. 2001. Operational space technology for global vegetation assessment. *Bulletin of the American Meteorological Society* 82 (9): 1949–1964.

———. 1997. Global drought watch from space. *Bulletin of the American Meteorological Society* 78: 621–636.

Kogan, F.N., J.T. Sullivan, and Pu Bu Ciren. 1996. Testing post-launch calibration for the AVHRR sensor on world desert targets during 1985–1993. *Advances in Space Research* 17 (1): 47–50.

Kogan, F.N. 1995a. Droughts of the late 1980s in the United States as derived from NOAA polar-orbiting satellite data. *Bulletin of the American Meteorological Society* 76 (5): 655–667.

———. 1995b. Application of vegetation index and brightness temperature for drought detection. *Advances in Space Research* 15 (11): 91–100. https://doi.org/10.1016/0273-1177(95)00079-T.

Kogan, F. 1989. Remote sensing of weather impacts on vegetation in non-homogeneous areas. *International Journal Remote Sensing* 11 (8): 1405–1419.

Kogan, F.N. 1987. Vegetation Health index for areal analysis of NDVI in monitoring crop conditions. In *Preprint 18th Conference on Agricultural and Forest Meteorology*, 103–114. Boston: AMS.

Kushnir, V., G. Korotaev, F. Kogan and A. Powell. 2009. *Consequences of land and marine ecosystems interaction for the black sea coastal zone*.

MAMC (Map of California with Main Crop). 2015. Where does your food come from? Available from: https://search. yahoo.com/yhs/search?p=map+of+california+with+main+crops&ei=UTF-8&hspart=mozilla&hsimp=yhs-001

Masih, I., S. Maskey, F.E.F. Musa, and P. Trambaner. 2014. A review of droughts on African continent: s geospatial and long-term perspective. *Hydrology and*

Earth System Sciences 18: 3635–3649. https://doi.org/10.5194/hess-18-3635-2014.
Mathews K. 2014. How California can survive a prolonged drought. [cited 2014 Oct 24]. Available from: http://www. care2.com/causes/how-california-can-survive-a-prolonged-drought.html
McKee, T.B., N.J. Doesken, and J. Kleist. 1993. The relationship of drought frequency and duration to time scale. In *Preprints Eighth conference on applied climatology*. Anaheim, CA: Janvier.
MercyCorps 2017. Annual Report. https://www.mercycorps.org/gallery/AR2017/2017-annual-report
NCDC (National Climatic Data Center). 2011. *Billion dollar U.S. weather disasters*. Accessed 2013. http://www.ncdc.noaa.gov/oa/reports/billionz.html.
NDMC (National Drought Mitigation Center). 2017. *Type of drought*. http://drought.unl.edu/DroughtBasics/TypesofDrought.aspx.
News24, 2017. https://www.news24.com/Green/News/spain-portugal-struggle-with-extreme-drought-20171121.
Newsgrid 2017. http://www.aljazeera.com/indepth/inpictures/2017/01/spain-drought-reveals-lost-worlds-170128072138481.html.
NIDIS 2017. *Global drought information system*. https://www.drought.gov/gdm/current-conditions.
NOAA/NESDIS 2018. Vegetation Health indices and products. http://www.star.nesdis.noaa.gov/smcd/emb/vci/VH/index.php
NOAA 2017. *U.S. billion-dollar weather & climate disasters 1980–2016*. https://www.ncdc.noaa.gov/billions/.
NOAA/NESDIS 2017. *JPSS has new name: NOAA-20, Nov 21*. https://www.nesdis.noaa.gov/JPSS-1.
Obasi, G.O.P. 1994. WMO's role in the international decade for natural disaster reduction. *Bulletin of the American Meteorological Society* 75 (1): 655–661.
Orians, G.H. 1990. Ecological sustainability. *Environment* 32: 10–15, 34–39. http://www.potashcorp.com/industry_overview/2011/agriculture/16.
Palmer, W.C.. 1965. *Meteorological drought*. U.S. Weather Bureau Pap. No.45, 58.
Ped, D.A. 1975. On indicators of drought and wet conditions (I Russian). *Procedure Hydrometeorological Center of the USSR* 156: 19–39.
PotashCorpo. 2013. *Agriculture: crop overview*. http://www.potashcorp.com/industry_overview/2011/agriculture/16.
Redorbit. 2012. http://www.redorbit.com/news/science/1112703681/climate-change-southwest-us-forests-100112/
Rojas, O., A. Vrieling, and F. Rembold. 2011. Assessing drought probability for agricultural areas in Africa with coarse resolution remote sensing imagery. *Remote Sensing of Environment* 115: 343–352.
Sasson A. 2012. Food security for Africa: an urgent global challenge. *Agriculture & Food Security*, 1:2. https://doi.org/10.1186/2048-7010-1-2
Seiler, R.A., F.N. Kogan, and G. Wei. 2000. Monitoring weather impact and crop yield from NOAA AVHRR data in Argentina. *Advances in Space Research* 26 (7): 1177–1185.
Selyaninov, G.T. 1937. *Mirovoy Agroklimaticheskiy Spravochnik (World Agro Climate Handbook)*, 150. Leningrad: Gidrometisdat.
Sheffield, J., E.F. Wood, and M.L. Roderick. 2012. Little change in global drought over the past 60 years. *Nature* 491: 435–438.
Solomon, S., D. Qin, M. Manning, Z. Chen, M. Marquis, K.B. Averyt, M. Tignor, and H.L. Miller, eds. 2007. *Contribution of working group I to the fourth assessment report of the intergovernmental panel on climate change*, 996. Cambridge: Cambridge University Press.
Svoboda, M., D. LeComte, M. Hayes, R. Heim, K. Gleason, J. Angel, B. Rippey, R. Tinker, Palecki, and M. Stooksbury. 2002. The drought monitor. *Bulletin of the American Meteorological Society* 83: 1181–1190.
The Guardian. 2017. *Extreme heat warnings issued in Europe as temperatures pass 40C*. https://www.theguardian.com/world/2017/aug/04/extreme-heat-warnings-issued-europe-temperatures-pass-40c.
Tsuberbiller, E.A. 1963. *Suhoveis and drought control*, 65–73. Washington D.C: US Department of Commerce and National Science Foundation.
Tucker, C.J. 1978. A comparison of satellite sensor bands for vegetation monitoring. *Photogrammetry Engineering and Remote Sensing* 44 (1): 369–380.
Unganai, L.S., and F.N. Kogan. 1998. Drought monitoring and corn yield estimation in southern Africa from AVHRR data. *Remote Sensing of Environment* 63: 219–232.
USDA/ERS (Economic Research Service) 2017. *California drought: farms*. https://www.ers.usda.gov/topics/in-the-news/california-drought-farm-and-food-impacts/california-drought-farms/.
USDA/FAS 2017. *Gain report. Global agricultural information network*. SP1714, Jun 2.
USDM, 2017. http://droughtmonitor.unl.edu/VH. http://www.star.nesdis.noaa.gov/smcd/emb/vci/VH/index.php.
UWCD (United Water Conservation District). 2017. December 2016 Hydrologic Conditions Report, 2017 Water Year January 3, 2017. https://www.unitedwater.org/images/stories/reports/GW-Conditions-Reports/Dec%202016%20Monthly%20Hydrologic%20Conditions%20Report.pdf.
Verigo, C.A., and L.A. Razumova. 1963. *Pochvennaiya Vlaga importance in agriculture (soil moisture importance in agriculture)*, 289. St. Petersburg: Hydrometizdat.
Vicente, A.. 2017. *Spain, Portugal struggle with extreme drought*. Terra Daily News. Madrid. http://www.terradaily.com/reports/Spain_Portugal_struggle_with_extreme_drought_999.html.
Wang P., X. Li, J.Y. Gongand, and C.H. Song 2007. Vegetation temperature condition index and its application for drought monitoring. *Proceeding of International*

Geoscience amd Remote Sensing Symposium. 141–143. Sidney, Australia.

WB (World Bank) 2017. *Agriculture & rural development.* https://data.worldbank.org/topic/agriculture-and-rural-development.

Wilhite, D.A. 2000. Drought as a natural disaster. In *Drought*, Routledge Hazards and Disasters Series, Routledge, ed. Donald Wilhite, 3–19. London, New York: Tailor and Francis Group.

Wilhite, and Glantz. 1985. Understanding the drought phenomenon: the role of definition. *Water International* 10: 111–120.

Yan, N.N., F. Wu, V.I. Boken, S. Chang, and I. Di Yang. 2014. BA drought monitoring operational system for China using satellite data design and evaluation. *Geomatics, Natural Hazards and Risk*. https://doi.org/10.1080/19475705.2014.895964.

Zhang, N., Y. Hong, Q. Qinand, and L. Lin. 2013. VSDI: a visible and shortwave infrared drought indexfor monitoring soil and vegetation moisture based on optical remote sensing. *International Journal Remote Sensing* 34 (13): 4585–4609. https://doi.org/10.1080/01431161.2013.779046.

6 Vegetation Health-Based Modeling Crop Yield and Food Security Prediction

6.1 Introduction

Weather-related crop losses have always been a concern for farmers, governments, traders, policy makers, and international organizations for the purpose of balanced food supply/demands, trade, and distribution of aid to the nations in need in cases of population hunger and starvation. Almost every other year of the twenty-first century has experienced large-scale crop losses due to bad weather, specifically droughts. Among the different types of agricultural production, grain is the major source of food supply. The amount of grain production depends on the planted area and yield. While grain crop area is relatively stable, yield fluctuates considerably with weather changes. Among weather disasters, drought plays a major role in large-scale crop losses. In the most recent three to four decades, drought-related crop losses have been reported to increase due to climate change (IPCC5 2014; IPCC4 2007). Therefore, it is very important to estimate crop yield in advance of harvest in order to anticipate food shortages and planning food aids to countries with weather-related shortage of crop yield and production to prevent hunger and starvation.

Over the years, a large-area assessment of crop production has been based on in situ observations such as weather, climate, soil, economic, and agricultural technology (e.g., cultivar changes, fertilizers, breeding, crops protection, mechanization, diseases/pest control, etc.), using different modeling methods: stochastic, deterministic (process-based), and their combination (Obuhov 1949). From all in situ observations, economic development and agricultural technology estimations provided predictions of long-term (multiyear) tendencies in crop yield and can be reasonably modeled by statistical approximations of yield time series. Short-term or annual crop yield fluctuations are normally controlled by weather, which are more problematic to estimate because it is difficult to predict strong yearly and seasonal weather variations. There are currently two known methods for estimation of weather impacts on agriculture for an advanced prediction of crop production. They are based on the development of crop-weather models when monthly weather parameters (mostly precipitation and temperature) or weather-based indices (for example, Palmer) were used, and in the recent three decades on satellite data.

One of the principal shortcomings of weather-based models is that weather variables are collected from low density weather station networks (around 15,000 stations (WMO 2018) for land surface of 510,100,000 km^2 or one weather station per 34,000 km^2) to adequately characterize very high spatial variability precipitation, soil moisture, and temperature, especially when they became extreme. This is very important in regions/countries with limited and/or strongly variable water supplies, where the majority of

F. Kogan, *Remote Sensing for Food Security*, Sustainable Development Goals Series,
https://doi.org/10.1007/978-3-319-96256-6_6

staple food crops are grown. For example, in Africa, the weather station network is so limited that one station on the average is available for every 40,000–45,000 km^2 of crop areas, which is not sufficient for development of an accurate crop-weather model. Even in the USA, a major world agricultural country, one weather station is available for every 500–1200 km^2 of crop areas. Besides, the use of weather parameters is often problematic following economic, political, and social situations, experienced often in developing and sometimes even in developed countries. Therefore, in the recent 20–30 years, satellite data, free of the indicated problem, have been successfully used to model crop and pasture conditions and using these models for assessment of agricultural production.

Satellite-based crop yield modeling advanced considerably in the 1990s with the introduction of vegetation health (VH) technology (Kogan 1987, 1989, 1997). Regional and country-based yield-VH models has been developed for such grain crops as spring and winter wheat, corn, rice, barley, sorghum, some non-grain crops (potatoes), and for pastures in the USA, Brazil, Argentina, Ukraine, Greece, Morocco, Zimbabwe, China, Mongolia, India, Poland, Bulgaria, and other countries and regions (Kogan et al. 2003a, b, 2013, 2015a, b, 2016, 2017; Liu and Kogan 2002; Salazar et al. 2008; Kogan 1995, 1997, 2001, 2002; Hayas and Decker 1996; Le Comte et al. 1988). They showed not only good results of independent verification and 2-month advance assessment of crop production but also potential for 3–4 months advanced prediction of crop yield. The VH method is working well because the parameters the models are composited of are biologically oriented, distributed frequently over space (0.5, 1, 4, and 16 km^2 resolution) and time (weekly) compared to weather data. This chapter introduces the approach for yield modeling with satellite-based VH indices and provides yield modeling results for a few grain crops in the major grain producers such as the USA, China, Russia, India, and Argentina, and also for several minor grain-producing countries, such as Ukraine and others. Some results are also shown for other crops and pastures.

6.2 Modeling Principles

In developing the modeling principles, the goal was to investigate the crop-indices relationship, determining the strength of the correlation and if the strongest correlation coincides with the critical period of crop development when the crop productivity is highly sensitive to environmental conditions. Therefore, the principal data used were crop yield and satellite-based vegetation health (VH) indices. The yield-VH modeling strategies were to: (a) have at least 20 year yield observation in order to have wide varieties of weather conditions; (b) extract the weather component from the annual crop yield multiyear time series; (c) extract the weather component from the NDVI and BT weekly time series values; (d) correlate the weather-related components of crop yield with satellite-based components; (e) develop yield-VH models, and (f) independently validate the results.

Yield time series. Agricultural production of any crop is created in the fields and its amount is controlled by two components: the area occupied by the crop and crop yield. The area of most of the crops is relatively stable over a certain period since it was set up over the years of crop economics and regional development. Therefore, variations in agricultural production from year to year depend almost completely on crop yield. On a large area (more than 1000 ha), multiyears yield values are controlled generally by the applied agricultural technology (fertilizers, machinery, pest/disease control, etc.) and weather/climate components. Agricultural technology is a slow changing process, leading to relatively slow changes in long-term production or yield time series. If agricultural technology is progressing intensively, crop yield is progressively increasing, showing an upward trend over time following the technology progress. In the opposite situation, if technology is not progressing, crop yield might experience declining trend. Following Fig. 6.1, there are several types of long-term yield trend in response to the applied agricultural technology. In Austria cereal (grain) yield is increasing stably, indicating some progressive contribution of the applied agricultural technol-

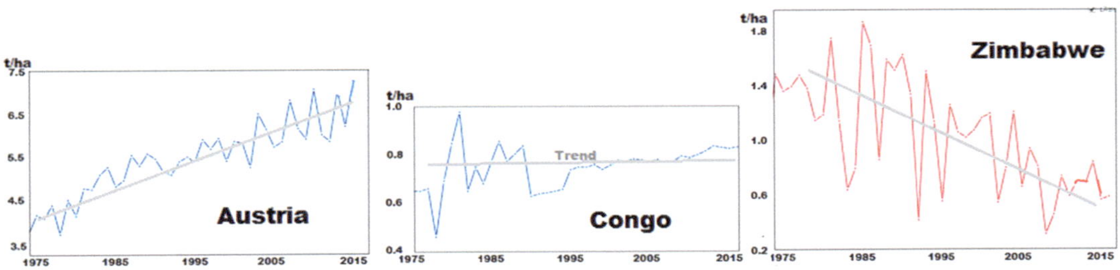

Fig. 6.1 Cereal yield time series and trend during 1975–2016 in Austria, Congo, and Zimbabwe (WB 2017)

ogy; in Zimbabwe, grain yield is declining following the contribution of the stagnated agricultural technology; and Congo shows no changes in cereal yield. These long-term changes or trend can be approximated (in most cases) by a linear Eq. (6.1) using statistical methods.

$$Yi = YTi + \mathrm{d}Yi, \tag{6.1}$$

where Y is yield (t/ha), T is a slowly changing function representing the deterministic component (trend) regulated by the intensity of agricultural technology changes, $\mathrm{d}Y$ is a random component regulated by weather fluctuations, and i is the year or coded year number. The random component ($\mathrm{d}Y$) can be approximated by either a difference (Eq. (6.3) or a relative difference's ratio of the actual yield and the trend yield (Eq. 6.2). In most cases of yield analysis, the ratio (expressed in %, Eq. 6.2) is a method of approximation (Obuhov 1949).

$$\mathrm{d}Yi = 100 \times \left(\left(Yi - YTi \right) / YTi \right) \tag{6.2}$$

$$\mathrm{d}Yi = Yi - YTi \tag{6.3}$$

Vegetation health indices. As described in Chaps. 3 and 4, the principle for constructing VH indices stems from the properties of green vegetation to reflect/emit solar radiation in the VIS (visible), NIR (near-infrared), and IR (infrared) wavelengths of solar spectrum, measured by the NOAA operational polar-orbiting environmental satellites. If the vegetation is healthy, it reflects little radiation in the VIS (due to high solar radiation absorption by chlorophyll contained in green biomass), much in the NIR (due to scattering of light by leaf internal tissues and a low light absorption by pigments, water and other leaf constituents), and emits less thermal radiation in the IR spectral bands (because most of the accepted energy is partitioned into the latent heat flux compared with the sensible heat flux resulting in a relatively cooler canopy compared with the case when the canopy is water stressed). As a result, healthy vegetation has a higher NDVI and a lower Brightnes temperature (BT). Conversely, unhealthy vegetation has a lower NDVI and a higher BT.

The VH indices were calculated from the VIS- and NIR-based normalized difference vegetation index (NDVI) and IR-based brightness temperature (BT) processed thoroughly to reduce temporal noise and calculating smoothed NDVI (SMN) and smoothed BT (SMT) and finally converting them to the vegetation condition index (VCI) from SMN, temperature condition index (TCI) from SMT and vegetation health index (VHI), combining SMN and SMT (Eqs. 6.4–6.6). The details of the algorithm are presented in Kogan (1989, 1987, 1997). Vegetation condition index (VCI) characterizes vegetation greenness and vigor and, through them, the chlorophyll and moisture contents. The temperature condition index (TCI) characterizes how hot the land surface, including the vegetation canopy, is. Besides, TCI characterizes the moisture availability through the near-surface radiation and aerodynamic conditions (Jensen 2000; McVicar and Bierwieth 2001; McVicar and Jupp 1998, 1999, 2002; Mo et al. 2005; Anderson et al. 2007), which control the surface energy budget and hence the BT. VHI characterizes both moisture and thermal conditions together.

$$\text{VCI} = (\text{SMN} - \text{SMN}_{\min}) / (\text{SMN}_{\max} - \text{SMN}_{\min}) \quad (6.4)$$

$$\text{TCI} = (\text{SMT} - \text{SMT}_{\min}) / (\text{SMT}_{\max} - \text{SMT}_{\min}) \quad (6.5)$$

$$\text{VHI} = a \times \text{VCI} + (1-a) \times \text{TCI}, \quad (6.6)$$

where a is a coefficient, changing from 0 to 1 and showing contribution of VCI and TCI into the total vegetation health (VHI).

Yield-vegetation health models. One of the principles used in the data analysis was to investigate the relationship between the weather-related yield fluctuation (anomaly) and satellite data characterizing the weather fluctuations components. Therefore, the data were expressed as a deviation from some standards: for yield—from the multiyear linear trend (expressed as dY) and for VH (VCI, TCI and VHI)—from NDVI-based SMN and BT-based SMT indices normalized by their 36 year climatology. Both correlation and regression analysis of yield-VH deviations (from trend and climatology, respectively) were performed to investigate the dependence of dY from VCI, TCI, and VHI. Annual crop dY was correlated with each week's VCI, TCI, and VHI during a few years of investigation. The goals were to test if dY–VH relationship is strong during the *critical period* (a period of the strongest crop response to changes in weather conditions) of the investigated crops and to find the differences and similarities in the investigated crops' response and their correspondence to each crop's growth and development of environmental criteria. In addition, since the critical period continues for a few weeks in a row, all these weeks might show a strong correlation between dY and VH indices. An inclusion of all these weeks in the statistical models is not advisable, since the independent weekly variables (VH indices) are highly collinear (correlate strongly with the neighboring weeks' VHs). Therefore, two types of dY models were investigated: (a) with the independent variables for the week (during the critical period) with the highest Pearson correlation coefficient (Eq. (6.7)) and several weeks' indices (during the critical period) with the Pearson correlation coefficients (PCC) greater than 0.5. In the second case, the mean indices values for the selected weeks were used as the independent variables (Eq. 6.8).

$$\text{d}Y = a_0 + b_i(\text{VCI})i + b_{i+1}(\text{VCI})_{i+1} + \cdots + b_j(\text{TCI})_j + b_{j+1}(\text{TCI})_{j+1} + \cdots + e \quad (6.7)$$

$$\text{d}Y = a_0 + b_i(\text{VCI})i + b_{i+1}(\text{VCI})_{i+1} + \cdots + b_j(\text{TCI})_j + b_{j+1}(\text{TCI})_{j+1} + \cdots + e \quad (6.8)$$

where a and b are regression coefficients, e is a model error, and n and m number of weeks during crop's critical period for VCI and TCI aggregation.

The number of highly correlated VH variables in the models is normally large (considering three indices and 30 weeks for the period of crop development) and they are inter-correlated (collinear). Models, which include collinear variables, become statistically unstable and less accurate. In order to improve the model's stability and performance, variables with the lowest PCC should be removed. Reduction in the number of VH variable is normally based either on partial correlation coefficients (Snedecor 1965) or on weighting variable contribution and aggregating them to one variable (Obuhov 1949). The partial correlation coefficients (pCC) estimate the contribution of each independent variable when other variables are fixed at the average level. Predictors with pCC < 0.6 are normally removed from the model. Weights are calculated based on dY-VH's weekly Pearson correlation coefficients (Eq. 6.9). Aggregation of the weighted variables follows the procedure shown by Eq. (6.10) for VCI example.

$$W_i = \text{PCC}_i^{\,2} / (\text{PCC}_i^{\,2} + \text{PCC}_{i+1}^{\,2} + \cdots + \text{PCC}_{i+n}^{\,2}) \quad (6.9)$$

$$W_{i+1} = \mathrm{PCC}_{i+1}{}^{2} / \left(\mathrm{PCC}_{i}{}^{2} + \mathrm{PCC}_{i+1}{}^{2} + \cdots + \mathrm{PCC}_{i+n}{}^{2}\right)$$

$$W_{i+n} = \mathrm{PCC}_{i+n}{}^{2} / \left(\mathrm{PCC}_{i}{}^{2} + \mathrm{PCC}_{i+1}{}^{2} + \cdots + \mathrm{PCC}_{i+n}{}^{2}\right)$$

$$\mathrm{VCI}_{k} = \mathrm{VCI}_{i} \times W_{i} + \mathrm{VCI}_{i+1} \times W_{i+1} + \cdots + \mathrm{VCI}_{i+n} \times W_{i+n} \quad (6.10)$$

In Eqs. (6.9) and (6.10), *W*—weight, PCC—Pearson correlation coefficient, squared to provide more contribution for the variables with a higher influence on d*Y*, *i* and *n*—first and last weeks of predictors' aggregation, respectively, *k*—the new aggregated variable.

6.3 Yield-Vegetation Health Models

6.3.1 Global Grain and Food Security

In the past 30 years, satellite-based crop yield models addressed very important problems of global food availability and food security (Holdren and Ehrlich 1974; Hardin 1986; Tucker and Sellers 1986; Le Comte et al. 1988; Kogan 1995, 1997, 2001, 2002; Unganai and Kogan 1998; Salazar et al. 2007; Kogan et al. 2013, 2015a, b, 2016, 2017). Among many crops, grain (wheat, rice, rye, corn, coarse grain) is the principal commodity used globally for food and feed. Availability of grain creates the most serious global food insecurity situation, especially in years when grain production was below consumption and whether such a situation strongly affects developing countries (*see* Chap. 2). In the past four to five decades, at the background of a steady growth of grain production and consumption (Fig. 2.2), in quite a few years, the difference (production minus consumption) becomes negative (grain production becomes below the consumption), indicating a potential situation for development of global food insecurity. Moreover, as seen in Fig. 2.6, this gap has been increasing since 1988 (shown by a declining trend), especially in the twenty-first century. This imbalance is gradually increasing and will complicate future problems with global food security. Moreover, anticipated drought intensification, due to climate change drought expansion, and penetration to new areas creates additional concerns about the food security situation. If grain production is larger than consumption food security concerns might be in a single county, several countries, and even a large region. However, it is very important to notice that global production–consumption differences have been gradually increasing in the twenty-first century (Fig. 2.6), indicating that the future situation is not likely to improve. Besides, the situation is further complicated because the Earth's population growth rate is increasing, while grain production rate is gradually leveling off. Therefore, crop yield-VH modeling is very important for pre-harvest grain production predictions and advanced projections of food security situation. The next subchapters provide examples of yield-VH modeling.

6.3.2 Corn in China

Having a quarter of the world's population, one of China's most important agricultural priorities is directed to provide enough domestic food and feed, and also commodity for trade. Although China has only 7% of the world's arable land, the country is one of the largest producers of agricultural commodities including such an important food source as grain. In recent years, nearly 20% of the global total grain, including the largest amount of rice and wheat, have been produced by China's farmers. For example, in 2002, China's contribution to the total world production of wheat, corn, and rice accounted for 16%, 20%, and 31%, respectively (FAO 2016; USDA 2006). Therefore, in addition to domestic use, the amount of grain produced in China annually is very important for global supply and demand and food security.

The Northern Plains (north-east) are the main agricultural areas of China. Thirty-eight percent of China's arable land and 35% of the total number of people are located there (Ning 1998). Although the lands are generally productive, a semiarid climate with frequent droughts limits agricultural output, including grain production. By the amount of natural water resources, China is ranked number six in the world. However, the distribution of water is unbalanced geographically. The Northern Plains makes up only 7.5% of China's total available water resources (Ning 1998). Therefore, irrigation can't offset dry climate in that area. In drought years, up to 25% of agricultural production might be lost in the Plains. Since climate controls a large portion of variation in China's agricultural production, weather data have been traditionally used for assessment of the production losses. Unfortunately, the weather-station network in China is not dense enough for a very large agricultural area, the number of farmers growing crops, and weather changes over space. Therefore, in the late 1990s, China began to use satellite-based vegetation health (VH) technology for these purposes. The recent advances in the application of operational satellites proved that observed radiances are an excellent data for monitoring the environment and agricultural activities. This subchapter provides an application of VH for modeling grain production in Jilin province, which is the breadbasket of China's Northern Plains.

Average Jilin province corn yield (t/ha) was obtained from the China's Central Statistical Administration, and average weekly VH (VCI and TCI) data were obtained from NOAA/NESDIS (2018) from 1980 through 2001. The VH data were aggregated over the main corn-cultivated area to match with corn yield from 1981 through 2001. Figure 6.2a displays corn yield time series during 1980–2001. Following the linear trend estimate (upward line) corn yield in Jilin province over 22-year period increased 30% (from 4.5 to 5.8 t/ha) following improvements in agricultural technology, which is a very strong contribution. However, weather-related yield deviation from the trend is very large in both directions up and down. For example, a very interesting example is corn yield deviation in 1997 and 1998, which were estimated at 0.79 and 1.23, respectively, indicating a 21% yield reduction in 1997 from unfavorable and a 23% increase due to favorable weather in 1998. Below-trend corn yield reduction can eliminate contribution of agricultural technology to increase yield, especially when corn yield in Jilin province drops below trend for several years in a row (1980, 1981, 1982 and 2000, 2001). In the last 2 years, yield reduction was very large (33% and 28%, respectively) when corn yield dropped to the level of the early 1980s because Jilin province was affected by large-area, intensive drought (NOAA/NESDIS 2018).

Although Jilin province is the largest producer of corn in China (USDA 1994), geographic location and the environmental conditions are not favorable there. The major area of corn cultivation in the central and western Jilin province receives 500–700 mm of annual precipitation (*P*), 80% of which falls during May to September (CIA 2017). This amount is generally marginal for corn growth. Moreover, because of the hot summer, potential evapotranspiration (*PET*) in this area is large, exceeding the amount of precipitation. As the result, the natural annual water balance ($P - PET$) in Jilin province is 200–400 mm short (Gol'tsberg 1972). Ninety per cent of this deficit occurred during warm season. Therefore, corn, which is planted in Jilin province during May through mid-June and harvested in September to October, is affected severely by shortages of water (USDA 1994).

Figure 6.2b shows dynamics of correlation coefficients for d*Y* versus VCI, TCI, and VHI between weeks 10 (mid-March) and 40 (end of October). These results clearly indicate that in the early period of the corn-growing season, which is pre-planting, planting, and emergence (weeks 15–20, April to early June), the correlation for all VH indices is very small (PCC = 0.0–0.15). It increases sharply during the corn's green mass accumulation (weeks 22–28), reaching maximum in late June for VCI (PCC is 0.63), mid-July for TCI (0.80), and mid-August for VHI (0.83). This maximum correlations coincide with

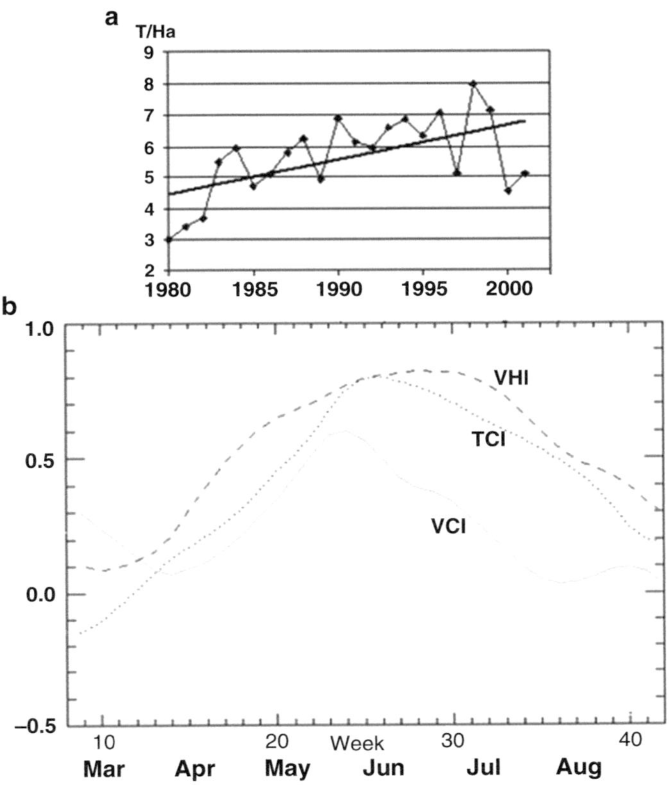

Fig. 6.2 1980–2001 corn yield time series for Jilin province, China (**a**) and dynamics of Pearson correlation coefficients (PCC) between annual corn yield departure from trend and weekly vegetation health indices (**b**)

the critical period in corn development, which starts 2–3 weeks prior to and ends 2–3 weeks after corn tassels. During this period, plentiful water supply and cooler temperatures stimulate larger corn production. The figure confirms that positive correlation with VCI, TCI, and VHI indicates that below-trend yield is associated with moisture and thermal stress (the three indices are below 40) and above-trend yield is associated with favorable (moist and cool) conditions (VCI, TCI, and VHI above 60). It is known that crop response to moisture and thermal conditions is not equal during the growing season. Corn is more moisture dependent in the period of green leaf formation and more temperature dependent during the reproductive stages. This is confirmed by Fig. 6.2b, which indicates that in the environment of the Jilin province, the pick of TCI correlation is shifted (relative to VCI) to corn tasseling period (weeks 26–36 or June–July) and correlation with thermal conditions is stronger. For example, for week 22–24 (leaf formation), the contribution of VCI and TCI are almost the same (PCC is nearly 0.5), while for week 32 (reproductive stage), TCI contribution to corn yield is stronger (PCC = 0.75) compared to VCI (PCC = 0.37).

These results were used to build regression models (Kogan et al. 2005a, b). Two equations were developed: the first one, using VCI and TCI combination as independent variables (Eq. 6.11) and the second, using VHI only (6.12). The variables in these equations were selected based on the largest values of PCC. These equations were used to calculate the estimated dY (called EdY) and build the scatter plot of EdY versus dY shown in Fig. 6.3a, which confirms that the relationship is very strong, explaining 75% and 81% of dY variance.

$$dY = 0.502 + 0.005 \times VCI27 + 0.034 \times TCI26 + 0.106 \times TCI27 + 0.107 \times TCI28$$
$$+0.03 \times TCI29; \qquad R^2 = 0.75 \tag{6.11}$$

$$dY = 0.501 + 0.114 \times VHI26 + 0.316 \times VHI27 + 0.289 \times VHI28 + 0.076 \times VHI29; \qquad R^2 = 0.81 \tag{6.12}$$

where R^2 is coefficient of determination for the models.

These equations were tested independently using a "Jacknife" (or "one-in-one-out") approach when years were eliminated one by one from the dataset and in each case, a new equation (with the same variables) was developed (without an eliminated year) and each equation was applied to predict the eliminated years. This procedure was performed 19 times and independently estimated EdY was calculated for each year not included in the development of the model. Finally, the Fig. 6.3 displays corn yield correlation of observed (dY) versus independently simulated EdY corn yield and annual time series of these parameters in Jilin province. The correlation is quite strong since PCC = 0.87 and the two time-series match quite well, except for years 1984 and 1991 due to some remaining VH error from two volcanoes aerosols.

6.3.3 Winter Wheat, Corn, and Sorghum in the USA

The United States is one of the most important among all grain-producing countries. Being the second (after China) largest producer of grain in the world (provides nearly 16% of global production (Table 5.3)), the USA is number one in grain exports (provides between 16 (in the worst year) and 24% (in the best year) of world total exports (Table 5.4)), which is more than any world countries, including the entire European Union. Therefore, some tests of VH-based modeling grain crops were investigated following the presented methodology. In Salazar et al. (2007) it was shown that AVHRR-based VH indices can be successfully used to predict winter wheat (WW) crop yields over Kansas state, including large-size several regions (around 209, 920 km^2), medium-size, Crop Reporting District (20,000–51,000 km^2) and small-size, county (500–1300 km^2). Statistical analysis of WW in Kansas indicated that WW growth and productivity can be monitored during the entire growing season and that yield can be estimated 1–3 months in advance of harvest. Specifically, the WW yield was shown to correlate strongly with AVHRR-based VH indices (Kogan 1997; Salazar et al. 2007) during the critical period of crop development and that VH indices can be used as yield predictors.

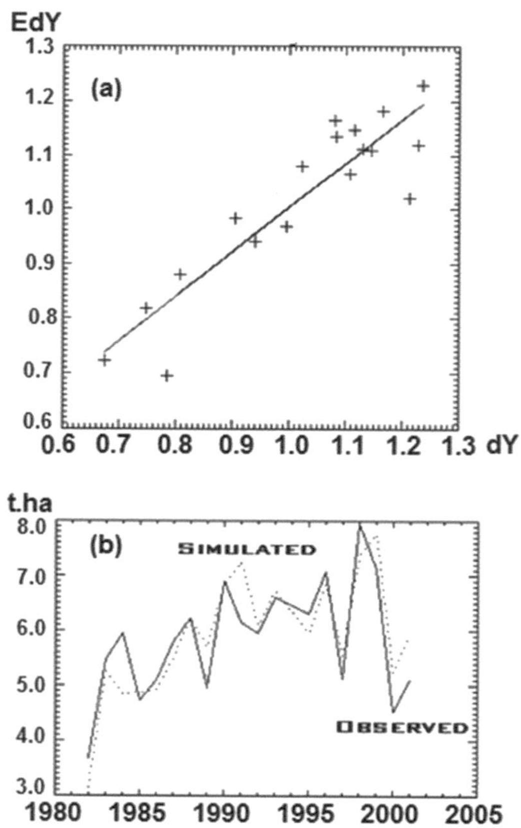

Fig. 6.3 Correlation between independently modeled and actual corn yield in China's Jilin province during 1981–2001 (**a**) yield deviation from trend and (**b**) actual yield (model 6.3)

Since the results of modeling a single crop were successful, the tests were expanded to two other major crops in Kansas, sorghum and corn, which have different crop calendars and different biophysical responses to weather impacts. Kansas, called the "granary" of the USA (Shroyer 2004), provides 24%, 39%, and nearly 4% of WW, sorghum, and corn, respectively, of the total value of crop production in the USA during 2003–2005 (Kogan et al. 2015a, b). Besides, these crops were interesting since they are fundamentally different, being C4 compared to WW's C3 photosynthetic type. The goals were to (a) investigate whether sorghum and corn yields, similar to WW yield, correlate strongly with VH indices during the critical period (when crop productivity is very sensitive to weather conditions, both moisture and temperature) of their development; (b) estimate if VH indices can be used as predictors of sorghum and corn yields; (c) build d*Y*-VH statistical models for these crops; (d) assess their performance; and (e) develop a scheme for early assessment of regional crop yields. Corn is also interesting to test because it is less than sorghum drought resistant.

The Kansas yield data (tons per hectare (t/ha)) for corn, winter wheat, and sorghum from 1980 through 2005 were obtained from the USDA's National Agricultural Statistics Service (NASS) database (http://www.nass.usda.gov). Weekly VH indices (VCI, TCI, and VHI) during 1981–2005 for the areas of the three crops cultivation were obtained for each 4 km^2 land surface from the NOAA-VH WEB (https://www.star.nesdis.noaa.gov/smcd/emb/vci/VH/vh_browse.php). The weekly Kansas's area-average VCI, TCI, and VHI for each crop were aggregated as mean values of all 4 km^2 pixel inside the crop-grown area. Some of the VH data were not available due to some technical problems, which included failure of NOAA-13 satellite after launch on 13 September 1994. Therefore, the 1994 and 1995 yield data were excluded from the analysis. Moreover, yield and VH data from June to September 1993 were removed from the statistical analysis because record flooding produced standing water and NDVI reduction (near zero) across North Dakota, South Dakota, Nebraska, Kansas, Minnesota, Iowa, Missouri, Wisconsin, and Illinois (Stallings 1994).

Time series of WW, sorghum, and corn yield during 1980–2005 are shown in Fig. 6.4. As seen, in a 25-year period, all three Kansas crops experienced general yield growth due to technology improvement. The upward trends were approximated by lineal regression line following Eq. (6.1). Following Table 6.1 corn has the largest and WW the smallest yield. However, WW had the highest increase (18.2%) and sorghum the smallest due to technological innovations. At a background of this stable yield increase,

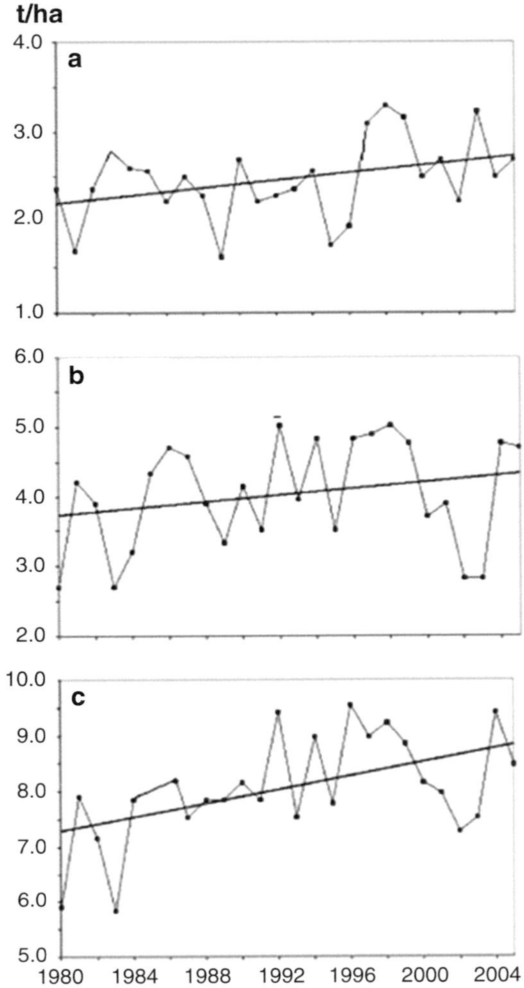

Fig. 6.4 Average crop yield in Kansas, USA, for winter wheat (**a**), sorghum (**b**), and corn (**c**) during 1980–2005

Table 6.1 Technology- and weather-related dynamics of WW, sorghum, and corn yields in Kansas, USA, during 1980–2005

	Yield trend (t/ha)			Yield deviation from trend (t/ha)			
				Above trend		Below trend	
Crop	1980	2005	Increase (%)	Max	Min	Max	Min
Winter wheat	2.2	2.6	18.2	0.80	0.05	0.84	0.07
Sorghum	3.8	4.2	10.5	1.10	0.23	1.60	0.08
Corn	7.3	8.5	16.4	1.60	0.04	1.62	0.09

weather-related yield variations around the trends changed from almost 0 to 0.8–1.6 t/ha with the smallest for WW. However, it is important to indicate that nearly 70% of weather-related variations are larger than 0.5 t/ha, with an almost equal number going up and down. Another important consideration is that most of the large variations (up and down from trend) for all three crops coincide (1980, 1983, 2000–2003—down, 1992, 1994, 1996–1999, 2003—up). This indicates that (a) similar weather impact on quite different crops and (b) if yield is strongly reduced drought was the major cause.

The 25-year yields deviations from the technological trend were correlated with weekly areal-mean values of VH indices. Figure 6.5 shows the dynamics of correlation coefficients for the end-of-season dY of each crop with each week's (from the first week in January to the last week in December) VCI (moisture conditions) and TCI (thermal conditions) during 1980–2005 (except of removed data). The goals were to investigate if (a) dY correlates strongly with VH indices during the "critical period" (a period starting from 2 to 3 weeks before flowering and ending with grain filling) of the investigated crops; (b) dY correlates weakly during noncritical and/or out of the growing season periods; and (c) to find the differences and similarities in crops' response to moisture (VCI) and thermal (TCI) conditions and the correspondence of this response to crops' growth and development.

The correlation results (Fig. 6.5) emphasize clearly a distinctive and right difference between spring (early season) WW, when flowering and grain filling occur during April to May, and summer crops (sorghum and corn), which reproduce during July to September. As expected, the correlations of WW's dY with VH indices are low (PCC is near zero) during January–March when WW is dormant and values of the indices are driven by soil and snow rather than by vegetation. WW's dY is highly correlated with VCI (0.82–0.88) during weeks 12–21 (April to May), when WW is heading, flowering, and filling grain. Similarly, WW is highly correlated with TCI (0.69–0.73) during weeks 16–18 (late April to early May). The indicated periods are the most critical for WW because optimal moisture and thermal conditions are needed for maximum WW productivity. Both VCI and TCI reflect this situation quite well, correlating strongly with dY. By early summer, when harvest begins and ends a few weeks later, correlations of WW's dY with VH indices drop very quickly.

Summer crops have the same bell-shaped correlation curve dynamics reflecting sorghum and corn responses to changes in weather impacts (Fig. 6.5). During July to September, when sorghum and corn biomass is developing, correlations of dY with VCI and TCI are very strong. These crops are very sensitive to moisture and thermal conditions, which are indirectly represented by VCI and TCI, respectively. For sorghum, the highest correlation of dY with VCI (0.76–0.88) occurs during weeks 28–37 (mid-July through mid-September), and with TCI (0.77–0.87) occurs during weeks 27–33 (July through mid-August). For corn, the strong correlation of dY with VCI (0.78–0.88) occurs during weeks 31–37 (August through mid-September), and with TCI (0.77–0.90) during weeks 28–34 (mid-July (0.77–0.90) during weeks 28–34 (mid-July through August). During the indicated periods, sorghum and corn are very sensitive to environmental stress such as a very high temperature and water deficit or excess, which may reduce the potential crop yield. High correlations of dY with

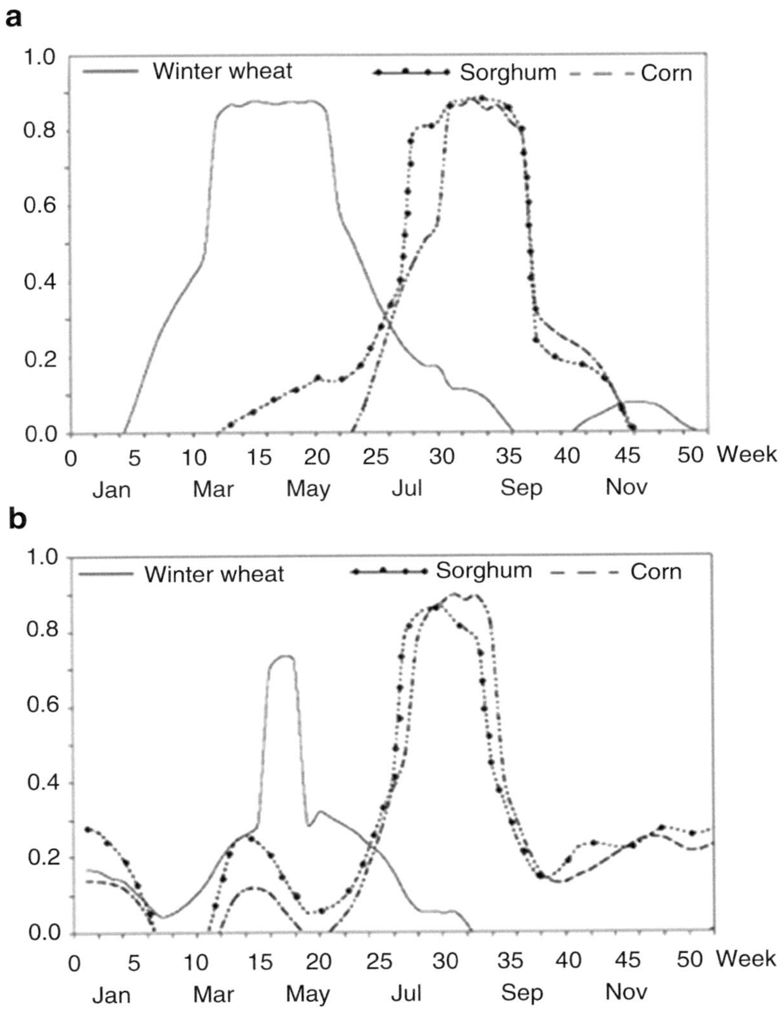

Fig. 6.5 Annual dynamics of the Pearson correlation coefficients (PCC) between dY of winter wheat, sorghum, and corn and VH indices: VCI (**a**) and TCI (**b**), Kansas, USA

VCI and TCI during the investigated crops' critical period show the ability of the indices to be used as the predictors of crop yield. It is important to note that the correlation of dY with VH indices, during the sorghum critical period, is slightly lower than during the corn critical period. This is because sorghum is more drought resistant compared to corn. It is important also to emphasize that VCI is an integral or cumulative indicator of crop growth, development, and productivity, whereas TCI is not. Therefore, the critical period characterized by TCI starts earlier and is shorter than that characterized by VCI (Fig. 6.3).

Figure 6.6 shows the scatter plots of dY with VH indices for the weeks with the highest PCC. Positive correlation of dY with VCI indicates that the transition from moisture stress (lower vegetation greenness) to favorable moisture conditions (higher greenness) corresponds to dY changes from below to above the trend line; positive correlation with TCI indicates that dY changes from below to above trend when TCI has similar to VCI stress-to-no stress changes (reduction of dY from hot and increase to cool weather). Except for a few cases, extreme deviations in dY are associated with VCI and TCI below 20 and

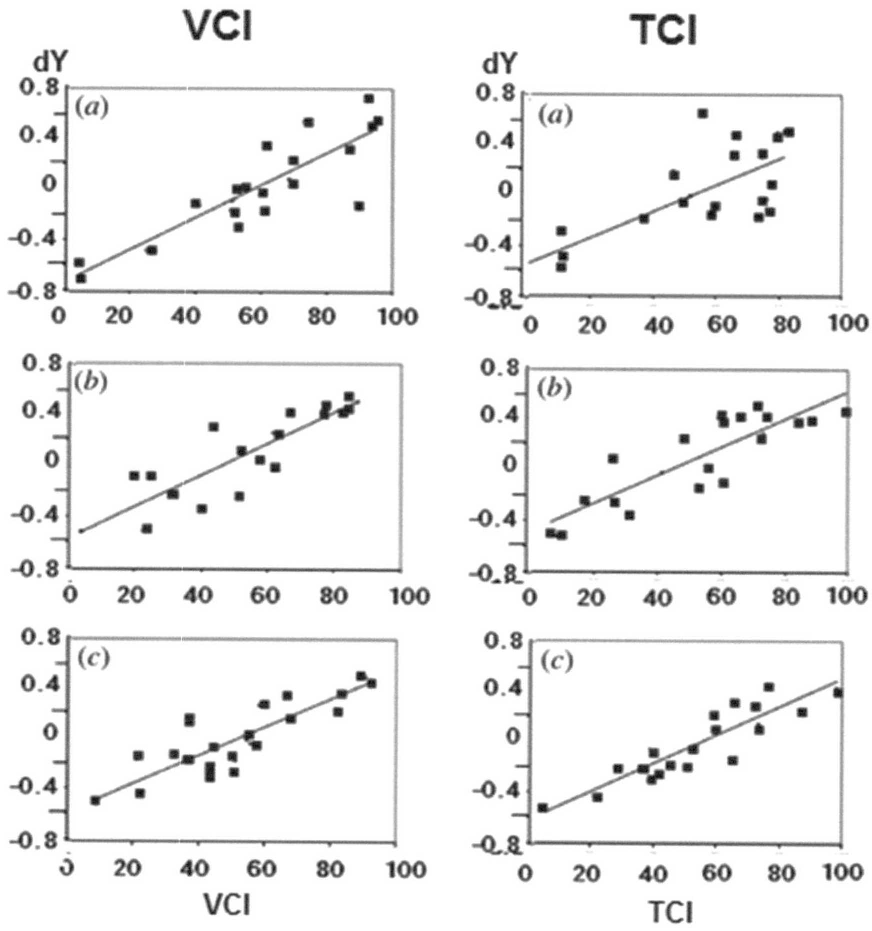

Fig. 6.6 Scatter plots of dY correlation with VCI and TCI for the weeks with the highest PCC during 1981–2005, Kansas, USA: (**a**) winter wheat, (**b**) sorghum, and (**c**) corn

above 80. It is important to note, that standing water and/or soil saturated with water reduce NDVI and VCI, producing a false signal in dY estimation. Such a situation occurred in 1993 when excessive early spring rainfall produced flooding in the US Midwest, including a part of Kansas. As a result, WW yield was reduced, which is observed in Fig. 6.6 (week 15), where one dY point located below the regression line although VCI is very high (90). Fortunately, such conditions are rare and cover a small area. The highest correlation of dY with VCI for WW, sorghum, and corn was for weeks 15, 33, and 33 and with TCI for weeks 17, 30, and 31, respectively. If VCI and TCI for these weeks (Fig. 6.5) are used as the only independent variables in a simple linear regression model, these would account for 77%, 78%, and 77% of VCI and 54%, 75%, and 80% of TCI variation in dY for WW, sorghum, and corn, respectively.

Although there was only a 1–2% difference between coefficients of determination (R^2) values, we developed two regression models (*a*) and (*b*) using dY as a dependent variable and VCI and TCI as independent variables. Model (*a*) uses VCI and TCI for 1 week only with the highest PCC between dependent and independent variables (Fig. 6.5) Eq. (6.9) and model (*b*) uses the average VCI and TCI for several weeks with the highest correlation (Eq. (6.10)). The regression results support our preliminary conclusions from the analyses of the Figs. 6.5, 6.6, and 6.7. There is a strong relationship between dY and the independent variables (weeks with the highest corre-

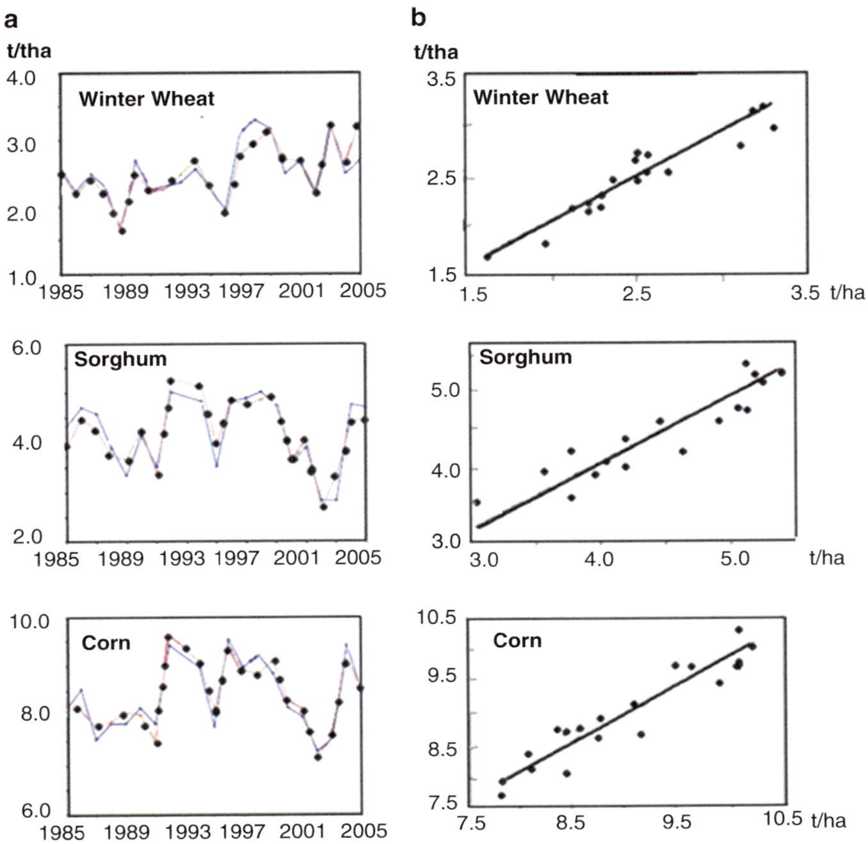

Fig. 6.7 Observed (Y) versus the independently simulated (predicted, PY) average Kansas (USA) crop yields (winter wheat, corn, and sorghum) using models approximated by equations: (**a**) 6.9 and (**b**) 6.10

lation coefficient of each crops' dY with VCI and TCI, Fig. 6.5). The R^2 values are 0.79, 0.90, and 0.90 for WW, corn, and sorghum, respectively. The tests of the composite hypothesis showed that all regression coefficients are statistically significant with F-statistic of 32.83, 74.59, and 77.68, respectively. There is also a strong relationship between dY and the independent variables (average VCI and TCI for several weeks with highest PCC). The R^2 values are 0.81, 0.89, and 0.92 for WW, corn, and sorghum, respectively. The tests of the composite hypothesis for case (b) showed that all regression coefficients are statistically significant with F-statistic of 37.19, 70.34, and 91.78, respectively. This is a very important result indicating that VH indices can be applied to many crops with different physiology and requirements to weather conditions.

The results of this modeling showed that satellite-derived moisture (VCI) and thermal (TCI) conditions are very good predictors of different crops in one of the main USA's grain-producing region. An independent models' testing showed that VH indices provide accurate prediction of WW, sorghum, and corn 2–4 months in advance of harvest. Such early predictions are giving farmers and others in the industry the opportunity to use this information to make more efficient crop management, planning, and market decisions. In addition, with such early predictions, international organizations will be able to make early decisions regarding food security situation and provide advanced planning with food assistance to avoid population hunger and starvation. Similar models can be developed for the entire USA (or other major grain producers) either by administrative region (states in the USA) or by the average yield and VH aggregated for the entire area of crop cultivation. This research also shows that there is potential for VH application in modeling other grain crops to address food security situation. Interestingly, the models that used independent variables for only 1 week (with the strongest impact on yield) showed similar results as the models that used independent variables for several weeks with a high correlation. This indicates that single-week models are preferable, since they have 1–3 weeks' lead time versus multi-week models. There is some potential for models' improvement through combining VH indices with weather parameters, especially those measuring winter conditions (snow cover, minimum temperature).

6.3.4 Winter Wheat in Ukraine

Weather-related crop losses are very serious problems for Ukraine which is a country where agriculture is the major sector of the economy. Every 2–3 years, weather vagaries, especially droughts in Ukraine, have caused shortfalls of agricultural production. In some years such as 2003 or 2007, the losses might be staggering. In these years, Ukraine reduces the amount of grain sold at the international market. Therefore, early predictions of crop losses in response to weather fluctuations is an important issue not only for estimation of country's food supply/demand but also for the international crop production, trade, and for global food security.

Ukraine has 24 administrative regions (oblast) with the area ranging from 8097 to 33,310 km^2 (average area is 24,000 km^2). Environmentally, Ukraine can be divided into the following agroclimate zones: Forest (Plane-Polissya) in the north (mixed forest zone, 26% of Ukrainian territory), Forest-Steppe in the center (34%), and Steppe in the south (40% the most intensively cultivated agricultural area). Wheat is one of the most important cultivated crops in Ukraine. For the past 5 years, wheat accounted for almost 45% of all grain produced, and 48% of all crops' area. Winter wheat constitutes 95% of the entire wheat production while the rest 5% are accounted by spring wheat. Winter wheat (WW) is sown in the autumn, and harvested in July next year. Ukraine has the best in the world chernozem soils to produce excellent harvest if the weather is supportive. Unfortunately, Ukraine's misfortune is to be located in a dry climate zone, where annual shortage of water (due to lack of precipitation) with excessive warm season's temperatures causes a strong evapotranspiration, which leads to losses of 200–400 mm of climatically available precipitation. Unfortunately, droughts affect Ukraine every 2–4 years. The largest agricultural losses occur when drought is preceded by a very cold winter with a lack of snow to cover winter wheat to protect the crops from damages (winterkill). Therefore, monitoring weather, especially drought impacts on agriculture in Ukraine, is a very important task for prediction of agricultural production (including the main WW crop), food supply/demand, potential trade, and food security in general (Kogan et al. 2013).

Weather data are traditionally used in Ukraine for agricultural assessments. One hundred eight operational weather stations provide good information about precipitation, temperature, snow, and other weather parameters used for assessments of the impacts on crops. However, for the 223,000 square miles of Ukrainian territory, (on the average one station per 2000 square miles) this number is not sufficient for the entire country and administrative regions (oblast) since each station covers (on the average) nearly one million acres of land. Therefore, an attempt was made to use the NOAA/NESDIS (2018) vegetation health (VH) technology, which provides environmental information for every 4 km^2 of Ukraine's territory, to predict crops yield and production. The VH indices were used to model winter wheat (WW) yield in Odessa oblast, one of the important southern oblast of Ukraine. Odesa's WW annual yield data were collected from the Ukrainian Statistical Administration for the period 1980–2006. VH indices (VCI, TCI, and VHI) at 4 km^2 resolution were collected inside Odessa oblast from NOAA (NOAA/NESDIS 2018). The values of the indices were averaged for the investigated region.

Figure 6.8 shows the result of modeling WW yield in Odessa administrative region (oblast) of Ukraine. The yield time series and their technological trend during 1980–2006 is shown in

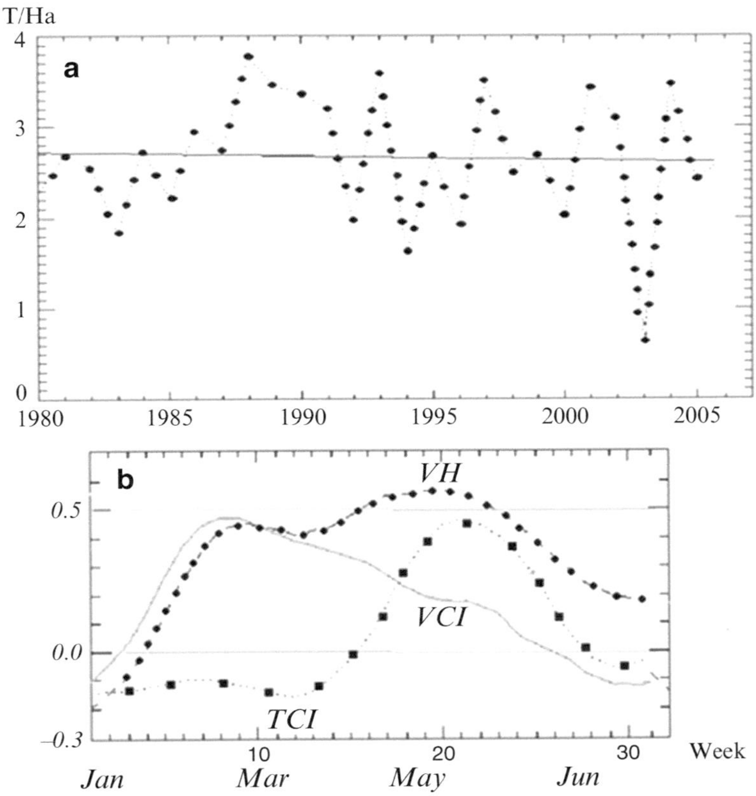

Fig. 6.8 Odessa oblast, Ukraine (**a**) winter wheat (WW) yield (*Y*) and trend during 1980–2006, (**b**) correlation dynamics of WW deviation from trend (d*Y*) versus VH indices (VCI, TCI, VHI)

Fig. 6.8a. Following the trend dynamics, it is possible to conclude that technology provides a strong contribution to WW yield stability over years in Odessa, which is 2.7 t/ha, and this yield value is slightly above the USA's Kansas winter wheat (2.5 t/ha) in 2004. More importantly, Odessa yield fluctuates considerably due to weather changes. Departure of yield below the trend is related mostly to drought conditions and are important to estimate for prediction of annual yield losses. In most of these years (1983, 1985, 1992, 1994, 1996), WW is losing 20–30% of its yield and production, but in the years of a very severe drought such as 2003 (NOAA/NESDIS 2018) WW yield in Odessa is reduced nearly 70%. Figure 6.8b displays the PCC dynamics for the end-of-season winter wheat d*Y* with weekly VH indices from the first week in January to July. As seen, during January and February the correlation is close to zero since WW is dormant. The correlation is increasing sharply at the beginning of the growing season in March (VCI's PCC = 0.5) due to strong dependence of WW on moisture conditions. Temperature is less important that time (TCI's PCC is near 0). During the critical period of winter wheat development (heading, flowering, and grain filling) in May to June, thermal conditions became very important (TCI's PCC = 0.5) and moisture conditions (VCI) are secondary factor affecting WW yield (PCC = 0.25). Very interesting results are presented with dY-VHI correlation dynamics (Fig. 6.8b), which shows two picks of environmental importance (reflecting combination of VCI and TCI) March (the highest PCC = 0.49) and May–June (PCC = 0.53). Following these results, Ed*Y* model (Eq. 6.13) was developed from several weeks of VHI during these two picks (beginning of the growing season and critical period) of the highest d*Y*-VHI correlation.

$$EdY = 0.286 - 0.057VHI5 + 0.067VHI6 - 0.041VHI18 + 0.044VHI19, \quad (6.13)$$

where EdY is modeled WW yield deviation from the trend and VHI is vegetation health index for the weeks indicated by the attached numbers.

Equation (6.13) was validated independently using "Jack-Knife" techniques. The two WW yield's time-series are well correlated ($R^2 = 0.75$) with the standard error of estimation 0.2 t/ha. Equation 6.13 predicts WW yield in the second half of April, which is 2 months in advance of harvest. Such early prediction is appropriate for planning the harvest, storage of production, and for marketing/economic decisions. Although the modeling results are appropriate, we should indicate that the model predicts yield based on spring's weather conditions. As was mentioned, some of the WW losses might occur in winter during below freezing temperatures and a lack of snow cover to protect WW plants from winter-kill. Therefore, this model should be improved modeling winter conditions.

6.3.5 Corn in Argentina

Argentina is one of the world's leading countries in grain production (contribute 3.7% to global production) and delivery of grain to international market (4–8% of global grain). Although Argentina is located in a moderately wet climatic zone (Gol'tsberg 1972), it has experienced huge losses of grain crops from year to year, which have grave consequent for the country's economy. Grain losses are caused by weather variability. Especially big losses occurred during droughts. Therefore, advanced predictions of weather-related grain losses are needed to improve the country's economy, provide advanced assessments of grain delivery to the global market, which will help to ease food security situation.

Weather data is a fairly good source of information used traditionally in Argentina for monitoring crop growth and assessing crop production. However, poor spatial distribution of weather stations often makes this task difficult to fulfill. Sometimes weather data is incomplete and/or not available early enough for timely assessments. In addition, weather observations are location specific and do not adequately represent diversity of weather variation over the large areas where grain crops are grown. Furthermore, prediction of final crop production requires estimation of cumulative environmental impact, which is only possible at the end of the harvest. Since an early grain yield prediction prior to the harvest is very important for monitoring agricultural losses and optimizing agricultural market, satellite data were tested to model grain production in Argentina. This work was stimulated by research and development done at NOAA in the 1990s using vegetation health methodology (Kogan 1989, 1990, 1995, 1997; Hayas and Decker 1996; Liu and Kogan 2002; Seiler et al. 2000, 2006). Accuracy of the satellite-based assessments of weather impacts on crops' yield and production is superior due to their much higher spatial and temporal resolutions. Also, the use of these data led to delineate more precisely the affected areas and the period of the highest crop's sensitivity. Besides, compared to weather data, satellite-based indices combine the contribution of moisture and thermal conditions based on cumulative vegetation responses during the season (Seiler et al. 2000, 2006).

The studied area covers Argentina's Cordoba province (29° to 35°S and 62° to 66°W), which is located in the center of the country and ranked fifth in size among all other Argentinian provinces. Cordoba is divided by administrative units called departments, where each department is used as the basic unit for the analysis to represent environmental characteristics in different areas of the province. Eighty-three percent of Cordoba's surface is dedicated to different agricultural activities developed under variable soil–climatic conditions. The surface of the province is flat except for a small range of mountains north–south oriented in the western part. Representative soils in the area are mollisoils with textures that follow a gradient from finer in the east to coarser in the west of the province. Cordoba is the second largest maize producer in the country contributing 32% to the total national production (SAGPYA 2004). The climate of the region is a combination of semiarid and subhumid with a monsoon type precipitation regime and moderated thermal conditions during summer and winter. The mild and short winters and moderate summers with extended length of the growing season allow for the normal maturity of summer crops such as corn, sunflower, and soybean, which are extensively cropped in the region. However, due to weather variability, the time of occurrence of the climate variables, their extreme values, frequencies, and their interaction with other physical variables of the environment become very important to determine the potential production of the region, the inter-annual yield variability of the crops and pastures, the quality of the natural resources, and finally the economy of the region. The annual crop season in Argentina, located in the Southern Hemisphere (SH), runs from the first week in July (week number 1) continuing to the last week of June (week number 52) of the next year. The first 8–10 weeks and the last 6 weeks (from 46 to 52) represent winter in SH; weeks 12 through 40 (September to March) represent the period when crops are grown.

Cordoba is located in central Argentina and has 26 departments (Fig. 6.9a)) growing corn (Seiler et al. 2006). Yearly corn yield, total production, and harvested area were used for the analysis in each department. Yield data were represented by average values for each of Cordoba's departments annually. The data were obtained from the Crop Reporting Division of the Secretary of Agriculture in Argentina, for the crop seasons 1981/1982 to 2002/2003. Corn yield trend was calculated for each department as a linear function obtained by correlating the observed yield to the calendar year number. Yield departures from the trend, used for the data analysis, were calculated as ratio (Eq. 6.2). Correlation and regression analysis were used to regress corn yield departures from technological trend (dY) with the VH (VCI, TCI, and VHI) indices and to develop corn dY-VH models. Correlation of corn dY with VH indices were done for each department and for every week during the crop-growing season. The highest Pearson correlation coefficient (PCC) of this relationship during the crop season is shown in Fig. 6.9b–d for each department and for each index. As seen, the correlation is very strong PCC = 0.70–0.75.

Figure 6.9b–d shows the correlation results of de-trended (from technological trend) corn yield with vegetation indices for those departments, which have significant corn production in Cordoba. Also, the week of the highest correlation between yield and each of the indices is shown for each department. In general, the correlation values were around 0.5 and above for those departments in the east and in the south half of the province, with the exception of some lower values for the departments D5 and D12. However, for the largest corn-producing departments, the correlation is more consistent exceeding 0.7. A regular curve (not shown) of weekly correlation values through the season shows an increase from pre-planting to approximately the time of critical period in corn development which is about 2–3 weeks before and after tasseling and then decreases rapidly. This period happens to occur for corn from the end of December (week 26) through mid-January (week 29), depending on the planting dates. The results show that the week of the highest correlation for any of the three indices is located in that period in most of the departments, indicating the time of the vegetation conditions of higher impact on yields. Positive correlation with VCI, TCI, and VTI also indicates that yields below the trend are associated with moisture and thermal stress while above-trend yields are associated with no weather stress.

Cordoba's dY-VH (VCI, TCI, VHI) models, for six of the largest corn-producing departments (1, 6, 9, 19, 23, 25), are shown below. The models predict deviation of yield from the trend for each department. Independent variables for the models are the vegetation indices selected according to the highest week's correlation, and the statistical significance from a stepwise procedure used to select variables which significantly contribute to the models. Except for Departments 1 and 23, the

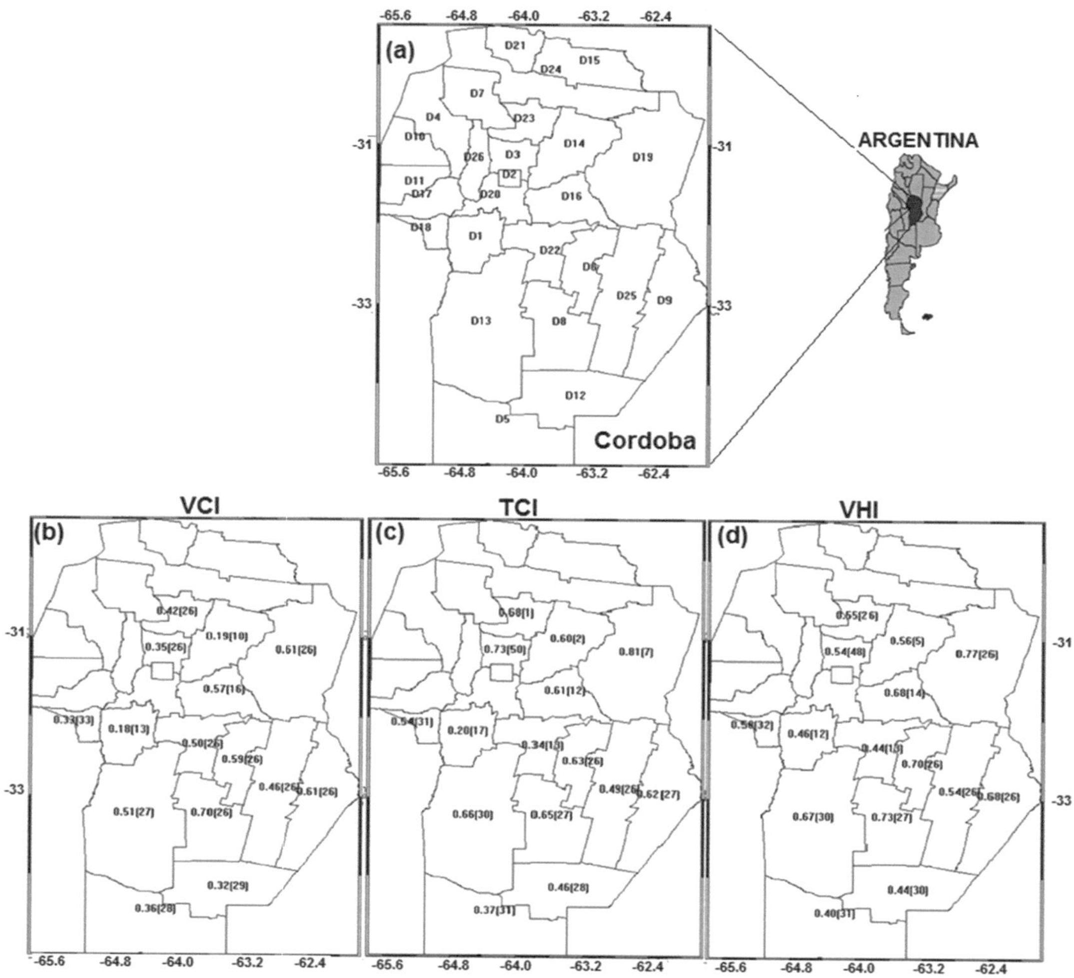

Fig. 6.9 Cordoba province (CP) in Argentina, CP's administrative departments (AD), their locations and numbers (**a**); (**b**)–(**d**) Correlation coefficients between corn dY in AD with VCI (**b**), TCI (**c**) and VHI (**d**) for the week (week number (in prentices) is counted from July 1 as week # 1) with the highest correlation

selected indices included in the model explain more than 60% ($R^2 > 0.60$) of the yield variability around the trend. The error of these models (MSE) was below 11% (below 18% for department 1 and 23) of the respective mean of the observed yield (Seiler et al. 2006). The testing of the models is summarized in Table 6.2. The number of samples N in the table is the number of years used for models' development, which is the same for models testing using the Jacknife independent testing. Some years in the models development were missing due to absence of or missing yield data. Good agreement between observed (O) and predicted (P) yield is shown by the simple correlation coefficient values and by the index of agreement for all department predictions. Prediction errors go from a minimum of 15% in department 9 to a maximum of 28% in department 1. Except for departments D6 and D25 where the yield estimates come close to the harvest, the rest of the departments used estimates that can be done about 2 months before harvest. Meanwhile, the index of agreement is close to one and R is 0.83–0.94, indicating a very good agreement between observed and independently predicted corn yield.

Table 6.2 Quantitative measures of VH-based independently predicted versus observed corn yield in the six administrative departments of Cordoba Province, Argentina

Indicator	Parameter	Dep 1	Dep 6	Dep 9	Dep 19	Dep 23	Dep 25
Mean observ. *Y*	MeanO (t/ha)	3.7	3.8	5.0	4.6	4.8	4.2
Mean predict. *Y*	MeanP (t/ha)	4.0	3.8	5.0	4.7	4.9	4.2
Sample number	*N*	16	20	20	16	16	19
Mean bias error	MBE (t/ha)	0.26	0.03	0.07	0.06	0.14	0.03
Root sq. error	RMSE (t/ha)	1.06	0.81	0.77	1.06	0.91	1.14
Systemat. error	RMSEs (t/ha)	0.36	0.36	0.08	0.25	0.28	0.35
UnSystem. error	RMSEu (t/ha)	1.00	0.73	0.76	1.04	0.85	1.09
Agreement	*D*	0.912	0.937	0.969	0.909	0.945	0.931
Correlat. coef.	*R*	0.874	0.884	0.940	0.834	0.918	0.870

$dY = 106.38 + 9.26 \times VHI\_22 – 9.47 \times VHI\_24 + 0.83 \times VHI\_31$	$R^2 = 0.52$	(1)
$dY = 81.57 + 1.18 \times VCI\_28 – 0.89 \times VCI\_36$	$R^2 = 0.63$	(6)
$dY = 27.08 + 1.28 \times VCI\_22 + 0.22T\ TCI\_28$	$R^2 = 0.80$	(9)
$dY = 47.72 + 0.33 \times VCI\_27 – 0.76 \times TCI\_27$	$R^2 = 0.63$	(19)
$dY = 79.11 + 4.42 \times VHI\_23 – 9.09 \times VHI\_24 + 5.38 \times VHI\_25$	$R^2 = 0.43$	(23)
$dY = 76.69 – 0.86 \times VCI\_15 + 2.14 \times VCI\_22 – 0.87 \times VCI\_39$	$R^2 = 0.75$	(25)

Results of presented analysis indicate that the seasonal and inter-annual variability of the vegetation indices serves as a good proxy for identifying the impacts of weather anomalies on vegetation and crops production in Argentina. The VH indices demonstrated to be good predictors of corn yield in semi-wet and wet climate conditions of Argentina. Moreover, seasonal and inter-annual variability of the vegetation indices serves as a good proxy for identifying environmental sources of variations mainly weather anomalies impacts on vegetation and crops production. In addition, the VH indices demonstrated to be good predictors of corn yield. The developed VH-based corn yield models explained up to 80% of the yield variation according to administrative departments. This integrated approach applied to the province of Cordoba using satellite data and statistical models suggests an effective mean to monitor seasonal and inter-annual vegetation responses and to estimate corn yield at regional scales. Also, it provides insights to economic decisions, marketing strategy, and providing help to nations with food security issues. In addition, they have to be used for further analysis into understanding the regional climate potential for crop production and for the effects of the climate variability on vegetation and corn yield.

6.3.6 Wheat in Australia

Agriculture in Australia provides significant contribution to the national economy. Grains and oilseed crops produce about 35–45 million tons of grain per annum from an area of slightly more than 20 million hectares, with an annual gross production value of around a $9–13 billion (AG 2017). Wheat is the major grain crop grown in Australia, which is sown in autumn and harvested in spring or summer (depending on weather). The main producing states are Western Australia, New South Wales, South Australia, Victoria, and Queensland. Wheat grown for domestic consumption and feedstock is predominately produced on the east coast. The majority of Australian wheat is sold overseas with Western Australia the largest exporting state. The major export markets are in the Asian and Middle East regions, which include Indonesia, Japan, South Korea, Malaysia, Vietnam, and Sudan (AG 2017).

Although Australia has nearly 800 million hectares of land, most of them are occupied by desert and some by grassland (Fig. 6.10a). The climate of Australia is mostly arid and a shortage of rainfall plus hot weather puts considerable constraint on water resources, which strongly affects strongly agriculture and other human activities

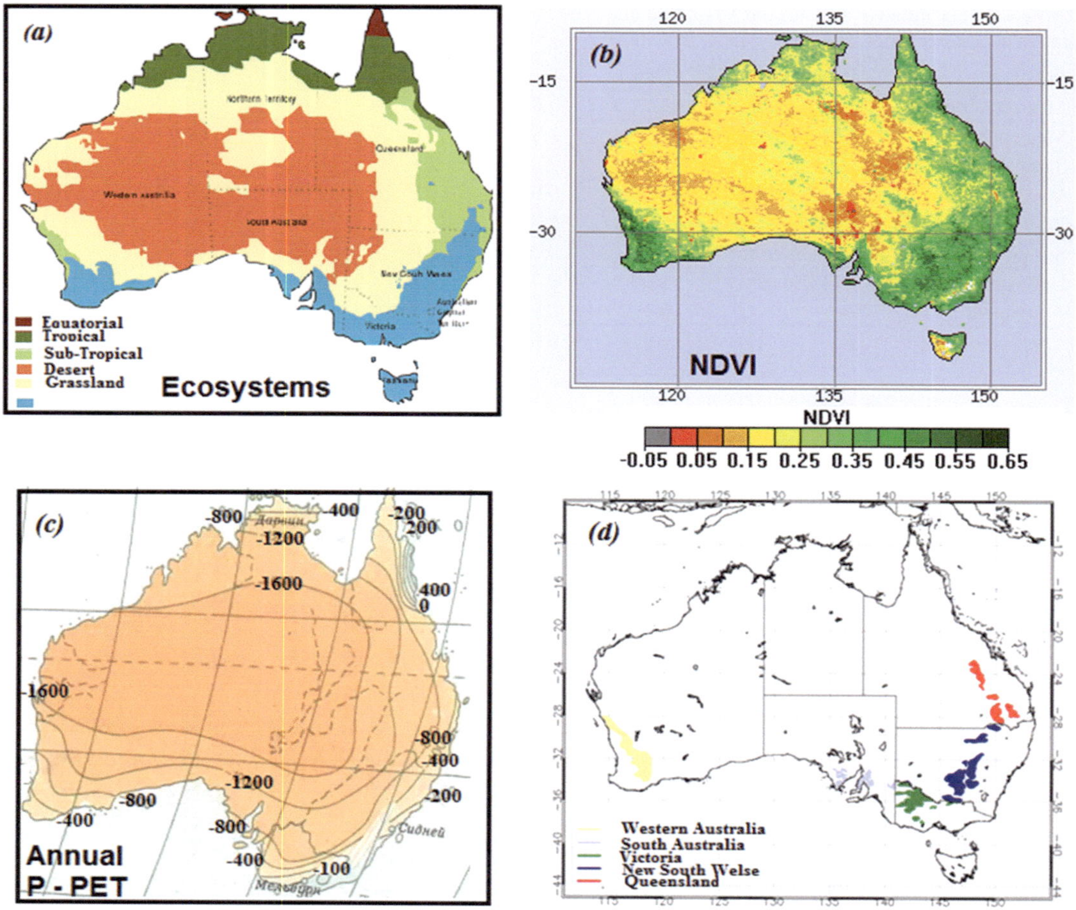

Fig. 6.10 Environmental resources and wheat in Australia: (**a**) ecosystems, (**b**) normalized difference vegetation index (NDVI), (**c**) shortage of water (difference between annual precipitation (*P*) and potential evapotranspiration (PET)), (**d**) wheat-growing area

(AG 2016, 2017). Annual precipitation in the two biggest ecosystems (desert and grassland) are greater than in the other world's big deserts (for example, Sahara) and grassland (China) areas, accounting for 50–250 mm in desert and 250–500 mm in grassland (Gol'tsberg 1972). Therefore, these areas have some vegetation and the normalized difference vegetation index (NDVI) is quite elevated (0.10–0.25, Fig. 6.10b), compared to similar ecosystems in the other areas (VH 2017). However, precipitation in nearly 90% Australia is much below potential evapotranspiration (the difference is −1200 to −1800 mm, Fig. 6.10c) due to extremely high temperatures during the growing season (Gol'tsberg 1972) limiting agricultural activities. Moreover, Australian soils have low fertility, especially compared to such extremely productive soils as chernozem in the southern European Russia and US prairies. Therefore, of the 800 million hectares of Australia land, only 10% is suitable for agriculture (AG 2017). Wheat is the major crop grown there. Australia contributes around 1.3% to global grain production (Table 5.3) and 4–8% to global grain in trade, which is substantial to address food security by providing help to countries in need. However, as seen in Fig. 6.10d (USDA 2016), its wheat area is very limited and patchy, which should be taken into consideration for environmental data collection and wheat yield modeling.

Matching wheat areas with climate and weather characteristics in Australia indicates that the wheat-growing areas still have negative balance (−300 to −600 mm, Fig. 6.10c) between annual precipitation and potential evapotranspiration (Gol'tsberg 1972) indicating a shortage of moisture for a successful wheat harvest. Moreover, at the background of precipitation shortage, the warmest month temperature is relatively high (24–26 °C) and up to 10 days during the growing season air temperature might be 40 °C and higher, which creates intensive drought conditions leading to a considerable reduction of wheat production. Therefore, wheat is completely dependent on precipitation and temperature. The situation with successful wheat growth has deteriorated considerably in drought years, when up to 55% of wheat production might be lost (USDA 2006). The weather station network in Australia, which provides data for estimation of wheat production, is not distributed uniformly to provide reliable information for a sketchy wheat area (Fig. 6.10d). Therefore, 4 km^2 satellite-based VH data collected specifically for the area of wheat cultivation were used to model Australian yield during 1980–2014. Wheat yield (*Y*, t/ha), production (*P*, tons, (t)), and area (*A*, hectares, (ha)) were collected for the country during 1961–2014 (FAO 2016). Yield was calculated as a ratio of *P/A*. Satellite-based 4 km^2 weekly VH indices data were collected from the NOAA Global Vegetation Health data set (in 8-bit counts format) from 1980 through 2014 (NOAA/NESDIS 2018; Kidwell 1997).

Between 1981 and 2014, wheat yields (*Y*) in Australia experienced upward trend (Fig. 6.11a) due to improvements in the applied agricultural technology. This trend (*Yt*) was approximated linearly as a function of year number by the following equation:

$$Yt = 0.1198 \times \text{year} - 222.84 \qquad (6.14)$$

This equation approximates technology-based yield level provided that weather is near normal (close to many-year mean). Following the equation, technology provided wheat yield increase

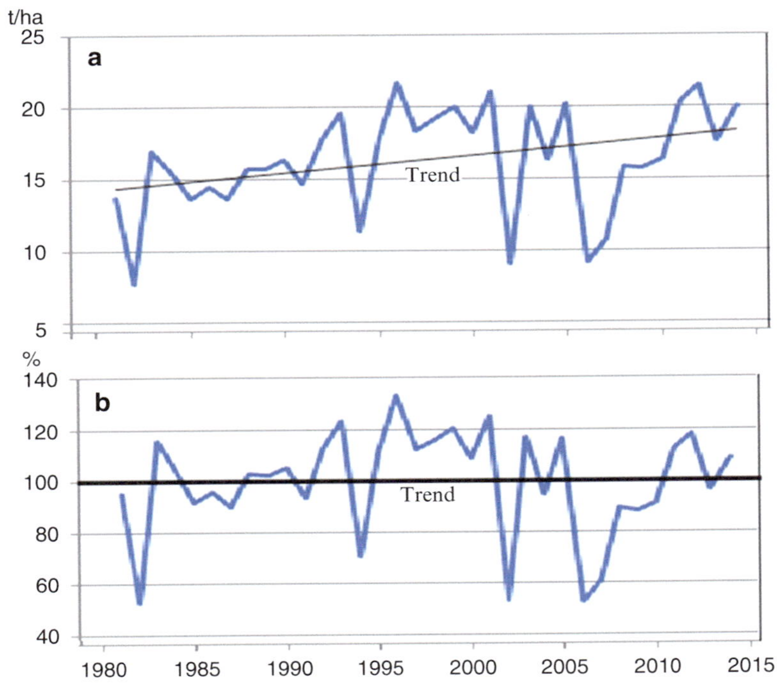

Fig. 6.11 Hard wheat yields in Australia during 1981–2014: (**a**) collected annually (*Y*) and linear trend (*Yt*), (**b**) deviation from trend (d*Y*%) from Eq. (6.2)

from 14 t/ha in 1981 to nearly 18 t/ha in 2014, adding nearly 22% during 25 years. In spite such a strong increase, weather-related variations around the trend, shown in Fig. 6.11b, are very large indicating strong weather contribution to yield variations from year to year. During the most unfavorable weather years associated with droughts, wheat yield might be reduced up to 40–50% below the trend (1982, 2001). These deviations (dY) were approximated following Eq. (6.3).

Figure 6.12 Scatter plots of hard wheat dY versus weekly VCI, TCI, VHI during the growing season in Australia

According to the modeling methodology (Sect. 4.2), dY-VH (VCI, TCI, and VHI) correlation analysis was performed. Wheat in Australia is typically sown in April to May and harvested in October to November. Correlation dynamics of dY with weekly VH (VCI, TCI, VHI) indices was performed (Kogan et al. 2018). The goals were to investigate (a) the strength of dY-VCI, dY-TCI, and dY-VHI relationship and (b) if the strongest correlation coincides with wheat's critical period (approximately, 2–3 weeks before and after heading). The correlation analysis indicated that in the early period of wheat season, which is pre-planting, planting, and emergence (March to May), the correlation for all three VH indices was low (around 0.35). It increases sharply during the green biomass accumulation (in June), reaching maximum at weeks 38–42 (July to August). Thermal conditions (dY correlation with TCI) are as important as moisture (dY-VCI) and their combination (dY-VHI). The importance of thermal conditions begins 2 weeks later as compared to moisture. During that period, good water supply and cooler temperatures stimulate larger wheat yield. Following dY-VH correlation indices indicates that when VH indices have values below 40, indicating moisture stress (lack of water) and thermal stress (hot weather) yield is below trend. Oppositely, above-trend yield occurs when VH indices are above 60 (favorable conditions, when weather is wet and cool).

$$\begin{aligned} \mathrm{d}Y_{\mathrm{VCI}} &= -1.06 + 0.0036 \times \mathrm{VCI}_{36} + 0.0036 \times \mathrm{VCI}_{37} + 0.0039 \times \mathrm{VCI}_{38} + 0.0039 \times \mathrm{VCI}_{39} \\ &+0.0038 \times \mathrm{VCI}_{40} \qquad R = 0.799; \qquad \mathrm{StDev} = 0.242; \end{aligned} \tag{6.15}$$

$$\begin{aligned} \mathrm{d}Y_{\mathrm{TCI}} &= -0.726 + 0.0029 \times \mathrm{TCI}_{38} + 0.003 \times \mathrm{TCI}_{39} + 0.003 \times \mathrm{TCI}_{40} + 0.003 \times \mathrm{TCI}_{41} \\ &+0.003 \times \mathrm{TCI}_{42} \qquad R = 0.684; \qquad \mathrm{StDev} = 0.300; \end{aligned} \tag{6.16}$$

$$\begin{aligned} \mathrm{d}Y_{\mathrm{VHI}} &= -0.982 + 0.0035 \times \mathrm{VHI}_{36} + 0.0038 \times \mathrm{VHI}_{37} + 0.0038 \times \mathrm{VHI}_{38} + 0.0038 \times \mathrm{VHI}_{39} \\ &+0.0038 \times \mathrm{VHI}_{40} \qquad R = 0.777; \qquad \mathrm{StDev} = 0.276; \end{aligned} \tag{6.17}$$

$$\mathrm{d}Y_{\mathrm{VcTc}} = 0.80\mathrm{VCI}_{38} + 0.69\mathrm{TCI}_{40} \qquad R = 0.785; \qquad \mathrm{StDev} = 0.256 \tag{6.18}$$

Finally, regression Eqs. (6.15–6.17) have been developed for each index. All three have high multiple correlation coefficient (R). Moisture conditions (VCI) are the most important and thermal conditions are slightly less. Meanwhile, moisture–thermal combination (VHI) has a very good predictive power. At the same time, it is important to mention that following regression coefficient values, contribution of different weeks' variables into final model prediction is almost equal. Therefore, an additional model was developed with two variables only VCI and TCI for the weeks of their highest correlation. The last Eq. (6.18) supposed to have similar (r = 0.785 and StDev = 0.256) to dY-VCI and dY-VHI equations. Since the independent variables in Eqs. (6.15–6.18) are collinear, they were combined together with the weight corresponding to the value of regression coefficients. These equations are written below:

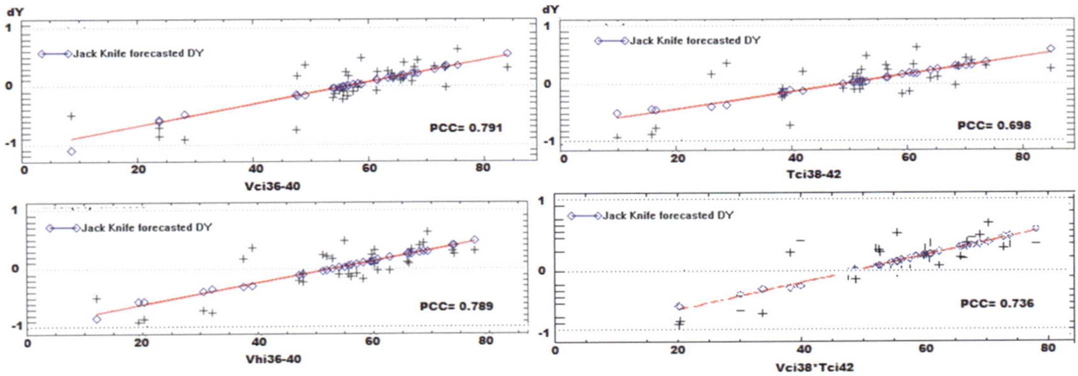

Fig. 6.12 Correlation between Australia hard wheat dY and combined multi-week VCI, TCI, VHI and VCI and TCI indices (Eqs. 6.19–6.22) from independently verified models and actual data

$$dY_{vci} = -1.063 + 0.0189 \times VCI_{36-40} \qquad R = 0.776 \tag{6.19}$$

$$dY_{tci} = -0.726 + 0.0148 \times TCI_{38-42} \qquad R = 0.685 \tag{6.20}$$

$$dY_{vhi} = -0.982 + 0.0186 \times VHI_{36-40} \qquad R = 0.779 \tag{6.21}$$

$$dY_{vctcI} = 1.175 + 0.0206 \times VCI_{38} \times TCI_{42} \qquad R = 0.755 \tag{6.22}$$

The last four equations were verified independently (using Jeck Knife techniques) and these results are shown in Fig. 6.12.

This globally universal satellite-based VH technique for monitoring vegetation health, including drought detection and impacts prediction, was used to model wheat yield in the principal growing areas of Australia. Correlation and regression analysis during 1981–2014 showed strong relationship between wheat yield anomaly (deviation from technological trend) and VCI, TCI, and VHI indices (weekly weather-related components of the environment) during the period of highest wheat yield sensitivity to weather (critical period). This period starts 2–3 weeks before and ends up 2–3 weeks after wheat heading. The developed models provide prediction of wheat yield 1–2 months in advance of harvest. This paper proved again that the VH methodology can be applied to many different environments. The advance yield prediction in Australia from NOAA operational satellites will help the Australian Government to plan in advance wheat delivery to the international market and also providing help the countries with food problem addressing food security.

6.3.7 Rice in Bangladesh

Rice is a very interesting crop since it requires unusual environmental conditions for a plentiful production. The major world rice-producing countries are China, India, Indonesia, Bangladesh, Myanmar, Vietnam, and Thailand (Abdus et al. 2003). Bangladesh is a densely populated agriculture-based country and rice is the largest and most widely cultivated food crop in Bangladesh. The 150 million people of the country obtain 60–70% of their calories from rice and its products. The introduction of new crop varieties, increased utilization of chemical fertilizer, pesticides, exploitation of ground water for irrigation have significantly increased the rice crop yield over the country though the rate is still much below the highest attainable value (Salam 1981). Nearly 50% of the cropland is double cropped and 13% is triple cropped rice. As a result, much of the country has areas where the fraction of sown rice is over 90% of the land area (BBS 2006). Rice ecosystems in Bangladesh are dominated by rain fed (over 50% of the rice area) and irrigation, although significant amounts of upland and deep water rice still exist. Bangladesh produces near 40 million ton and imports 1 million ton of rice annually (Rahman et al. 2009). The country has three rice crop seasons within a year: Boro crop, Aman, and Aus, which account for 50%, 43%, and 7% of total annual rice production, respectively. Monsoon rains are most important for the Aman crop, harvested in November to December. Two varieties of Aman rice (local and hybrid) are cultivated in Bangladesh. It also categorized as two plantings type namely broadcast and transplant Aman. This rice requires a special habitat of prolonged flooding. Most Aman is transplanted between the end-of-June and mid-August (in some years in mid-September) and harvested in November to December.

Bangladesh lies in the Indo-Gangetic plain of South Asia, bordered by India to the west, north and north-east, Myanmar to the south-east and Bay of Bengal to the south. The territory is situated on one of the largest, complex deltas in the world where the Ganges, Brahamaputra, and Meghna rivers enter the Bay of Bengal. The climate of Bangladesh is subtropical warm, wet, and humid (Rahman et al. 2006). The country experiences natural disasters such as excessive rainfall (floods), droughts, and tropical cyclones, which can reduce rice yield. There are three seasons: hot in summer (March to June), warm and humid monsoon (June to September), and a cold and dry (October to February). The annual average rainfall varies from 1500 mm to 5000 mm, temperature and humidity range is 12–30 °C and 65–90%, respectively. Aman rice is mainly grown on the floodplains and deltas of the Meghna, Jamuna, and Ganges river systems (Anwar 1999) in the six administrative divisions (Barisal, Chittagong, Dhaka, Khulna, Rajshahi, and Sylhet).

The Aman crop experiences extremes of dry and wet conditions over its long growing season. Temperature has positive effect on the yields of all three rice seasons in Bangladesh. Rainfall impact is complicated: at the early stages of Aman rice it is negatively correlated with yield, during the biomass development and flowering stages, the rainfall is positively correlated. Following the Bangladesh climate, nearly 80–90% of annual rainfall occurs within June to September but December to mid-March is almost rainless. For rain-fed Aman rice, during panicle initiation and flowering stages water shortage normally leads to a considerable reduction of fertilizers' transfer from soil to leaves for uninterrupted photosynthesis during limited daytime. As a result, more empty grains are produced and crop yield is reduced. Monsoon rains are the most important for the Aman crop, harvested in November to December (Abdus et al. 2003).

Quite a few models have been developed to estimate rice yield before harvest. However, most of them require field data on crop conditions that are not easily available during rice-growing season. The models complexity and the ground data requirements render them impractical for large-scale applications. Another approach, which uses weather data also, has some problems connected to low density weather stations, which do not adequately represent the diversity of weather over the large areas where crops are grown (Salam 1981). A rice monitoring system that provided timely and accurate information on crop status and yield well before harvest and without requiring a dense ground-based observation network would be of great value.

Therefore, similar to other countries, satellite-based VH indices were used in Bangladesh for early drought and flood detection and monitoring their impacts on Aman rice (Rahman et al. 2009). Such system would be especially interesting to test, since rice responds differently to environmental conditions compared to other crops discussed earlier. Aman rice yield, production and sown area, estimated by sampling surveys for each administrative division and aggregated for the entire Bangladesh, were obtained from the Bangladesh Bureau of Statistics for the years 1991–2005 (BBS 2006). Two Aman rice types were modeled: local and hybrid. The NOAA 4 km^2 weekly VH indices (VCI and TCI) were collected from NOAA/NESDIS (2018) and aggregated for the same six administrative divisions and averaged as the mean values for the entire Bangladesh during 1991–2005.

Following the methodology (Sect. 6.2), the research strategy was to: extract the weather component (dY) from the Aman rice annual yield, correlate dY with VCI, TCI, and VHI (investigate the strength of the relationship and determine if the strongest correlation coincides with rice criti-

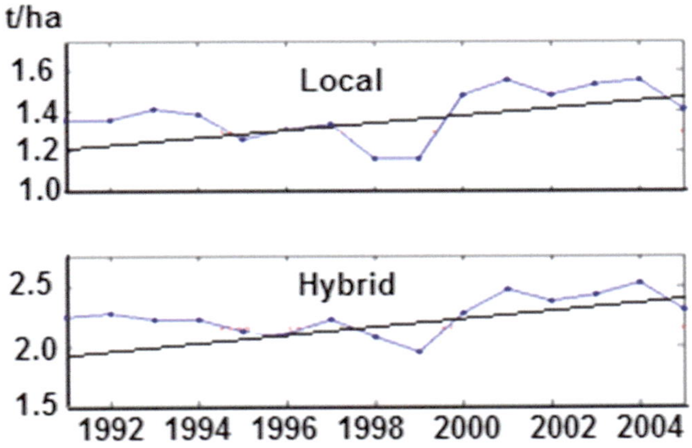

Fig. 6.13 Aman rice (local and hybrid) yield and linear technological trend in Bangladesh during 1991–2005

cal period), and develop regression models. Two Aman rice types were investigated; Local (Loc) and Hybrid (Hyb, more productive). Linear regression approximation (Obuhov 1949; Brockwell and Davis 2000) was applied to Loc and Hyb time series (Fig. 6.13). First, what is noticeable from the figure, the Hyb yield is 54% more productive than the Loc. Second, technology improvements stimulated both yield increase with the same speed. Third, although Hyb is advanced compared to Loc both technologies provided almost equal contribution to yield grows (13% and 11%, respectively, in 15 years) and fourth, weather-related yield fluctuations (above- and below-trend line) are similar for both varieties. For example, yield deviation from the trend in 1999 and 2001 were estimated at 83 and 109 for Loc (17% yield reduction in 1999 due to unfavorable and 9% increase due to favorable weather in 2001) and 85 and 105 for Hyb, respectively.

Dynamics of correlation coefficients for d*Y* versus moisture (VCI) and thermal (TCI) conditions between weeks 20 (end of May) and 45 (early November) is shown in Fig. 6.14a. As seen, d*Y*s for both types of Aman rice are highly correlated with VCI and TCI from early October through early November (weeks 40–43). Pearson correlation coefficients (PCC) of Loc yield with VCI are −0.61 to −0.69 and with TCI are −0.39 to −0.40; for Hybrid correlation with VCI (−0.53 to −0.59) and TCI (−0.46 to −0.53). October to November period is known as a very critical for both types of Aman yield varieties, because Aman rice yield goes through reproductive phase starting at panicle initiation and ending at flowering and ripening phases or ends of maturity. Interestingly, it is observed that a negative correlation of yield anomaly with VH indices indicate that below-trend (losses) Aman rice yield are associated with higher TCI and VCI, which indicate cooler thermal conditions, and wetter moisture conditions. Oppositely, above average yield is associated with lower both TCI (hotter weather) and VCI (lower rainfall) during those months. These results clearly indicate that in a wet climate of Bangladesh, excessive rainfall during the monsoon season negatively affects Aman rice yield of both varieties highly technological and regular. Similar results were obtained by Xiao et al. (2006).

$$\begin{aligned}&\mathrm{d}Y-\mathrm{Loc}=147.54-3.97\mathrm{TCI}_{40}+4.32\mathrm{TCI}_{41}-2.34\mathrm{TCI}_{42}+1.22\mathrm{TCI}_{43}-0.98\mathrm{VCI}_{40}-1.04\mathrm{VCI}_{41}\\&+3.95\mathrm{VCI}_{42}-2.41\mathrm{VCI}_{43}\quad R^2=0.98;\mathrm{RMSE}=2.19;\quad F=34.58;\quad P<0.002\end{aligned} \tag{6.23}$$

$$\begin{aligned}&\mathrm{d}Y-\mathrm{Hyb}=136.62-2.55\mathrm{TCI}_{40}+2.96\mathrm{TCI}_{41}-1.92\mathrm{TI}_{42}+0.90\mathrm{TCI}_{43}-0.03\mathrm{VCI}_{40}-1.44\mathrm{VCI}_{41}\\&+1.91\mathrm{VCI}_{42}-077\mathrm{VCI}_{43}\quad R^2=0.97;\quad \mathrm{RMSE}=1.73;\quad F=28.15;\quad P<0.003\end{aligned} \tag{6.24}$$

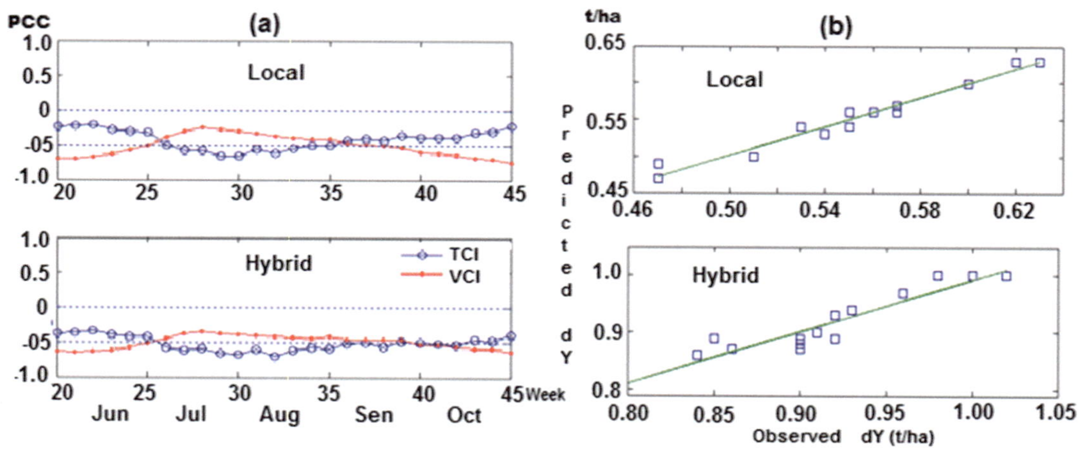

Fig. 6.14 Aman rice in Bangladesh (local and hybrid) during 1991–2005: (**a**) dynamics of correlation coefficients for d*Y* versus VCI and TCI and (**b**) independently predicted d*Y* against observed d*Y*

In the multiple regression analysis, d*Y* was regressed on the linear combination of VCI and TCI values for weeks 40–43 with the highest PCC. The overall models (Eqs. 6.23 and 6.24) are highly significant with R^2 nearly 1.0, Root Mean Square Errors (RMSE) <5.0, high F values of 34.58 (Loc), 28.15 (Hyb) and p values much smaller than 0.05 (Obuhov 1949). Scatter plots of observed versus independently predicted yield are shown in Fig. 6.14b. Independently validated results confirm the reliability of the models since R^2 values are very high, 0.97 for Loc and 0.89 for Hyb (Willmott 1982). Aman rice yield can be predicted from VH indices at approximately 4–8 weeks prior to harvest time. This information will improve the harvesting process, storage of production, marketing and transportation. These results are complementary to crop modeling in the other countries.

6.3.8 Cereals in Russia

Russia is the fourth largest grain-producing country in the world after China, India, and the USA. Grain is a very important agricultural commodity for Russia determining availability of food and feed, marketing strategies and tools for political decisions. Since the demand for grain is higher than supply, Russia purchases on the average nearly three million metric tons (MMT) of grain annually from the international market. In some years, when large-scale drought causes huge crop losses, this amount is doubled, as it was in 1999 and 2000, when purchases reached 7.8 and 5.8 MMT, respectively (FAO 2016). Therefore, advanced knowledge about Russia's amount of grain grown, collected and anticipated purchases in a particular year are very important not only for estimation of domestic grain production and use, but also projections of global supply and demand, international trade and consequently global food security.

Since the time of the USSR, after World War II, non-advertised government policies about annual requirements for total grain production, called cereals (includes winter and spring wheat, barley, oats, buckwheat, and pulses), was to produce one ton of cereal per one person per year (Verigo and Razumova 1963). Unfortunately, this target between 1950 and 1990 has never been attained. Maximum amount of grain was collected in 1978 (209 MMT), for nearly 260 million population. In most years, grain production was much below obtained in 1978, especially in the years of intensive droughts. For example, in 1972, USSR collected 162 MMT cereals (Kogan 1986). After the disintegration of the USSR, the situation with a lack of cereals production has not changed, especially due to grain losses following frequent droughts in Russia's principal agricultural areas. From the most recent events, the 2010 droughts in Russia reduced total cereal production to 75 MMT compared to 97 MMM in a no drought 2011. The 2010 cereals losses in Russia led to a shortage of grain supply compared to demands, a reduction of global grain, lack of grain on the international market, and the corresponding price increase (Goldenberg 2012; U.S. Drought 2012).

Weather in Russia plays a major role in determining production of cereals, individual grains, and other crops. Weather forecasts in the former USSR and currently Russia are issued by the Hydrometeorological Center (HCR 2017; Kogan 1986). The Center has a group of scientists, who develop crop-weather models and forecast cereals and individual grain crops yield and production every year based on weather observations (mostly precipitation (P) and temperature (T)) obtained from Russia's meteorological stations (Kogan 1986). Unfortunately, the number of weather stations in Russia, similar to other countries, has low density and the provided weather information is not comprehensive to estimate accurately grain production. Therefore, satellite-based VH indices were used to model cereal yield (Kogan et al. 2015a, b). The goal was to investigate potential for VH-base multiple grain crops (cereals) estimation and prediction of drought-related cereals losses in advance of harvest.

Southern Russia (mostly European) is the main grain basket in the country, producing nearly 70% of high-quality grain. The Saratov Administrative region (oblast, SO) in the Lower Volga ecosystem located in the southeastern European Russia was selected for this test since SO is one of the principal grain-growing areas of Russia. Agricultural land of SO covers nearly 9 million (mil) hectares (ha), with 6.1 mil ha occupied primarily with grain crops, which fields are traditionally huge and have not changed over years in order to receive maximum production with the best grain quality (Voeikov and Gortsev 2014; Zernoimport 2014). This region has fertile, well-drained and nutritional chernozem soils, appropriate for a successful cultivation of grain crops such as wheat, barley, oats, and corn. However, climate of this area is semiarid, and the shortage of water puts considerable constraint on grain production Gol'tsberg 1972). The situation is complicated by hot weather accompanied by desiccative winds (suhovei). Moreover, the unusually light snow cover does not make up for summer moisture deficit. Although this area has a few rivers, the area of commercial grain crops is so large that the amount of available water is not sufficient to offset dry climate. Therefore, grain crops are generally not irrigated. In drought years, up to 40% of grain production might be lost in that area.

Annual cereal yield data (production in tons per area in hectares) were obtained from the Russia's Central Statistical Administration for the period 1972–2001. The data were presented by spring-planted cereals, which included wheat, barley, oats, corn, and pulses. Spring wheat, barley, and oats occupy 90% of the area. All crops are planted in late April early May and are harvested in August September (USDA 1994). SO cereal production and area from farms were summarized from their reports to the total SO level at the end of agricultural year; yield (Y) was calculated by dividing SO production by area. Weekly VH indices (VCI, TCI, and VHI) with 4 km^2 resolution were collected during 1982–2001 from NOAA/NESDIS (2018) and aggregated to SO mean values. Since grain crops occupy slightly over six million ha in SO, VH indices' data were aggregated from slightly over 17,000 4 km^2 pixels, which number sixteen hundred times exceeded the number of weather stations used for spatial aggregation of SO mean P and T. Further modeling procedure included calculation of SO $\mathrm{d}Y$ (Eq. 6.2) and correlation of annual $\mathrm{d}Y$ with the weekly VH indices (VCI, TCI, and VHI) and with SO area mean monthly total precipitation anomaly ($\mathrm{d}P$) and temperature ($\mathrm{d}T$) anomaly (departure from normal) collected from Russian weather data archive. Figure 6.16 demonstrates dynamics of Pearson correlation coefficient (PCC) for these relationships.

Figure 6.15a shows mean SO cereal yield (t/ha) and its linear trend during 1971–2001. The trend was approximated from the 10 year longer than the VH data set in order to avoid errors at the

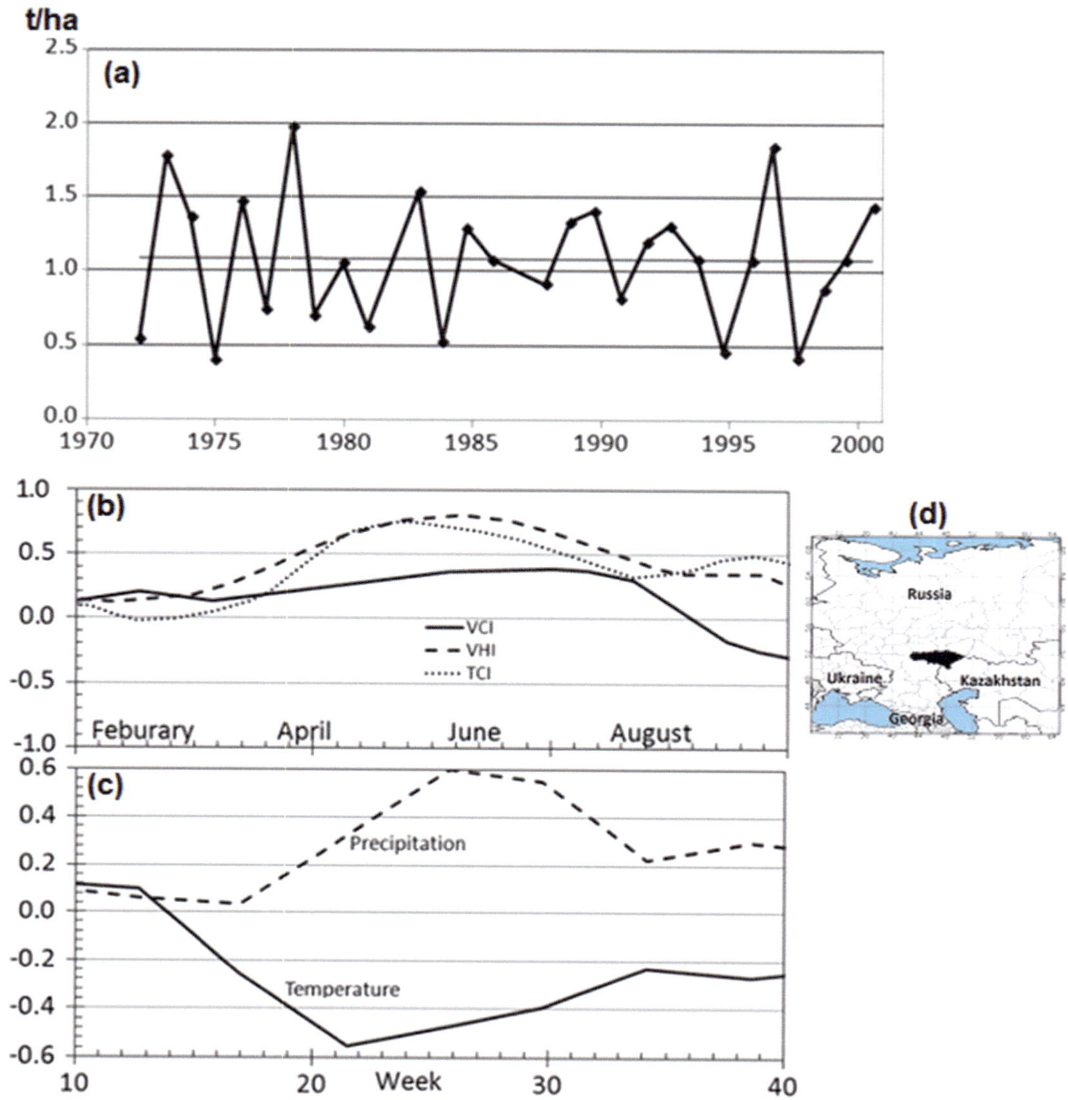

Fig. 6.15 Russia's Saratov administrative region (oblast): (**a**) cereal yield (*Y*) and trend time series during 1972–2001; and dynamics of correlation coefficient for SO cereal yield anomaly (d*Y*) with (**b**) weekly VCI, TCI, VHI and (**c**) with monthly mean precipitation and temperature anomalies (d*P* and d*T*, respectively); (**d**) Location of SO

ends of short-period time series (Snedecor 1965). Following Eq. 6.1, trend was approximated as $YT = 1.08 + 0.000013 \times \text{(year)}$. Trend analysis indicates that agricultural technology does not increase the yield during the indicated 30 year period. More important that variations of yield around the trend are strong and frequent. During severe droughts, such as 1975 or 1998, cereal yields dropped below the trend up to 66%. However, in good-weather years (1978, 1997), yield exceeded the trend up to 85%, restoring slightly multiyear yield balance. As seen, the correlation dynamics of d*Y* with the three indices and weather parameters is smoothed (Fig. 6.15b). More importantly, in the early spring, which is cereals' pre-planting, planting, and emergence (weeks 10–15, March to April), the correlation for all indices is near zero. It increases sharply during the green biomass accumulation in April and May (weeks 17–22). Thermal conditions during this period show much stronger correlation, playing a greater role because at the general background of year-to-year moisture deficit in SO, productivity of cereals depends much on extreme air temperature. If weather is hot, crops become stressed thermally and the end-of-season production is reduced. Oppositely, cooler weather stimulates higher production. The maximum correlation in the figure occurs in June to July (weeks 26–30) for VCI and early June (weeks 23–25) for TCI. The timing of these maximums coincides with the critical period of spring-planted cereal development when they experience enhanced sensitivity to weather conditions.

It is important to emphasize, correlation with monthly *P* and *T* data (Kogan 1986) shows similar to satellite PCC dynamics (Fig. 6.15c). Correlation coefficients' time series of d*Y* with SO regional mean total monthly precipitation anomaly (d*P*) and mean monthly temperature anomaly (d*T*) shows identical d*Y* versus VH indices correlation dynamics. Differences in the sign of the correlation for temperature anomaly (negative) and TCI (positive) is because TCI was designed to reflect a reverse temperature scale (from high-to-low, Eq. (6.4)) to match with VCI, while in situ temperature reflects changes from low to high. Strong similarity between VH-based satellite and in situ data in their impacts on cereal yield confirms again that VH-based methodology is highly reliable in prediction of agricultural production and advanced monitoring global, continental, country and regional food security issues.

Following the analysis of cereal yield's relationships with VH indices, it is possible to conclude that cereals, similar to individual crops in the other countries, can be predicted from space-based VH indices. Specifically, yield becomes below trend if VH indices are below 40 and above trend if VH > 60. However, it should be emphasized that even for the weeks of the highest correlation, VCI explained only 15–30% of d*Y* variance (PCC = 0.36–0.41), while TCI is more predictive index, explaining 35–50% of d*Y* variance (PCC = 0.47–0.75). Moreover, the relationship of SO d*Y* with VHI was the strongest explaining up to 64% of d*Y* variance (PCC = 0.68–0.80). The largest correlation of d*Y* with VHI stress the importance of combining moisture (VCI) and thermal (TCI) indices together even if their contribution to VHI was approximated as equal (Kogan 1997). Following the modeling principal (Eqs. 6.1, 6.2, 6.7 and 6.8), d*Y*-VH models were developed (Eqs. 6.25–6.28).

$$\mathrm{d}Y = 0.502 + 0.034^{*}\mathrm{TCI}_{26} - 0.106\times\mathrm{TCI}_{27} + 0.107\times\mathrm{TCI}_{28} - 0.03\times\mathrm{TCI}_{29} \qquad R^2 = 0.69; \quad (6.25)$$

$$\mathrm{d}Y = 0.501 + 0.114\times\mathrm{VHI}_{26} - 0.316\times\mathrm{VHI}_{27} + 0.289\times\mathrm{VHI}_{28} - 0.076\times\mathrm{VHI}_{29} \qquad R^2 = 0.74; \quad (6.26)$$

$$\mathrm{d}Y = 0.462 + 0.006\times\mathrm{TCI}_{26-29} \qquad R^2 = 0.74; \quad (6.27)$$

$$\mathrm{TCI}_{26-29} = 0.269\times\mathrm{TCI}_{26} + 0.257\times\mathrm{TCI} + 0.246\times\mathrm{TCI}_{28} + 0.229\times\mathrm{TCI}_{29} \quad (6.27a)$$

$$\mathrm{d}Y = 0.496 + 0.006\times\mathrm{VHI}_{26-29} \qquad R^2 = 0.81; \quad (6.28)$$

$$\mathrm{VHI}_{26-29} = 0:246\times\mathrm{VHI}_{26} + 0.249\times\mathrm{VHI}_{27} + 0.253\times\mathrm{VHI}_{28} + 0.252\times\mathrm{VHI}_{29}; \quad (6.28a)$$

Equations 6.25 and 6.26 provide d*Y* modeling using weekly TCI and VHI, respectively. Based on R^2 values, the estimates are statistically significant, explaining 69% and 74% of the d*Y* variance, respectively. However, it is important to indicate that all four variables of the considered equations represent the neighboring weeks' TCI and VHI (weeks 26–29). Unfortunately, these variables are highly collinear (each variable correlate strongly with the weeks before and after). Since collinearity might reduce the accuracy of a model, two other Eqs. 6.27 and 6.28 were developed where TCI and VHI variable were converted to one cumulative variable, summarizing TCI and VHI values for weeks 26 through 29 with the weights derived from PCC following the formula (6.7) and shown by (6.27a) and (6.28a). The new one-variable models show slightly better performance explaining 74 (compared to 69% in (6.25)) and 81% (compared to 74% in (6.26)) of d*Y* variance.

Figure 6.16 present the result of SO cereal yield validation and drought area and intensity estimates during the last 3 years. Losses of cereals were independently estimated quite accurately (Fig. 6.16a), especially in 2010, when drought reduced average SO cereal yield to 0.4 t/ha, while the general trend level for that year was 1.2 t/ha (FAO 2016); the cereal losses were 66%. These losses were extremely high and are supported by a huge drought area with the highest intensity (Fig. 6.16b), especially compared to the other 2 years (2011 and 2012). As seen in Fig. 6.16c, d, in the mid-summer of all 3 years, SO experienced thermal stress (TCI < 40). This stress was extreme in 2010 when TCI was zero (exceptional stress) for 1.5 months and below 20 (severe-to-exceptional stress) for 3–4 months. In 2011 and 2012, the stress was weaker (TCI around 25) and shorter, continued less than 2 weeks (Fig. 6.16c). Based on Fig. 6.16d, the 2010 thermal stress of severe-to-exceptional and exceptional (the strongest) intensity affected 100 and nearly 50% of SO area, respectively. In 2011 and 2012, the area of similar intensity thermal stress was much smaller.

The presented SO cereal yield models (especially (6.27) and (6.28)) predict the yield at the end of July or early August, approximately 2 months in advance of harvest and 6 months in advance of official grain production statistic is released. The error of production loss prediction is 7–10%. The error of prediction drops to 3–5% in the years of intensive droughts. Such a high accuracy and a very advanced time of predictions are very useful for assessments of food availability, making marketing decision and also for providing help to the nations in need helping to address food security issues. The new vegetation health method can be successfully used for drought early detection (6–8 weeks earlier than before) in any corner of the globe, outlined more accurately drought area, and its impacts on crops well in advance of harvest. This is a vital step for prediction of global food security and trade. The current developments in satellite technology and sensor design, along with great achievements in numerical methods, speed, and capacity of computers, and available hardware and software, will bring more progress in the application of operational satellites. With accumulation of satellite data, we will continue to enhance the accuracy of hazard detection, monitor the environment, increase the lead time of estimates, and better diagnose impact assessment. New products geared toward early warning of insect development, diagnosis of epidemics, and other human health problems will be more aggressively pursued. New operational VIIRS sensor is currently widening our abilities to detect problems earlier and with higher accuracy and spatial resolution. New high-resolution capabilities of satellite sensors will help with precision agriculture and the ability to move from the problem of detection to prediction and the mitigation of consequences. A new satellite era will provide the ability to estimate economic effectiveness of applied technologies for profit optimization.

6.3.9 Spring Wheat in Kazakhstan

In the early 1990s, the United States Agency for International Development (US AID) provided funds to investigate possible applications of the VH tools in Kazakhstan. This work included quantitative validation of the new AVHRR-based VH indices in dry ecological environment of this country and their calibration against ground data (Gitelson et al. 1995, 1998). These objectives were set up because Kazakhstan is a huge agricultural country (wheat and cattle producer), where agriculture accounts for a large share in the total country's budget. Kazakhstan's dry climate with

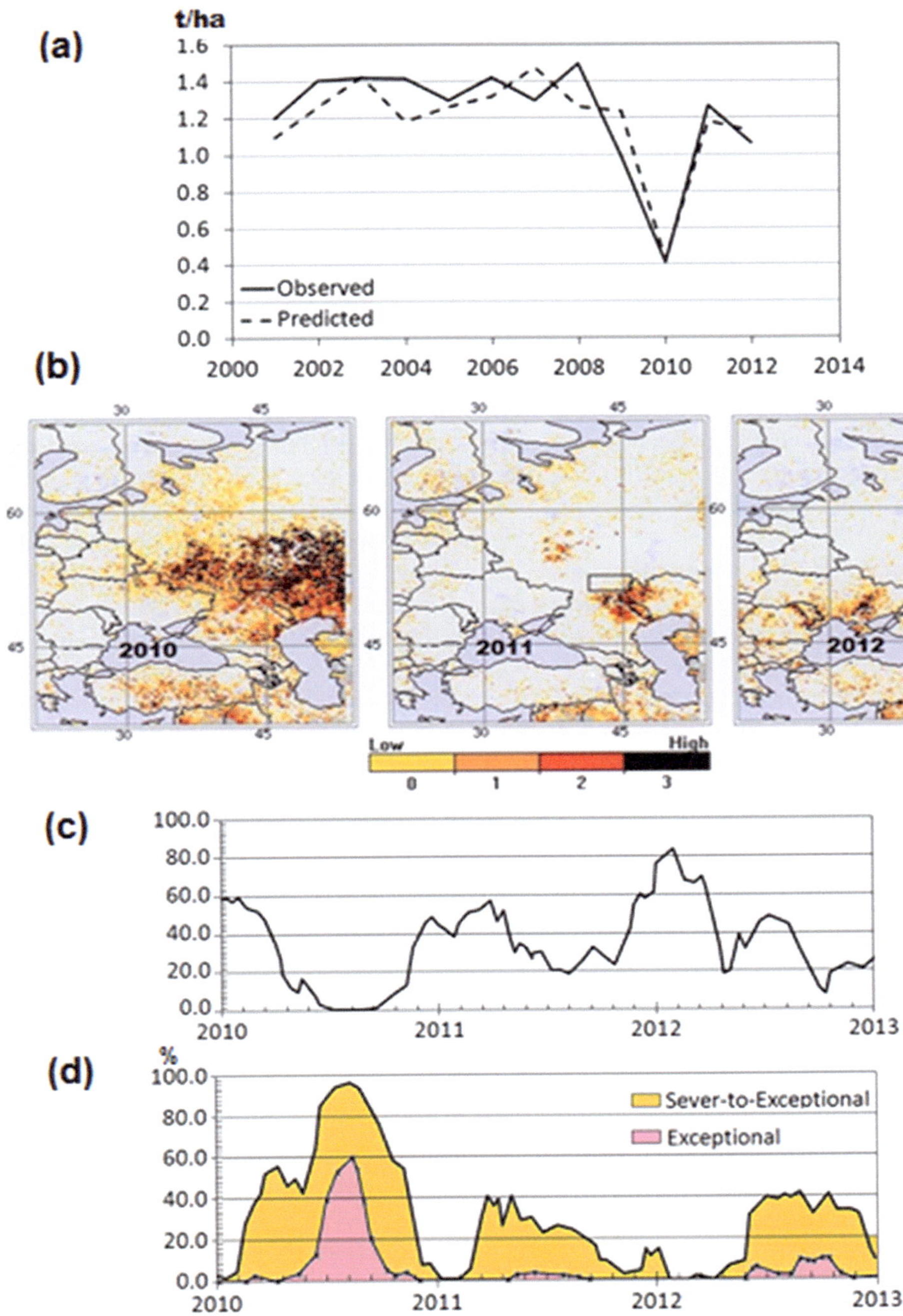

Fig. 6.16 Russia's Saratov oblast (**a**) cereal yield observed and independently modeled (2001–2012); (**b**) 2010–2012 drought area and intensity in European Russia, (**c**) dynamics of TCI-based thermal stress in 2010–2013 and (**d**) VHI-based drought area and intensity (%) in 2010–2012

frequent droughts is incompatible with agricultural goals, limiting agricultural production. Moreover, in recent years, the weather network shrank considerably and the quality of observations deteriorated, which jeopardized the environmental services for crop production prediction. Pastures occupy 87% and crops 13% of Kazakhstan's total area. Following USDA/FAS (2010) report, Kazakhstan is an important producer and exporter of high-quality wheat. Average annual wheat production is around 13 million tons. However, due to weather variations wheat output might change from 10 and 17 million tons. Following the production changes, wheat exports (mostly to Europe, Russia, northern Africa and central Asia) fluctuates between two and eight million tons (WB 2017).

The climate of Kazakhstan is arid and semiarid. Annual precipitation changes from nearly 50 mm in southern desert ecosystem to 300–450 mm in the northern steppe zone; only a small area in the southeastern mountain receives around 1000 mm of rains. Annual potential evapotranspiration in the northern principal agricultural areas of Kazakhstan is 500–700 mm, exceeding the amount of precipitation almost twofold (Gol'tsberg 1972). Most of Kazakhstan's crops and pastureland are located in arid and semiarid zones. Rainfall fluctuates strongly during the growing season and from year to year, putting additional constraints on agriculture. Drought is the most typical phenomenon of the Kazakhstan climate, occurring every 2–3 years. Extreme droughts occur every 7–10 years, leading to considerable losses (more than 60%) in agricultural production. Therefore, such dry climate conditions are incompatible with Kazakhstan's agricultural goals and there is a strong need for an advanced prediction of wheat production for planning exports, grain availability in order to avoid food insecurity. Annual assessments of spring wheat production have been done in the past based on weather data (Ulanova 1975). However, in the past 20 years, the weather network, which has provided data for assessment of weather impacts on wheat and pastures in Kazakhstan, shrank considerably. Currently one weather station is available for 37,000 km^2 of Kazakhstan land (CIA 2017). Besides, the quality of weather observations deteriorated, jeopardizing crop-weather assessments (USID 2015). Therefore, satellite-based VH indices were tested in Kazakhstan as the predictors of wheat conditions and yield.

The ground data for VH indices testing included plant density (PD) measured in spring wheat fields at six weather stations from 1985 through 1994. The PD measurements included a calculation of a number of stems per 1 m^2 area in four places of the middle of the selected field, every 10 days during the growing season (WMO 1972). The PD for the field was calculated from the four 1-m^2 samples. When wheat plants appear after planting, each plant consists of one stem, but density increases when the plants grow additional stems (shooting phase), more plants emerge in moist conditions and decreases with dry weather. Following the shooting (or tillering stage), each plant is potentially able to produce up to six additional stems (grown from the nodal buds), and each of the shoots is potentially able to produce an ear with grains. The number of additional stems and their productivity depend on moisture and thermal conditions; good moisture supply and warm temperature stimulate stems' appearance, survival and higher PD. Therefore, the PD is an indirect measure of vegetation productivity. Since PD depends on moisture and thermal conditions prior to and during the shooting stage, this parameter is an appropriate indicator of weather impacts on wheat biomass and production (Paulsen 1978). During the 1985 through 1994 growing period, wheat's PD was measured each 10 days as the mean number of plants per square meter, characterizing wheat density of the field. Since the goal was to compare PD with VH indices, the PD was normalized using the approach similar to VCI and TCI approximation (equation below). We used multiyear median values of density for each 10 day period of the growing season (PDmed) and also multiyear minimum (PDmin), and maximum (PDmax) values (Eq. 6.29).

$$\mathrm{PDcond} = \left(\mathrm{PD} - \mathrm{PDmed}\right) / \left(\mathrm{PDmax} - \mathrm{PDmin}\right), \tag{6.29}$$

where PDcond—PD condition, PDmed—1985–1994 PD medium for each 10 days of measurements, PDmax and PDmin—1985–1994 PD maximum and minimum for each 10 day, respectively.

Figure 6.17 shows seasonal variation of wheat's PDcond and VCI for six stations in 1991 and 1992. As seen, there is a good match between the temporal behavior of the VCI and PDcond for all ecosystems and for both favorable (VCI > 60) and stressed (VCI < 40) vegetation conditions. Station 1 shows the closest correspondence between VCI and PDcond for both years. It should be noted that these years had quite different weather. The year 1991 had severe drought, which was reflected when VCI was below 40 and PDcond below 20, while during mostly favorable weather in 1992, both the VCI and PDcond were very high, 70–90 and 30–60, respectively. It is notable that the dynamics of the VCI and PDcond were similar, changing from a reduction during weeks 18–22 to an increase thereafter. In general, the dynamics of the VCI and PDcond matched quite well. The only two stations where some shifts occurred in 1991 were stations 2 (weeks 22–28) and 3 (weeks 25–33). Correlation analysis of the diagrams confirms this relationship with individual stations $R^2 = 0.7–0.8$ with exception of station 5 ($R^2 = 0.47$), where both VCI and PDcond were high with low variation inside and between the years. Following these relationships, VCI values around 50 characterize near-normal vegetation conditions. They corresponded to a near-zero deviation of PDcond from the multi-year median values. VCI below 30, which specify strong drought conditions (Kogan 1995), corresponded to a reduction of PDcond below 20. The highest PDcond recorded in this study was around 60, which corresponded to a VCI around 15, indicating exceptional drought. For VCI over 60, the density of the vegetation exceeds the median value, indicating that conditions are

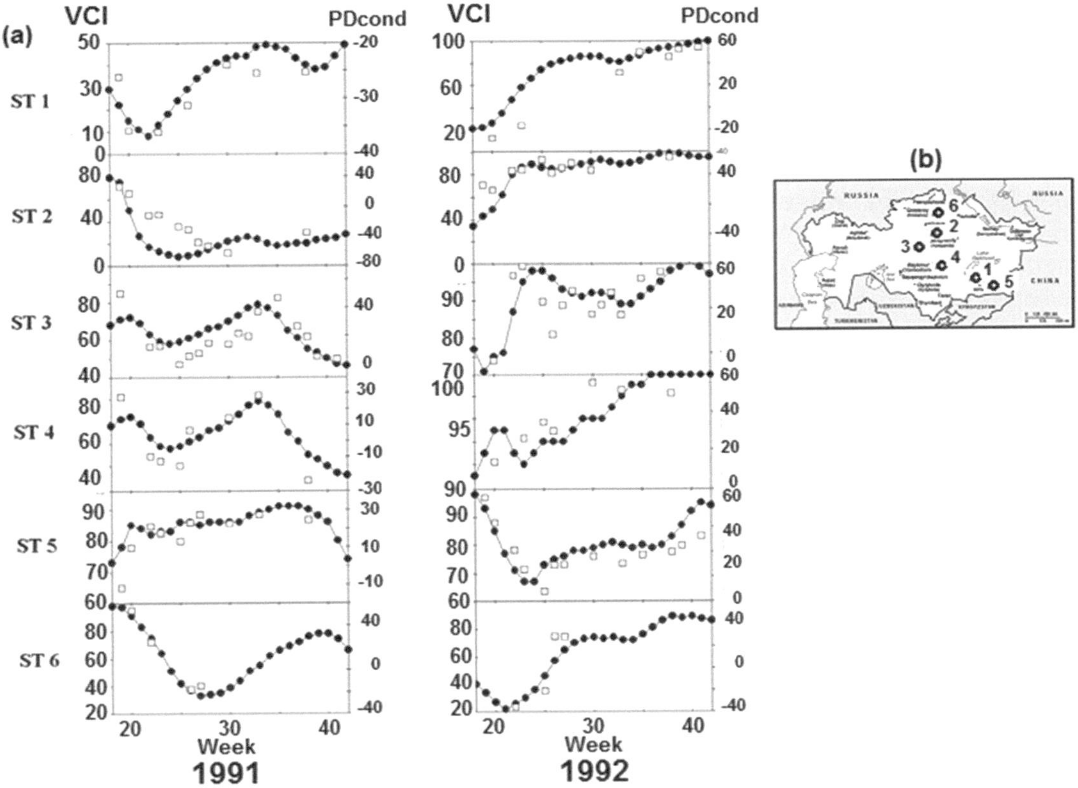

Fig. 6.17 Seasonal dynamics of 10 day PDcond (squares) and weekly VCI in 1991 and 1992 (black line with circles) for six measuring stations in Kazakhstan

favorable for the development of healthy vegetation. Finally, it should be emphasized that selected stations were located in different climatic and ecological zones and had considerable variation in elevation (200–700 m) and NDVI values (0.05–0.47), the conditions, especially those related to drought impact.

Although selected stations were located in different climatic and ecological zones, with different elevations (200–700 m) and considerable difference in NDVI values (0.05–0.47), the all-stations' VCI versus PDcond correlation was quite strong (Fig. 6.18a) with $R^2 = 0.76$ a Standard Error of PDcond estimation less than 16% (considering that the variation in vegetation density was large (between −60 and +70). The RMSE was smaller for low density crops, indicating a higher accuracy in estimation of VCI-derived drought conditions. Moreover, mean Kazakhstan wheat yield (Y) collected from the Central Statistical Administration was well correlated with area mean summer VCI during 1990–2005 (Fig. 6.18b). A big difference between Y and VCI in 1998 was due to cooler temperatures (TCI = 70). Therefore, at the background of considerable moisture deficit (VCI = 22), wheat yield reduction was not large enough.

6.3.10 Corn in Zimbabwe

In the recent three decades, frequent and severe droughts have become major climate disasters through many countries in Africa affecting strongly regional and global food security. For example, very severe and widespread droughts affected the region in 1980s and 1990s (1982–1984, 1991–1992, 1994–1995). Beyond the strongest drought episodes, there were a number of moderate droughts (1986–1987 and others). Following SADS (1992), one of the worst droughts affecting southern Africa in the past 30–40 years occurred in 1991–1992, when 18 million people (16% of the region) were at risk of starvation in 1992 (WMO 1994; Wilhite 1993; US Government 2002). Moreover, the damage to the environment was extensive and the death toll of livestock and wildlife was unprecedented. Increasing cost of droughts in the twentieth century, hunger and malnutrition in African and Asian countries, attracted unusually huge attention of international organizations and governments of developed nations. How to predict, and if not, at least to monitor an early droughts detection, estimate area, intensity, and duration and provide 1–2 months advanced warning of drought-related reduction in agriculture, predict food shortages, and finally assess food security situation in order to prevent malnutrition and hunger (Goldenberg 2012; WMO 1994). Because of some military, political, and economic conflicts in Africa, the weather network for monitoring drought and impact assessment on agriculture and specifically grain production (the major foodstuff) has been steadily declining making drought early detection and monitoring from the measured rainfall and temperature data not

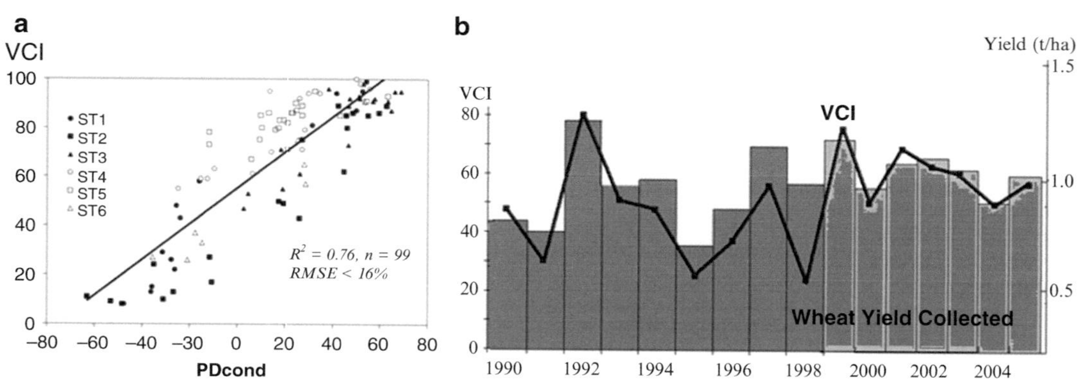

Fig. 6.18 Kazakhstan, spring wheat (**a**) Correlation of 10 day PDcond for all six stations with weekly VCI (collected in the area around the stations) in 1991 and 1992; (**b**) Time series of Kazakhstan wheat yield and mean summer (June to August) area-average VCI for the wheat-growing area

always reliable. Therefore, satellite-based VH data were tested in Zimbabwe, which is one of the main grain producers in southern Africa (USDA/SCS 2002).

Corn and wheat are the two most important grain crops in Zimbabwe. District average corn yield data for the country from 1985/1986 through 1993/1994 were used for VH-based modeling production losses from droughts in advanced of harvest. The corn yield data were collected from small scale communal farming sector of 58 country's districts. Corn accounts for nearly 60% of all cultivated land each year. However, what requires special emphasis for food security is that corn production accounts less than 60% of its requirements in Zimbabwe. The average size of a corn field is typically less than a hectare and fields are scattered throughout the district. Therefore, a very high-resolution operational data is required for drought detection and its impact on corn production. Unfortunately, most of the major corn planted areas are situated in low rainfall and low soil fertility locations, which are considered as semiarid and/or prone to drought. The rainy season in the country spans from October through April of the following year, although the major crop season in terms of crop response to the environmental conditions is between November and March of the next year (Kogan 1997). The vegetation and crops in the region follow the climate pattern quite closely unless human activity has altered the plant cover (USDA/SCS 1993).

Zimbabwe's corn yield for the administrative regions during 1985/1986–1993/1994 was investigated statistically to derive first, an existence of technological trend. This investigation showed that the majority of Zimbabwe's districts multi-year corn yield have a flat technological trend (no increase-decline). Therefore, weather-related yield fluctuation (yield anomaly, d*Y*) were derived as a deviation of annual corn yield from the 1986 to 1994 mean corn yield values. Second, weekly 4 km^2 VH indices (VCI and TCI) were averaged over each of the 54 administrative province and third, correlated d*Y* with 9 year corn yield anomalies. The critical week in the corn yield highest response to the weather in Zimbabwe, was identified from the highest value of d*Y*-VCI and d*Y*-TCI correlation. An example of such correlation time series is shown for one of the main corn districts, Chegutu in Fig. 6.19. This district has a very typical corn growth period (early November, week 19 through mid-June of the next year, week 50). It is important to indicate that for VH assessments in agriculture, annual year for the Southern Hemisphere starts from the 1st week in July, as week number one, (the year of corn planting) and ends up in week 52 at the last week in June of the next year, when corn has been harvesting. Following the figure, two critically important periods in the corn life cycle are identified when corn d*Y* correlates strongly with VCI and TCI. The first period is January to February (weeks 28–34) when corn is extremely sensitive to thermal conditions (correlation d*Y*-TCI is 0.96–0.98) during the period of an intensive photosynthetically active biomass accumulation. The second phase of the highest correlation between d*Y* and VCI is in late March (week 38), when sensitivity of corn to water stress during the reproductive phase increases considerably. Weather conditions in the district #5 during these years varied from favorable in 1988–1989 to extremely unfavorable (a very intensive and long drought) in 1991–1992. In most of the 54 provinces (Fig. 6.20), PCC

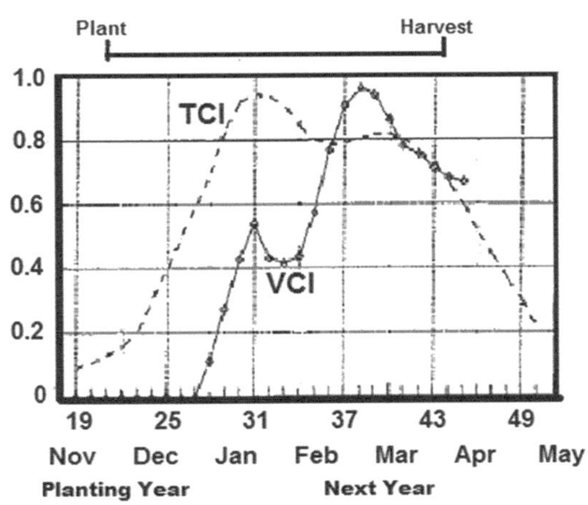

Fig. 6.19 Correlation of corn d*Y* (% deviation from the 1986 to 1994 regional mean yield) with weekly VCI and TCI for Chegutu, (#5) district during the growing season in Zimbabwe

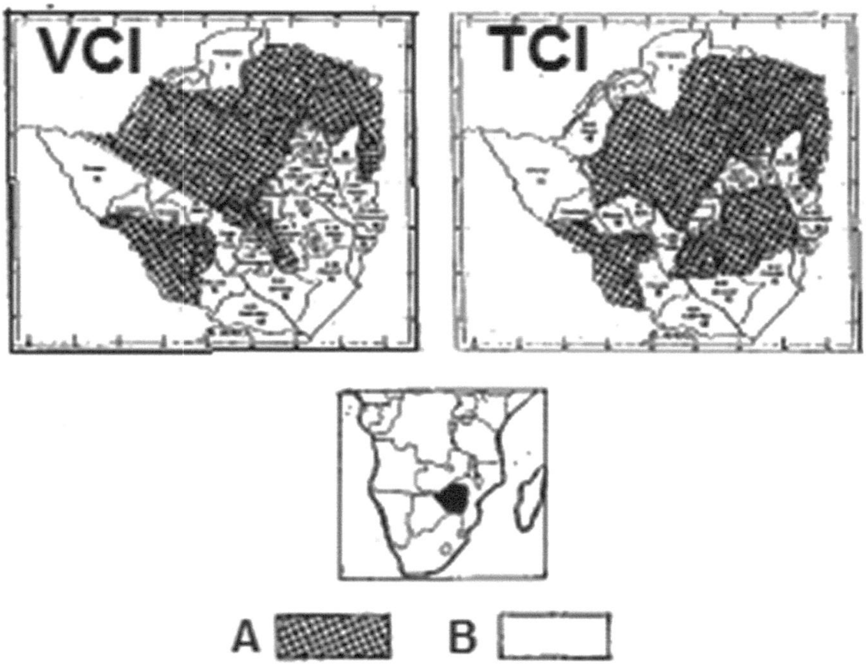

Fig. 6.20 Correlation between regional corn dY (% deviation from the 1986 to 1994 regional mean yield) and weekly VCI and TCI for three provinces of Zimbabwe, Africa (**a**) PCC = 0.75–0.95 and (**b**) PCC = 0.40–0.74)

between dY and the indices during the week of the highest (critical) requirements of corn yield to moisture and thermal conditions is very strong (PCC = 0.75–0.95). The correlation degrades in regions where corn area is small and/or environmental resource is very limited for successful corn farming in Zimbabwe. Both VH indices were highly correlated (0.72–0.95) with corn's dY in districts 5, 13, and 23. However, for the southern districts, where tea and coffee are major crops the correlation turned to be much smaller (0.3–0.5).

Zimbabwe models presented in this subchapter were developed for the four major communal districts (Chegutu (#5), Guruve (#11), Mt. Darwin (# 13), and Gokwe (#23)) to verify how satellite-based VH data can be used for an advanced (6–8 weeks) prediction of corn yield, especially for drought-related corn yield losses, and for assessments of predictions reliability. Significant drought fluctuations in Zimbabwe usually results in drastic changes of vegetation greenness, vigor, biomass, and finally yield. Following correlation analysis, in 19 out of 41 districts of Zimbabwe, dY correlated stronger with TCI, in 10 districts—with VCI and for the remaining 12 the statistical differences were insignificant (Unganai and Kogan 1998) (Fig. 6.21). Using regression approximations, corn yield models were developed and shown below for the four major regions (5, 11, 13, and 23). Two model variants were investigated: (a) dY was expressed as a function of the indices for the week of the highest weekly correlation; and (b) dY was a function of the indices for 2–3 the highest correlated weeks, when these weeks were combined into one variable (for each index) based on PCC-derived weights. Equations 6.30–6.33 and 6.34–6.37 represent these cases, respectively. For the other districts the models are in Unganai and Kogan (1998) (Fig. 6.21).

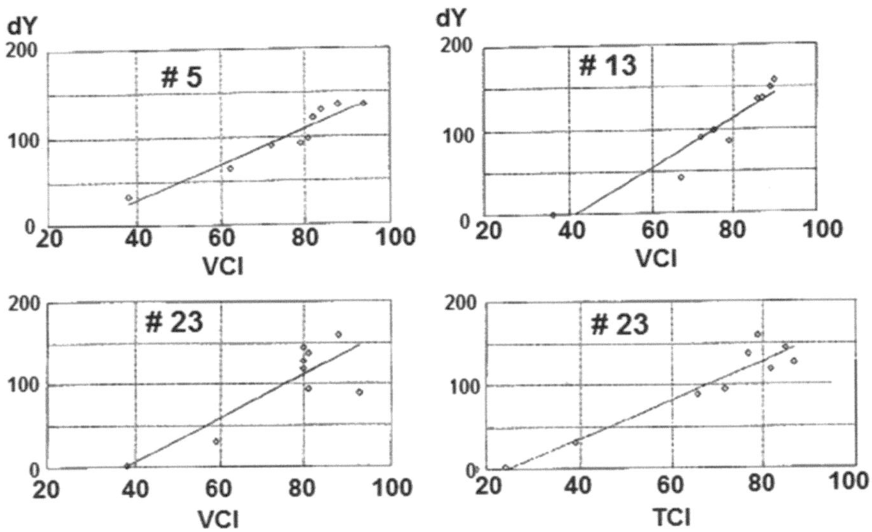

Fig. 6.21 Scattered plot and best fit of corn d*Y* (% of 1986–1994 mean yield) versus VCI and TCI at the weeks of peak correlation in Zimbabwe

$$\text{Chegutu}(5) \qquad dY = -70.16 + 1.41 \times \text{TCI}_{31} + 0.96 \times \text{VCI}_{37} \qquad R^2 = 0.92 \tag{6.30}$$

$$\text{Goruve}(11) \qquad dY = -22.75 - 0.81 \times \text{TCI}_{31} + 2.12 \times \text{VCI}_{39} \qquad R^2 = 0.75 \tag{6.31}$$

$$\text{Mt.Darvin}(13) \qquad dY = -39.57 + 0.53 \times \text{TCI}_{31} + 1.06 \times \text{VCI}_{38} \qquad R^2 = 0.91 \tag{6.32}$$

$$\text{Gokwe}(23) \qquad dY = -38.06 + 1.24 \times \text{TCI}_{30} + 0.96 \times \text{VCI}_{37} \qquad R^2 = 0.74 \tag{6.33}$$

$$\text{Chegutu}(5) \qquad dY = -31.65 + 0.45 \times \text{TCI}_{30-32} + 1.44 \times \text{VCI}_{36-38} \qquad R^2 = 0.92 \tag{6.34}$$

$$\text{Goruve}(11) \qquad dY = -29.74 - 0.11 \times \text{TCI}_{31-32} + 1.96 \times \text{VCI}_{37-38} \qquad R^2 = 0.72 \tag{6.35}$$

$$\text{Mt.Darvin}(13) \qquad dY = -84.57 + 0.52 \times \text{TCI}_{29-31} + 1.40 \times \text{VCI}_{35-37} \qquad R^2 = 0.91 \tag{6.36}$$

$$\text{Gokwe}(23) \qquad dY = -66.65 - 1.04 \times \text{TCI}_{29-31} + 1.96 \times \text{VCI}_{35-37} \qquad R^2 = 0.67 \tag{6.37}$$

Comparison of Eqs. 6.30–6.33 with the corresponding 6.34–6.37 indicates no advantage in integrating the indices for the neighboring weeks of the highest correlation. Following models 6.30–6.33 it should be concluded that multiple regression models that use TCI and VCI at their respective weeks of the highest correlation with d*Y* explains most of weather-related corn yield variations in Zimbabwe. These models can be used for prediction of corn yield in response to weather variations, which are especially important for frequent cases drought-triggered yield losses. The predictions might be done 6–8 weeks in advance of corn harvest. Such an early prediction allows for estimation of food shortages early enough, to arrange delivery and prevent malnutrition and hunger. These predictions provide estimates of food security situation for international organization and some ideas on the amount of food needed and timing.

6.3.11 Other Countries' Crops

Crop modeling research has started at NOAA in the mid-1980s (Van Dijk 1986; Kogan 1989, Le Comte et al. 1988). The NDVI was used during the critical period of crop growth to estimate crop yield. The regression models correlated yield in a country, with NDVI. First it was noticed that during wheat-growing season NDVI is normally reaches the maximum values at about the beginning of crops' reproductive stage and starts to decline after this stage when the crops accumulated the largest portion of dry matter (more than 50%). When nearly 90% of dry matter is accumulated in the plants, NDVI declines to less than half of its maximum values (Van Dijk 1986). Temporal NDVI coverage is an important determinant of the relationship between biomass and yield. Therefore, NDVI during the critical period of crop development can be used as yield indicator.

Burkina Faso (*Africa*). The work was done with *sorghum and millet*. Figure 6.22a shows correlation of these two crops' yields in 1983–1984 with 16 km² NDVI, collected for ten administrative regions (AR) in the area of each crop's location and averaged over the region. The timing of NDVI collection was 9 weeks after sorghum and 7 weeks after millet green-up, which is the critical period of these crops response to weather. The ARs' yields (*Y*) were correlated versus ARs'

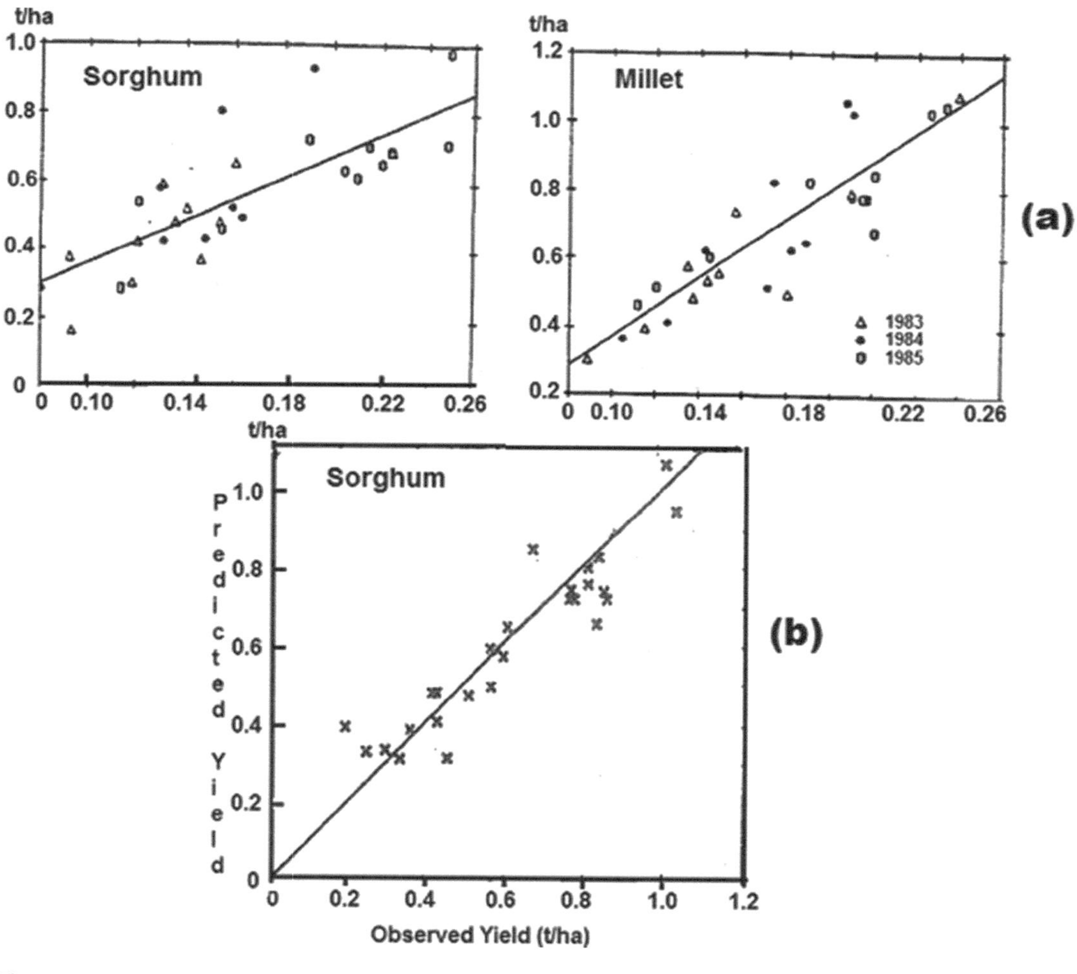

Fig. 6.22 Correlation of (**a**) sorghum and millet yield against NDVI in Burkina Faso during 1983–1984 and (**b**) model-predicted versus observed sorghum yield in 1985 (based on 1983–1984 model)

NDVI (Fig. 6.22) and the results were approximated by the following equations: for sorghum $Y = 0.046 + 0.0032 \times (\text{NDVI})$, $R^2 = 0.63$; for millet $Y = 0.042 + 0.0042 \times (\text{NDVI})$, $R^2 = 0.79$. An independent evaluation of these results (observed versus predicted) for sorghum in 1985 are shown in Fig. 6.22b (Le Comte et al. 1988).

Bulgaria (*Europe*). The demand for cost-effective tools to reduce weather-related yield losses is increasing presently in Bulgaria's agriculture. In 2016, Bulgarian scientists, together with NOAA, investigated the potential of early forecasting *corn* yield using datasets from Bulgarian long-term field experiments and satellite-based VH indices (Kogan et al. 2016). This research was interesting since crop data were experimental obtained from the Bulgaria Institute of Soil Science. The experiments were carried out with two typical corn varieties in eight representative agroclimatic regions of Bulgaria over the period 1975–1991 (Stoyanov 2008). The fields were fully irrigated at five levels of fertilization, ranging from 0 to 125% of the optimum rate, with names: ***B1*** (0), ***B2*** (125%), ***B3*** (100%), ***B4*** (75%), and ***B5*** (50%). The experiments provided crop data time series on annual corn planted *semi early* ***A1*** and *late* ***A2***. The yield and VH data from NOAA/NESDIS (2018) were collected for Gorny Dabnik (Pleven Administrative region) experimental field only (northern Bulgaria) during 1982–1991. Two technological experiments were tested *A1B1* (early planting, no fertilizers) and *A2B1* (late planting, no fertilizers). The standard d*Y*-VH procedure (Sect. 6.3) was used for modeling. The models below were developed for *A1B1* ((6.38) and *A2B1* (6.39) experiments.

$$dY = 0.440 - 0.0102 \times \text{VCI}_{17} - 0.0013 \times \text{VCI}_{23} + 0.0223 \times \text{VHI}_{16} - 0.0002 \times \text{TCI}_{21} \tag{6.38}$$

$$dY = 0.933 + 0.005 \times \text{VCI}_{22} - 0.003 \times \text{VCI}_{32} + 0.007 \times \text{VCI}_{41} - 0.003 \times \text{TCI}_{17} \tag{6.39}$$

Figure 6.23 displays correlation of predicted (pY) versus observed (oY) yields in dependent test (Fig. 6.23a) and 1981–1991 time-series in independent test for two technological treatments (*A1B1* and *A2B1*). The pY versus oY relationship for dependent and independent testing shows strong correlation (PCC = 0.95 and 0.84, respectively) for both experiments. However, correlation for *A1B1* experiment (an early planting) is stronger compared to late planting (*A1B2*) for independent test (PCC = 0.68 compared to 0.63, respectively).

Mongolia (*Asia*). Mongolia is a country of nomadic livestock husbandry. Its economy depends much on livestock products. In the past 20 years, livestock husbandry has expanded following a 38% increase in the number of domestic animals (FAO 2016). This trend is expected to continue. Mongolia's livestock production depends much on the productivity of natural grassland. The major obstacles limiting grassland productivity are dry climate with frequent droughts, which suppress the development of grass biomass (NAM 1990; Suttie 1999). When drought occurs, farmers move their livestock to less drought-affected areas. This is not a trivial task since information about grass condition in non-affected areas is not available. Therefore, early drought detection and assessment of its impact on biomass reduction are important tasks for livestock survival in Mongolia. Meteorological data are used sometimes for biomass estimation. However, there are only 35 weather-observing stations for an area of 150 million ha of land (on the average 4.3 million ha per station). Considering the dry climate and large spatial variability of summer rainfall in Mongolia, estimation of pasture conditions and biomass availability from meteorological data cannot be accurate for such a large area. Since the 1990s, scientists have been investigating the application of the AVHRR-based NDVI for land classification and monitoring dry/wet conditions in Mongolia (Adyasuren and Bayarjargal 1992; Bayarjargal et al. 2000). In parallel, grassland biomass has been measured from 1985. Since VH indices were successfully used as proxy for estimation of crop yield and pastures

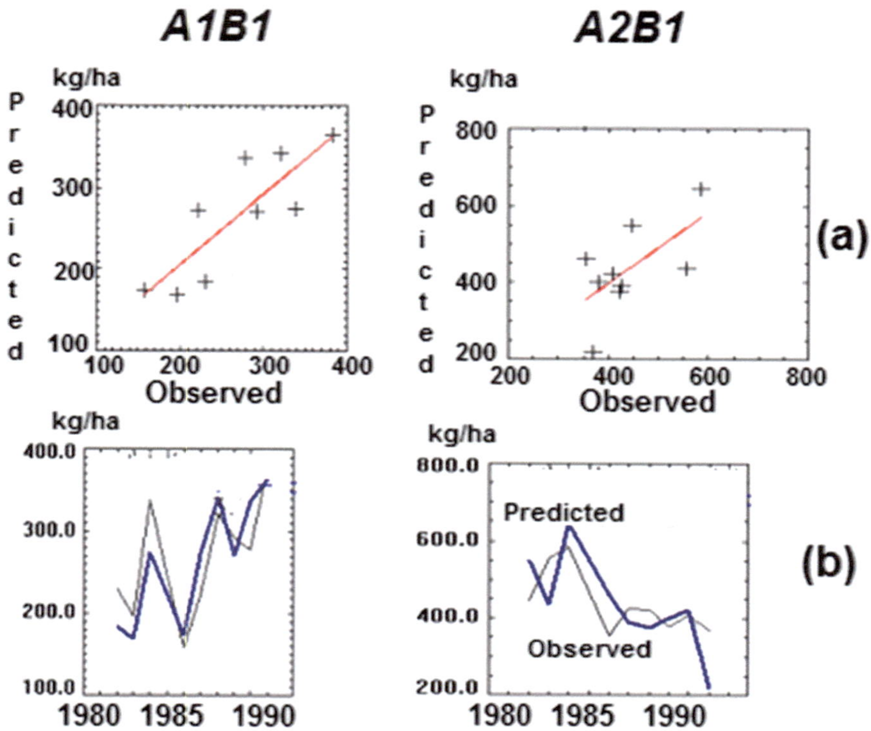

Fig. 6.23 Validation for experiments *A1B1* and *A2B1* (**a**) Correlation of predicted (p*Y*) versus observed (o*Y*) yield in dependent test and (**b**) time series of independent p*Y* and o*Y*

biomass in other parts of the world (Kogan 2001, 2002; Kogan et al. 2003a, b) this method have been applied in Mongolian environment (Kogan 2001).

Pasture biomass has been measured since 1985 at the Tumentsogt Research Station (TRS) located in eastern Mongolia (47.4° N and 112.5° E). The measurements were made at a 1 ha (2.4 acres) experimental field (EF) typical for the TRS location. This EF was fenced to preserve natural biomass growth. The procedure included cutting and weighing above-ground live biomass on the area of 1 m^2 in four corners of the EF every 10 days. The average weight from four 1 m^2 plots was recalculated for the entire 1 ha area and expressed in tons per ha (t/ha). For the following 10 days, measurements were moved to the neighboring locations of the EF. Final biomass for the field was expressed as a percent deviation from the calculated normal. The location of TRS has a typical steppe ecosystem with the dominant *Stipa grandis* species (Adyasuren and Bayarjargal 1992). Mean annual precipitation amount in the area is 250 mm with large variation (160–380 mm) from year to year. Following rainfall seasonal cycle, biomass at the station increases from 0.3 to 1.4 t/ha between May and August, declining thereafter. Amount of biomass fluctuates from year to year coherently with rainfall variation. Similar to other arid areas, above/below mean biomass in Mongolia was associated with wetter/drier conditions. Two cases in Fig. 6.24a indicate lower mean biomass (60–90) in 1986 and higher (110–130) in 1996 were associated with VHI below 40 (drought) and above 60 (wet conditions), respectively. Besides, VHI data correlate with biomass measurements quite well (Fig. 6.24b) with R^2 = 0.63 and standard error 9.85%. From all investigated years, biomass anomaly and VHI mismatched in 1990 because May to August period was extremely wet (rainfall 150–200% of normal) and cool (1–2 °C below normal), which resulted in excessive soil wetness and correspondingly smaller vegetation

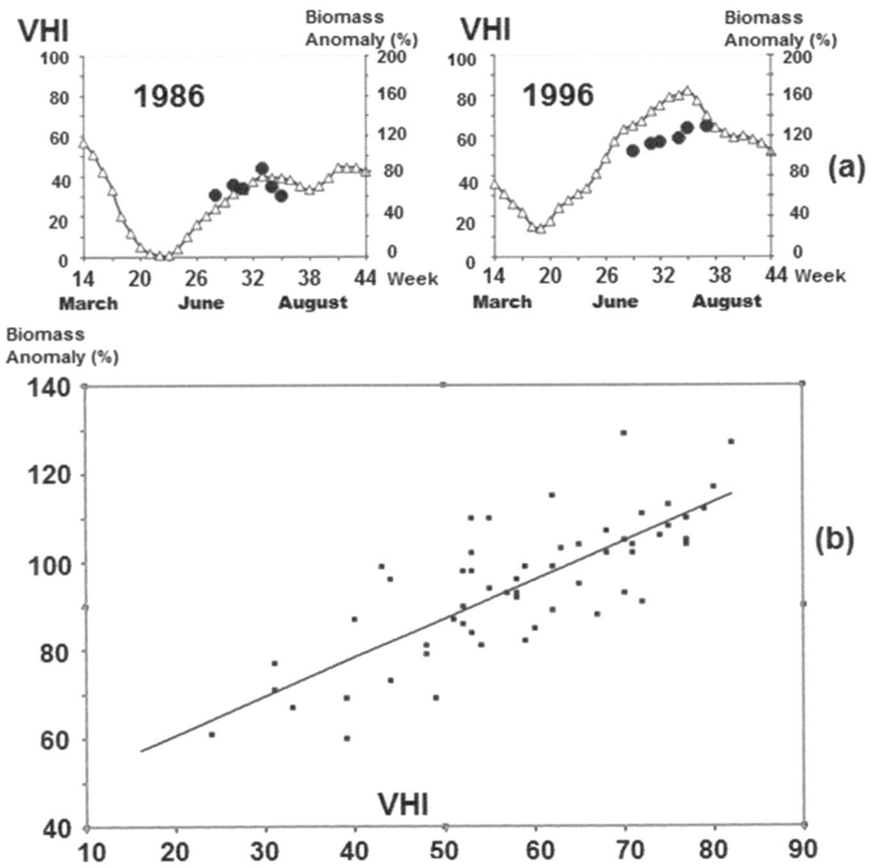

Fig. 6.24 Biomass and satellite data analysis at the TRS experimental field in Mongolia (**a**) Dynamics of 10 day biomass anomaly (filled circles) and vegetation health index (VHI) in dry 1986 (dry) and wet 1996; (**b**) correlation of biomass anomaly and VHI during 1985–1992

biomass (Unganai and Kogan 1998). Considering Mongolia's vast lands, harsh climate, limited weather network, spatial distribution of herds and lack of communication, these results can be used for providing advice to farmers on the quantity of biomass when they search for better grazing places.

India (*Asia*) is the third global largest producer of grain after China and the USA. During the years with favorable weather, India provides 10.5% of world grain production; however, in drought years this contribution might be reduced considerably (FAO 2016). Poor precipitation due to delayed arrival and/or early retreat of the southwest monsoon is considered the chief reason of drought in India. In addition, heat waves also play a crucial role, especially for drought intensification. In the Indian subcontinent, heat waves during the pre-monsoon and high air temperature in the subsequent monsoon season produce vegetation thermal stress, causing deterioration of vegetation health and reduction of grain production (Bandyopadhyay et al. 2015). So, moisture and thermal stress are the two primary parameters reducing grain and other crops production in India (Bandyopadhyay et al. 2016). Therefore, VH indices were tested as the indicators of drought, its impact on crops and prediction of crop losses (Bhuiyan et al. 2017). This test was interesting since kharif yield was calculated not only for grain crops (rice, corn, wheat, etc.) but also for other crops (soybean, groundnut, sugarcane, etc.) planted after monsoon arrival. The research was done in the Gujarat state of

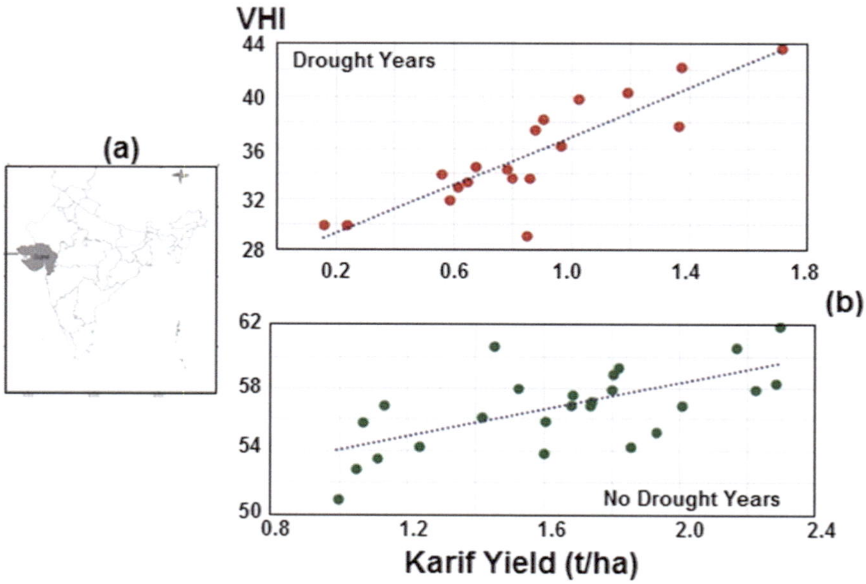

Fig. 6.25 Guitar Province, India (**a**) Location and (**b**) Correlation between mean district level kharif yield and vegetation health index (VHI) for drought and no drought years

India located on the west coast (Fig. 6.25a), covering an area of 196,000 km^2. The climate of the area is arid and semiarid owing to its proximity to Thar, the Great Indian Desert to the north. Rainfall in Gujarat is governed by the southwest monsoon, and occurs mainly during late June to the end of September, which accounts for more than 90% of annual rainfall. Rainfall is unevenly distributed in the state varying from 300–350 mm in the northwest to 600–700 mm in the South Central and northeastern and to more than 1500 mm in the southern Gujarat. Mean annual temperature of the state is 28–33 °C, mean summer (April to June) temperature is 40 °C, and the maximum temperature at many places rises to 48 °C, particularly during heat-wave events. The eastern and southern parts of Gujarat are covered by natural vegetation and agricultural crops (ACD 2012). Annual district statistical data of total kharif rice production (P) and planted area (A) were retrieved from the WEB portal of the Government of India for all the districts of Gujarat for the years 2001–2010 (GOI 2014); yield (t/ha) was obtained as a ratio of P/A.

Kharif yield was correlated with VHI index (Fig. 6.25b) to reflect not only moisture (VCI) but also thermal (TCI) conditions. More importantly, the data were separated to drought and no drought events. As seen in the figure, the correlation is very strong ($R^2 = 0.75$) but only for drought years, for no drought years, correlation also exists but much weaker ($R^2 = 0.47$). This has a reasonable explanation: agricultural practices in India (sowing, harvesting, crop types, cropping patterns, etc.) vary from year to year since it is adjusted by commencement and termination of the monsoon, rainfall spatial variation, and even average gap between 2 consecutive rainy days. These factors affect not only moisture and thermal conditions of vegetation but also VHI values. Therefore, the correlation between yield and average VHI for the cropland area gets attenuated for the non-drought years. It is important to indicate that although precipitation is the chief architect of drought, all the blames cannot be put on deficient rainfall alone as the role of temperature is crucial in India, especially during agricultural (or vegetative) drought, both in positive (cool) and negative (hot) sense. Therefore, monitoring, assessing and forecasting droughts in India solely on the basis of precipitation will be a gamble. In Gujarat, the dynamics of vegetative (and agricultural) drought did not always follow the spatiotemporal patterns of rainfall as is evident from the results of this study. Out of these 9 years of vegetative drought (Fig. 6.25b), thermal stress was acute and widespread in 8 years which was supported by TCI values. Therefore, low crop productivity in India's Gujarat province is associated with both moisture deficit and thermal stress in almost equal proportion. These conditions are derived accurately by VCI, TCI and especially VHI 1–2 months in advance of crops harvest. Such and early predictions of crops losses might be very helpful in determining food security problem and development of measures to reduce malnutrition and even hunger.

Brazil (*South America*) is the second (after the USA) leading soybean world producer country. So, soybean plays an important role in Brazil's economy. Commercial cropping of the soybean area in Brazil has increased substantially in the past 5–10 years reaching currently slightly over 13 million hectares. In the past decade, Brazil has recently emerged also as a major world soybean exporter (the third after the USA and European Union, Jeandson et al. 2017). Brazilian national average soybean yield is about 2.5 t/ha, although there are some areas, where under optimal conditions, yields might be 4.7–5.4 t/ha. In recent years, Brazilian soybean yield is gradually increasing from improved seed, increased fertilizers application and an advanced pest control management. Projections to the year 2020 indicate that Brazilian soybean production will increase to 105 million tons, largely due to Brazil's large expanse of potential arable land to the north that includes the states of Maranhão, Piauí, Tocantins, and Bahia (Jeandson et al. 2017). Soybean is normally planted during end of October to December and is harvested in March to May of the next year (USDA 2017). The tropical climate of Brazil is generally favorable for soybean since warm nights are suited seeds germination and mean daily temperatures around 25 °C provide good conditions to grow crop. Soil moisture is generally sufficient for soybean's seeds germination and for further establishment of seedlings. On a negative side, Brazilian tropical climate does not always provide optimal plant growth, since warm weather increases the spread of plant diseases (USDA 1994). In order to produce dry seeds at harvest time, dry weather is needed, which is not frequent in Brazil. Soybean also suffers if soil is waterlogged, what occurs frequently in some regions of Brazil (Jeandson et al. 2017).

Unfortunately, soybean yield data were available for a short period, 1986–1991, for states Mato Grosso, Goias Mato Grosso do Sul, Pararana, Canta Catarina, Sao Paulo, and Minas Gerais. Since the period was short, instead of trend analysis, weather-related yield fluctuations were estimated as annual deviations of yield from 6 year mean values. Weekly VH indices at 4 km^2 (NOAA/NESDIS 2018) resolution were collected from the first week in July (week # 1) 1986 through the last week in June (week # 52) 1992. All 4 km^2 pixels inside every state were averaged for every week. Finally, statistical soybean dY-VH models for each selected states were developed (Liu and Kogan 2002). From the two VH indices characterizing moisture (VCI) and thermal (TCI) conditions, soybean dY (annual yield deviation from 1986 to 1991 mean yield) was the most sensitive to TCI during the critical period of plant development, growth and formation of the productive organs. January to February is the most critical period of soybean reaction to thermal condition in Brazil when soybean go through flowering and filling stages. This period might shift to December or even to March depending on the state, climate zone and agricultural practice. The models are not shown here, since the period was short (6 years). They are presented in Liu and Kogan (2002). However, statistical estimates for the comparison of simulated and actual yields for the years of extreme losses in soybean production are shown in Table 6.3. In five out of seven administrative districts, R^2 explains a large portion of yield variance (71–92%) with root mean square error (RMSE) less than 10%. Only two regions Mato Grosso do Sul and Parana had smaller R^2 and larger RMSE (although still less than 10%). The independent tests of the largest soybeans losses during the period of this investigation were 1–6% except for Mato Grosso do Sul district in 1986 and 1990.

Table 6.3 Models' statistics and relative error of independent testing soybean yield losses during 1986–1991 in Brazil

Administrative region	R^2	RMSE (%)	Relative error (%)		
			1986	1990	1991
Mato Grosso	0.71	4		−2.9	
Goias	0.86	6		−7.0	
Mato Grosso do Sul	0.54	8	−7.4	−9.4	3.3
Parana	0.59	9	−4.1		5.3
Canta Catarina	0.88	6	6.5		−1.0
Sao Paulo	0.86	4		1.0	−2.5
Minas Gerais	0.92	4	−5.2	1.0	−1.5

6.3.12 VH-Crop Modeling for Food Security: Concluding Remarks

One of the major world tasks is to produce enough food to feed a fast-growing Earth population. Following the current multi-year tendencies, the world population is increasing faster than agriculture can produce. A lack of agricultural production compared to the demands for food is normally intensified in the years with weather-related disasters, specifically droughts. Therefore, crop losses have always been a big concern for farmers, traders, governments, policy makers and international organizations for the purpose of balanced food supply/demands, trade, and distribution of aid to the nations in need, especially in the years with serious crop losses and food security problems. In the twenty-first century, every 2–3 years many countries experienced large-scale crop losses due to a bad weather, accompanied by droughts in most years. Among different types of agricultural products, grain is the major source of food and feed supplies. The amount of grain produced annually depends on yield and planted area. While crop area is relatively stable, yield fluctuates considerably with weather changes from year to year. Among weather disasters, drought plays a major role in large-scale grain losses. In the recent two–three decade, drought-related crop losses have been frequent and they increased in some countries due to climate change. Therefore, it is very important to predict grain yield and production in advance of harvest, helping stabilize food security situation.

In the past three decades, two methods for modeling and prediction of grain and other crops yield have been known. They were based on (a) weather variables and their indices (precipitation, temperature, Palmer drought index, and others), and (b) satellite data and indices, mostly NDVI. In the mid-1980s, a new satellite-based vegetation health (VH) method was introduced to model grain, pasture and other crop yield. The VH method was tested in around 35 countries and showed very good results not only in monitoring crop yield but also in prediction of agricultural production well in advance of harvest. These results have been applied to food security predictions, helping to avoid lack of food and hunger and enhancing food security situation. This chapter demonstrated the application of the VH method for yield modeling using data from polar-orbiting operational satellites. VH-based modeling results will be shown for grain crops in the major grain-producing countries: the USA, China, Russia, Argentina, Australia, Ukraine, and India. Some examples are also shown for other countries, crops, and pasture. The VH method is working quite well because the parameters, this method composited on, are based on biological laws, characterizing vegetation condition and health. Besides, these parameters are distributed frequently over space (0.5, 1, 4, and 16 km^2 pixel resolution) compared to a very low frequency special distribution of weather data.

This chapter shows models, modeling results, and, most importantly, independent model verification. A detailed model analysis for six of the most important grain crops included: corn in China's Jilin province (the largest corn producer), winter wheat, corn and sorghum in the USA's Kansas state (the principal agricultural area), winter wheat (main grain crop) in Ukraine's

Odessa oblast (largest producer), corn in Argentina, wheat in the entire Australia, total cereals in Russia's Saratov oblast, rice in Bangladesh, spring wheat in Kazakhstan, and corn in Zimbabwe. In addition, corn in Bulgaria, rice in India, pasture biomass in Mongolia, sorghum and millet in Burkina Faso, and soybeans in Brazil were also modeled and predictions were investigated. The results of this modeling and independent verification showed that satellite-derived moisture (VCI), thermal (TCI), and combined (moisture–thermal, VHI) conditions are universal predictors of different crops in different grain-producing regions with different climate and ecosystem. Most interesting is that different crops in the same location (for example, winter wheat, corn, and sorghum in the USA) showed similar results of modeling with some differences following biological crops' characteristics. Finally, it is important to emphasize that VH-based crop models provide a very reliable forecast of crop yields and a production of 1–2 months ahead of harvest, which is very important for forecasting food security situation and helping to avoid a lack of food and even hunger.

6.4 Short Summary

One of the most important tasks helping to improve food security situation is an advanced estimation of crop yield and production well ahead of harvest. There are currently two methods for modeling an advanced estimation of crop yield. They are based on weather data and satellite indices. Satellite-based crop yield modeling advanced considerably in the 1990s with the introduction of vegetation health (VH) technology (Kogan 1989, 1997). Regional and country-based yield-VH models have been developed for such grain crops as spring and winter wheat, corn, rice, barley, sorghum, some non-grain crops (potatoes), and for pastures in the USA, Brazil, Argentina, Ukraine, Greece, Morocco, Zimbabwe, China, Mongolia, India, Poland, Bulgaria, and other regions. They showed not only good results of independent verification and 2-month advance assessment of crop production, but also potential for 3–4 months advanced prediction of crop yield. The VH method is working well because the parameters the models are composited of are biologically oriented, distributed frequently over space (0.5, 1, 4 and 16 km^2 resolution) and time (weekly) compared to weather data. This chapter introduces the method for yield modeling with satellite-based VH indices and provides yield modeling results for major grain crops in the principal grain-producing countries such as the USA, China, Russia, India, Argentina, Brazil, Ukraine, and some countries of Africa. Some results are also shown for other crops and pastures. The VH method is working well because the parameters of this method composited are based on biological laws of characterizing vegetation condition. Besides, these parameters are distributed frequently over space compared to weather data.

References

Abdus, S., G. Lindsay, M. Ross, and B. Mlcolm. 2003. A comparison of development opportunities for crab and shrimp aquaculture in southwestern Bangladesh, using GIS modeling. *Aquaculture* 220: 477–494.

ACD (Agriculture and Cooperation Department, Government of Gujarat). 2012. Annual Report 2011–12. http://agricoop.nic.in/imagedefault/ARE2012.pdf.

Adyasuren, T., and Y. Bayarjargal. 1992. Study of vegetation change on the territory of Mongolia using AVHRR and meteorological ground data. *Proceedings of the 13th Asian Conference on Remote Sensing*, Ulaanbaatar, Mongolia, D-9.

AG (Australia Government). 2016. Australia's Reference climate Station Network. http://www.bom.gov.au/climate/change/reference.shtml.

———. 2017. http://www.agriculture.gov.au/ag-farm-food/crops/wheat.

Anderson, M.C., J.M. Norman, J.R. Mecikalski, J.A. Otkin, and W.P. Kusttas. 2007. A climatological study of evapotranspiration and moisture stress across the continental United States based on thermal remote sensing: 2. Surface moisture climatology. *Journal of Geophysical Research* 112: D11112. https://doi.org/10.1029/2006JD007507.

Anwar, A. 1999. Climate change impacts and adaptation assessment in Bangladesh. *Climate Research* 12: 109–116.

Bandyopadhyay, N., C. Bhuiyan, and A.K. Saha. 2015. Temperature extremes, moisture deficiency and their impacts on Dryland agriculture in Gujarat, India. In *Drought: Research and Science-Policy Interfacing*, ed.

J. Andreu, A. Solera, J. Paredes-Arquiola, D. Haro-Monteagudo, and H. van Lanen. London: CRS Press.

———. 2016. Heat waves, temperature extremes and their impacts on monsoon-rainfall and meteorological droughts in Gujarat, India. *Natural Hazards*. https://doi.org/10.1007/s11069-016-2205-4.

Bayarjargal Y., T. Adyasuren, and S. Munkhtuya. 2000. Drought and vegetation monitoring in the arid and semi-arid regions of Mongolia using remote sensing and ground data. http://www.gisdevelopment.net/acrs/2000/ts8/ hami0004pf.

BBS (Bangladesh Bureau of Statistics). 2006. Yearbook of Agricultural Statistics of Bangladesh. Government of the People's Republic of Bangladesh. https://www.unicef.org/bangladesh/2006-08__MICS_2006_keyb_findings.pdf.

Bhuiyan, C., A.K. Saha, N. Bandyopadhyay, and F.N. Kogan. 2017. Analyzing the impact of thermal stress on vegetation health and agricultural drought—A case study from Gujarat, India. *GIScience & Remote Sensing*. https://doi.org/10.1080/15481603.2017.1309737.

Brockwell, P.J., and R.A. Davis. 2000. *Introduction to time series and forecasting*. 2nd ed, 238. New York: Springer.

CIA. 2017. The World Factbook. https://www.cia.gov/library/publications/the-world-factbook/geos/kz.html.

FAO. 2016, Crop production. Rome, Italy. http://fao.org.

Gitelson, A., F. Kogan, E. Zakarin, L. Spivak, and L. Lebed. 1995. *Estimation of Seasonal Dynamics of Arid Zone Pasture and Crop Production Using NOAA/AVHRR Data*. Final Report submitted to the U.S. Agency for International Development, Grant No. TA-MOU- CA13-056, p. 67. Beer Sheva: Ben-Gurion University of the Negev.

———. 1998. Using AVHRR data for quantitative estimation of vegetation conditions: Calibration and validation. *Advances in Space Research* 22: 673–676.

GOI (Government of India). 2014. District Season Crop Production Statistics from 1998. https://data.gov.in/catalog/district-wise-season-wise-crop-productionstatistics#web_catalog_tabs_block_10.

Gol'tsberg, I.A. ed. 1972. *Agroclimaticheskii Atlas Mira (World Agroclimatic Atlas)*. Moscow-Lemingrad: Gidrometizdat. 145 p.

Goldenberg, S. 2012. US drought could trigger repeat of global food crisis. *Guardian*, April 11. http://www.theguardian.com/environment/ 2012/jul/23/us-drought-global-food-crisis.

Hardin, G. 1986. Cultural carrying capacity: A biological approach to human problems. *Biological Science* 36: 599–606.

Hayas, M.J., and W.L. Decker. 1996. Using NOAA AVHRR data to estimate maize production in the United States Corn Belt. *International Journal of Remote Sensing* 17: 3189–3200.

HCR (Hydrometeorological Center of Russia). 2017. http://www.meteoinfo.ru/.

Holdren, J.P., and P.R. Ehrlich. 1974. Human population and the global environment. *American Scientist* 62: 282–292.

IPCC4. 2007. *Climate Change 2007: Synthesis Report*, 104. Geneva: Intergovernmental Panel on Climate Change https://www.ipcc.ch/publications_and_data/publications_ipcc_fourth_assessment_report_synthesis_report.htm.

IPCC5. 2014. In *Climate Change 2014, Synthesis Report. 5th Assessment*, ed. R.K. Pachauri and L. Meyer, 151. Geneva: Intergovernmental Panel on Climate Change https://www.ipcc.ch/pdf/assessment-report/ar5/syr/SYR_AR5_FINAL_full_wcover.pdf.

Jeandson, S.V., P.G. Edilma, C.S. Abraão, and P.M. Valderez. 2017. Climatic Conditions and Production of Soybean in Northeastern Brazil. https://www.intechopen.com/books/a-comprehensive-survey-of-international-soybean-research-genetics-physiology-agronomy-and-nitrogen-relationships/climatic-conditions-and-production-of-soybean-in-northeastern-brazil.

Jensen, J.R. 2000. *Remote sensing of the environment: An earth resource perspective*, 154. Upper Saddle River, NJ: Prentice Hall.

Kidwell, K.B. 1997. *Global Vegetation Index User's Guide*. Camp Springs, MD: US Dep. of Commerce, NOAA, National Environmental Satellite Data and Information Service.

Kogan, F. 1986. *Impact of Climate and Technology on Soviet Grain Production*, DELPHIC, 178 pp.

Kogan, F.N. 1987. Vegetation health index for areal analysis of NDVI in monitoring crop conditions. In *Preprint 18th Conference on Agricultural and Forest Meteorology*, 103–114. Boston: AMS.

———. 1989. Remote sensing of weather impacts on vegetation in non-homogeneous areas. *International Journal Remote Sensing* 11 (8): 1405–1419.

Kogan, F.N. 1990. Remote sensing of weather impacts on vegetation in non-homogeneous areas. *International Journal of Remote Sensing* 11: 1405–1419.

———. 1995. Droughts of the late 1980s in the United States as Derived from NOAA polar orbiting satellite data. *Bulletin of the American Meteorological Society* 76: 655–668.

Kogan, F. 1997. Global drought watch from space. *Bulletin of the American Meteorological Society* 78: 621–636.

Kogan, F.N. 2001. Operational space technology for global vegetation assessment. *Bulletin of the American Meteorological Society* 82 (9): 1949–1964.

———. 2002. World droughts in the new millennium from AVHRR-based vegetation health indices. *Eos Transaction of American Geophysical Union* 83 (48): 562–563.

Kogan, F., A. Gitelson, E. Zakarin, L. Spivak, and L. Lebed. 2003a. AVHRR-based spectral vegetation indices for quantitative assessment of vegetation state and productivity: Calibration and validation. In

Photogrammetric Engineering & Remote Sensing, vol. 69, 899–906.

Kogan, F., A. Gitelson, E. Zaharin, L. Spivak, and L. Lebed. 2003b. AVHRR-Based Spectral Vegetation Index for Quantitative Assessment of Vegetation State and Productivity: Calibration and Validation. Natural Resources. 278. Tampa, FL: American Society for Photogrammetry and Remote Sensing. http://digitalcommons.unl.edu/natrespapers/278.

Kogan, F., Y. Banjie, W. Guo, Z. Pei, and X. Jiao. 2005a. Modelling corn production in China using AVHRR-based vegetation health indices. *International Journal of Remote Sensing* 26: 2325–2336.

Kogan, F., B. Yang, W. Guo, P. Zhiyuan, and J. Xianfeng. 2005b. Modelling corn production in China using AVHRR-based vegetation health indices. *International Journal of Remote Sensing* 26: 1–12.

Kogan, F., N. Kussul, T. Adamenko, S. Skakun, O. Kravchenko, O. Kryvobok, A. Shelestov, A. Kolotii, O. Kussul, and A. Lavrenyuk. 2013. Crop modeling in Ukraine based on earth observation, meteorological data and biophysical models. *International Journal of Applied Earth Observation and Geoinformation* 23: 192–203. https://doi.org/10.1016/j.jag.2013.01.002.

Kogan, F., M. Goldberg, T. Schott, and W. Guo. 2015a. SUOMI NPP/VIIRS: Improve drought watch, crop losses prediction and food security. *International Journal Remote Sensing*. https://doi.org/10.1080/01431161.2015.1095370.

Kogan, F., W. Guo, A. Strashnaia, A. Kleshenko, O. Chub, and O. Virchenko. 2015b. Modelling and prediction of crop losses from NOAA polar-orbiting operational satellites. *Geomatics, Natural Hazards and Risk*. https://doi.org/10.1080/19475705.2015.1009178.

Kogan, F., Z. Popova, and P. Alexandrov. 2016. Early forecasting corn yield using field experiment dataset and vegetation health indices in Pleven region, North Bulgaria. *Ecologia i Industria (Ecology and Industry)* 9 (1): 76–80.

Kogan, F., W. Guo, and W. Yang. 2017. SNPP/VIIRS vegetation health to assess 500 California drought. *Geomatics, Natural Hazards and Risk*. https://doi.org/10.1080/19475705.2017.1337654.

Kogan, F., W. Guo, W. Yang, and H. Shennon. 2018. Space-based vegetation health for wheat yield modeling and prediction in Australia. *Journal of Applied Remote Sensing* 12 (2). https://doi.org/10.1117/1.JRC12.026002.

Le Comte, D.M., F.N. Kogan, C.A. Steiborn, and L. Lambert. 1988. *Assessment of Crop Conditions in Africa*. NOAA Technical Memorandum NESDIS AISC 13, Washington, DC. p. 66.

Liu, W.T., and F. Kogan. 2002. Monitoring Brazilian soybean production using NOAA/AVHRR based vegetation condition indices. *International Journal of Remote Sensing* 23: 1161–1179.

McVicar, T.R., and D.L.B. Jupp. 1998. The current and potential operational uses of remote sensing to aid decisions on drought exceptional circumstances in Australia: A review. *Agricultural Systems* 57: 399–468.

———. 1999. Estimating one-time-of-day meteorological data from standard daily data as inputs to thermal remote sensing based energy balance models. *Agricultural and Forest Meteorology* 96: 219–238.

———. 2002. Using covariates to spatially interpolate moisture availability in the Murray-Darling Basin: A novel use of remotely sensed data. *Remote Sensing of Environment* 79: 199–212.

Mo, X.G., S.X. Liu, Z.H. Lin, Y.Q. Xu, and T.R. McVicar. 2005. Prediction of crop production, water consumption and water use efficiency with a SVAT-crop growth model using remotely sensed data on the North China Plain. *Ecological Modeling* 183: 301–322.

NAM. 1990. *National Atlas of Mongolia*. USSR: Academy Press.

Ning, S. 1998. Scientists: water transfer feasibility. *China Daily*, 20 July 1998. Beijing, pp. 9.

NOAA/NESDIS. 2018. Vegetation health indices and products. https://www.star.nesdis.noaa.gov/smcd/emb/vci/VH/index.php

Obuhov, V.M. 1949. *Urozhainost I Meteorologicheskie Factoru (in Russian): Yield and Meteorological Factors*, 314. Moscow: Gosplanisdat.

Paulsen, G. 1978. *Growth and development, Wheat Production Handbook*. Vol. C-529, 1–8. Manhattan, KS: Cooperative Extension Service, Kansas State University.

Rahman, A., F. Kogan, and L. Roytman. 2006. Analysis of malaria cases in Bangladesh with remote sensing data. *American Journal of Tropical Medicine and Hygiene* 74: 17–19.

Rahman, A., L. Roytman, N.Y. Krakauer, M. Nizamuddin, and M. Goldberg. 2009. Use of vegetation health data for estimation of aus rice yield in Bangladesh. *Sensors* 9: 2968–2975.

SADS (Southern Africa Development Community). 1992. *Quarterly Food Security Bulletin*. Regional Early Warning Unit. Harare Zimbabwe.

SAGPYA. 2004. Secretaria de Agricultura, Ganaderia, Pescay Alimentos. Estimaciones Agricolas.

Salam, M.A. 1981. Application of remote sensing and geographic information system (GIS) Techniques for monitoring boro rice area extension in Bangladesh. In *Proceedings of International Seminar on Satellite Technology Applications in Communications and Remote Sensing*. New York: Wiley.

Salazar, L., F. Kogan, and L. Roytman. 2007. Use of remote sensing data for estimation of winter wheat yield in the United States. *International Journal of Remote Sensing* 28: 3795–3811.

Seiler, R.A., F.N. Kogan, and G. Wei. 2000. Monitoring weather impact and crop yield from NOAA AVHRR data in Argentina. *Advances in Space Research* 26 (7): 1177–1185.

Seiler, R.A., F.N. Kogan, G. Wei, and M. Vinocur. 2006. Seasonal and inter-annual responses the vegetation and production of crops in Cordoba Argentina assessed by

AVHRR derived vegetation indices. *Advances in Space Research*. https://doi.org/10.1016/j.asr.2006.05.024.

Shroyer, J.P., D. Whitney, and D. Paterson. 2004. *Wheat production handbook*. Manhattan, KS: K-State Research & Extension.

Snedecor, G.W. 1965. *Statistical Methods*. Ames, IA: The Iowa State University, 360 pp.

Stallings, E.A. 1994. *Hydrometeorlogical Analysis of the Great Flood of 1993*. Silver Spring, MD: Department of Commerce, NOAA, National Weather Service.

Stoyanov, P. 2008. Agroecological potential of maize cultivated on typical soils in the conditions of Bulgaria. Habilitation Thesis, "N. Poushkarov" Institute of Soil Science, Sofia, Bulgaria. p. 168.

Suttie, J.M. 1999. *Grassland and Pasture Crops: Country Pasture/Forage Resource Profile–Mongolia*. http://www.fao.org/ag/AGP/AGPC/doc/Counprof/Mongolia.

Tucker, C.J., and P.J. Sellers. 1986. Satellite remote sensing of primary production. *International Journal of Remote Sensing* 7: 1395–1416.

U.S. Drought. 2012. *The New York Times, Science*. http://topics.nytimes.com/top/news/science/topics/drought/index.html

Ulanova, E.S. 1975. *Agrometeorological Conditions and Wheat Yields*, 298. Arlington, VA: National Technical Information Service, U.S. Department of Commerce, Joint Publication Research Service-66418.

Unganai, L.S., and F.N. Kogan. 1998. Drought monitoring and corn yield estimation in southern Africa from AVHRR data. *Remote Sensing of Environment* 63: 219–232.

US Government. 2002. *Zimbabwe: Grain Production Shortfall Will Continue*. https://reliefweb.int/report/zimbabwe/zimbabwe-grain-production-shortfall-will-continue

USDA. 1994. *Major World Crop Areas and Climate Profile*. Agricultural Handbook No. 664, p. 279. Washington DC: U.S. Dept. of Agriculture, World Agricultural Outlook Board, Joint Agricultural Weather Facility.

———. 2006. *Outlook for U.S. Agricultural Trade*. Washington, DC: USDA. http://www.ers.usda.gov/publications.

———. 2017. *Brazilian Soybean*. https://www.usda.gov/oce/weather/pubs/Other/MWCACP/Graphs/Brazil/BrzSoyProd_0509.pdf.

USDA/FAS. 2010. *Kazakhstan Agriculture Overview*. Commodity Intelligence Report. https://ipad.fas.usda.gov/highlights/2010/01/kaz_19jan2010/.

USDA/SCS (US Dep. Agriculture/Soil Conservation Service). 1993. *Survey Manual U.S. Department of Agriculture Handbook No. 18*. Washington DC: United States Department of Agriculture.

USID. 2015. *Improving the Climate Resiliency of Kazakhstan Wheat and Central Asian Food Security Project*. Climate Change Resilient Development (CCRD). https://www.climatelinks.org/sites/default/files/asset/document/CRW%20Evaluation%20Report%2012-22-15.pdf.

Van Dijk, A. 1986. *Crop Conditions and Yield Estimation Method Based on NOAA/AVHRR Satellite Data*. PhD Thesis. University of Missouri, p. 203.

Verigo, C.A., and L.A. Razumova. 1963. *Pochvennaiya Vlaga Importance in Agriculture (Soil Moisture Importance in Agriculture)*. Leningrad: Hydrometizdat. 289 p.

Voeikov, V.I., and A.I. Gortsev. 2014. *Saratov Oblast*. http://encyclopedia2.thefreedictionary.com/SaratovOblast.

WB (The World Bank). 2017. *Agriculture and Rural Development: Cereal Yield*. https://data.worldbank.org/topic/agriculture-and-rural-development

Wilhite, D.A. 1993. Planning for drought: A methodology for drought assessment, management and planning theory and case study. In *Drought Assessment, Management and Planning Theory and Case Study*, ed. D.A. Wilhite, 87–109. Boston: Kluwer Academy.

Willmott, C.J. 1982. Some comments on the evaluation of model performance. *Bulletin of the American Meteorological Society* 63: 1309–1313.

WMO. 1972. *Twenty Years of WMO Assistance*. WMO Technical Note No. 338. World Meteorological Organization, Geneva. 188 p.

———. 1994. *Quarterly Bulletin*, 98. Geneva: World Meteorological Organization.

———. 2018. Observations. https://public.wmo.int/en/our-mandate/what-we-do/observations.

Xiao, X., S. Boles, S. Frolking, C. Li, J.Y. Babu, W. Salas, and B. Moore. 2006. Mapping paddy rice agriculture in south and Southeast Asia using multi–temporal MODIS images. *Remote Sensing of Environment* 100: 95–113.

Zernoimport. 2014. *Agriculture of Saratov Region*. http://zerno-import.ru/saratov_eng.php.

7 Vegetation Health for Insuring Drought-Related Yield Losses and Food Security Enhancement

7.1 Introduction

Agricultural production is a risky economic activity. In addition to market, institutional, political, and farmers' individual risks, strong production reduction in some years due to adverse weather presents an important source of risk in agriculture. Drought, floods, frost, hurricanes, etc. can partially or completely destroy farmers' harvests endangering farmers' livelihoods. Although farmers generally have a few options for reducing their exposure to adverse weather on its farm, such as using irrigation, drought resistant crops, crops with shorter cycles in order to avoid stressed weather, and others, the risk of reducing capacity of farm might be not sufficient to cope with extreme weather events. In such situations, farmers can effectively replace the damage with insurance payments having a crop insurance contract.

Extreme weather events usually have a systemic occurrence pattern covering large area and affecting simultaneously quite a few agricultural producers. This feature of extreme weather events presents an important prerequisite for the introduction of so-called index-based insurance. In contrast to the traditional yield insurance, index insurance is based not on the actual farm yields, but on an index, characterizing environmental conditions. Accordingly, an indemnity is paid whenever the realized value of the index exceeds or falls below a threshold index value (Skees et al. 1997). In case of weather-based insurance, the indemnification (insurance payment to compensate the farmer's yield loss) is conditioned on the actual realization of a selected weather index. An aggregate yield index is used in the case of area-based yield insurance to detect crop failure. As in the case of index-based insurance yield losses are assessed on the basis of an index (not on the farmer's actual yield), this type of crop insurance allows a substantial reduction of insurance administrative and other costs (because in this case there is no need to monitor single farm yields). In the last decade, several studies have been conducted to evaluate the potential effectiveness of insurance contracts based on weather and area yield-based indices (Skees et al. 1997; Turvey 2001; Vedenov and Barnett 2004; Breustedt et al. 2008; Kellner and Musshoff 2012; Bokusheva et al. 2016; Bokusheva 2011, 2014; Bokusheva and Breustedt 2012).

The effectiveness of index-based insurance products is, however, highly dependent on the availability of reliable and cost-effective information sources. In case of weather-based insurance, a rather sparse meteorological network is often considered a handicap for the development of this type of crop insurance. Additionally, in contrast to the data from a low density of ground weather stations, satellite digital data and images can be used to derive independent and in some

F. Kogan, *Remote Sensing for Food Security*, Sustainable Development Goals Series,
https://doi.org/10.1007/978-3-319-96256-6_7

cases more reliable information about plant-growing conditions because satellite indices are biophysically grounded and high-resolution (0.5 and 1 km^2) indicators. In recent decades, satellite data and images have been increasingly used for agricultural purposes. In the late 2000s, Barnett et al. (2008) discussed the potential for the application of remote sensing data for crop insurance purposes. They suggest that high-resolution remote-sensing data can seriously reduce problems of a high basis risk, i.e., the risk that occurs because of an imperfect correlation between low-resolution weather-based index and farmers' yields, which might result in a mismatch between farmers' actual yield losses and insurance payments because weather station is located far from the field where crop losses are evaluated.

In recent years, good results were obtained when satellite-based vegetation health (VH) indices were applied for insurance purposes. The VH index including the vegetation condition index (VCI), characterizing moisture condition, temperature condition index (TCI), characterizing thermal conditions, and the vegetation health index (VHI), characterizing both moisture and thermal conditions (Kogan 1995, 1998; Unganai and Kogan 1998; Liu and Kogan 1996; Kogan et al. 2003, 2012, 2016), which have been developed in 1981 and used to predict crop yields in different regions with high variations in climatic and ecosystem conditions. Several pilot projects have been implemented to evaluate the effectiveness of index-based insurance by employing satellite-based VH indices in Canada, India, Mali, and Kenya (Hazell et al. 2010; Miranda and Farrin 2012). However, satellite-based VH indices have been used to design insurance primarily for pasture vegetation, which yield is directly related to the quantity of green biomass generated during the growing period. A more recent study by Makaudze and Miranda (2010) developed index-based insurance employing satellite data for maize and cotton. Their study focused on Zimbabwe and was based on time series of the normalized difference vegetation index (NDVI) and maize and cotton yields aggregated at the district level. The authors found that insurance contracts based on the NDVI allowed a higher risk reduction than rainfall-based insurance contracts.

Although the results of many studies have shown that VHI, VCI, and TCI show considerable correlation with observed crop yields, the transition from satellite to ground (yield) data are dependent on data resolution. Low-resolution weather data are not sufficient for crop insurance since weather station might locate far from a farm. Moreover, low-resolution (more than 16 km^2) weather data can differ from farm to farm, as well as for different periods of crop planting and agricultural technology applied (Kogan et al. 2003). In addition, empirical research has shown that the use of aggregated yields might cause serious undervaluation of the farm yield variability and an underestimation of the farm risk (Claassen and Just 2010). Therefore, applicability of satellite-based indices for designing insurance products should be evaluated by using long time series of satellite data and farm yields. The novelty of the present study compared to previous analyses on the use of remote-sensing data for insurance purposes is that for the first time it conducts the analysis of the specially designed satellite-based VH indices in combination with farm yield data. In addition, to model the dependence between crop yields and VH indices, this study employs the copula approach, which, compared to linear correlation, presents a more adequate methodological framework for modeling dependence in the tails of multivariate distributions (Embrechts et al. 2002).

This chapter analyzes the effectiveness of two VH indices for insuring wheat yield losses in Kazakhstan's grain-producing farms. Kazakhstan is an important grain-producing nation. In the period from 2003 to 2012 the national average spring wheat production was 13.8 million metric tons (National Agency of Statistics of the Republic of Kazakhstan 2012). Both the high level of production and a relatively low domestic consumption of 6.2 million metric tons of spring wheat in 2010 (USDA/FAS 2012) make Kazakhstan a key player in the world grain market and are important factors assuring the country's contribution to global food security. However, the national spring wheat production can vary substantially from year to year subject to weather conditions during the vegetation period. Drought presents the main natural hazard to Kazakhstan's grain producers and is an important source of fluctuation in farm income. For example, in 2010, due to an extensive drought, the national spring wheat yield dropped to 0.73 tons per hectare (t/ha), while in 2011, a year with very favorable weather conditions, it rose more than two times, to 1.66 t/ha.

7.2 Copula Approach

The copula approach has been used extensively for modeling multivariate dependence structures in finance and actuarial mathematics. Recently, several applications of the copula approach have been carried out in the field of agricultural economics (Vedenov 2008; Bokusheva 2011; Goodwin and Hungerford 2015). According to Sklar's theorem (Sklar 1959), if F is a joint distribution function with marginal distributions $F_1, \ldots, F_d$, then there exists a copula C: $[0,1]^d \rightarrow [0,1]$ such that for all $x_1, \ldots, x_d$ in $\bar{R} = [-\infty, \infty]$, (eq. 7.1)

$$F(x_1, \ldots, x_d) = C(F_1(x_1), \ldots, F_d(x_d)) \quad (7.1)$$

where $x_1, \ldots, x_d$ is a vector of d random variables forming a joint distribution, and $F_1, \ldots, F_d$ are respective marginal distributions.

Therefore, according to Sklar's theorem, any continuous multivariate distribution can be uniquely described by two parts: the marginal distributions F_i and the multivariate dependence structure captured by the copula C. Sklar's theorem makes the copula approach a powerful instrument for modeling empirical joint distributions. In particular, it allows to reduce an empirical problem of modeling a multivariate dependence structure to the selecting appropriate marginal distributions and fitting an adequate copula. In general, there are many different copula families. In empirical research, however, the most commonly applied copulas are the Gaussian copula, Student's t copula, and three Archimedean copulas: the Clayton, Gumbel, and Frank copulas (Nelsen 1999; McNeil et al. 2005). To design and rate an index-based insurance contract, the approach, presented in Mainik and Schaanning (2012), was employed. In particular, it was assumed that the dependence between crop yields and VH indices might be stronger in the left tail of the distribution than if it is measured for the entire range of the distribution. Thus, in our analysis it was suggested focusing on the expected value of the yield μ (eq. 7.2) given that the satellite-based VH index W falls below its critical level, and in particular, its value at risk VaR at the confidence interval α, VaRα (W), i.e.: insurance contract parameters or potential indemnity payments and fair insurance premium (Bokusheva 2011, 2014; Bokusheva and Breustedt 2012).

$$\tilde{\mu}^* = \tilde{\mu} | w \leq \text{VaR}_{1-\alpha}(w) = E(Y) | W \leq \text{VaR}_{1-\alpha}(W) \quad (7.2)$$

where E is expectation operator.

The expression in (7.2) can be defined in terms of a copula as follows from Mainik and Schaanning (2012), shown in eq. (7.3).

$$H_{\not{y} W=w}^{(Y)} = C_{G(y)F(W=w)}^{(\upsilon)} | G(Y) = \upsilon \quad (7.3)$$

$$F(W) = u$$

where $F(W)$ and $G(Y)$ are marginal distribution of the index and crop yield, respectively. $H_{y|W=w}^{(Y)}$ is the conditional distribution of the yield variable, which can be derived by taking the first derivative of cupola with respect to u, i.e., (eq. 7.4)

$$C_{G(y)F(W=w)}^{(\upsilon)} = \partial / \partial u \times C(u, \upsilon) \Big|_{F(W)=u}, \quad (7.44)$$

To consider all realizations of the index variable below its VaR (7.2) has to be rewritten as in the eq. (7.5).

$$C_{G(y)F(W) \leq 1-\alpha}^{(\upsilon)} = 1 / P(F(W) \leq 1-\alpha) \int_{l=0}^{1-\alpha} \partial / \partial u \times C(u, \upsilon) | F(w) = l^{\partial l} \quad (7.5)$$

$$G(Y) = \upsilon.$$

The expression in (7.5) allows to determine the conditional distribution of the yield variable in terms of a copula and the marginal distribution of the VH indices. By integrating the expression in (7.5) over the interval [0,1] of the yield marginal distribution allows to obtain the yield distribution quintile corresponding to μ . Thus, μ can be determined and successively applied for the derivation of insurance contract parameters (Bokusheva 2014). Fair insurance premium does not incorporate any loading related to administrative costs of insurance companies, etc. and is calculated as the expected indemnity value.

7.3 Empirical Procedure

Empirical analysis has started by identifying for each county of the study area of Kazakhstan the most critical period of spring wheat vegetation, or the period that is the most important for grain formation and spring wheat yield productivity. This period was determined by identifying the weeks with the highest level of correlation between county-level spring wheat yields with VCI and TCI indices. The investigated period was between the beginning of June to the mid of September. In northern Kazakhstan climate, the period around heading has the highest correlation (0.6–0.8) between wheat yield and weather, which values drop thereafter.

To estimate dependence between the county average spring wheat yields and satellite-based VH indices, four different copula models: Gaussian, Clayton, Frank, and Gumbel survival copulas were employed. While the Clayton and the Gumbel survival copulas assume a strong dependence in the left tail of the joint multivariate distribution, the Gaussian and Frank copulas require radial symmetry of joint distribution, i.e., the same level of dependence in opposite corners of the copula. Since the Gaussian copula is based on near linear correlation concept, the Pearson correlation coefficient was used to model the dependence analysis for this copula. In the second step of the analysis, the indices selected in the first step, which estimated farm level wheat yield-VH indices dependence, were used. Subsequently, the Cramer-von-Mises criterion to select the most appropriate copulas for modeling dependence between farm yield and the selected indices was used. Hypothetical insurance contracts represented by indemnity payments and the insurance premium were specified on the basis of expressions (7.2) and (7.5). In the final step the method evaluated risk reduction effectiveness of insurance contracts, by comparing the risk of wheat yield revenues with and without insurance using the conditional value at risk or a downside risk measure (McNeil et al. 2005).

In our empirical application we define downside risk as probability of the index realization below the third decile ($p = 0.3$) of index marginal distribution. Consequently, the conditional value at risk is computed as:

$$\text{ES} = 1/0.3 \int_{p=0}^{0.3} q_p dp, \text{with} q_p \text{defined as yield distribution} p \text{ quantile}$$

7.4 Data

Satellite data were extracted from NOAA's Global Vegetation Health indices data set (NOAA/NESDIS 2018). NOAA/AVHRR-based weekly VCI (moisture characteristic) and TCI (thermal characteristic) at 16 km^2 were collected for the period 1986–2010. The study was performed in the administrative units of northern Kazakhstan shown in Fig. 7.1. Two oblast (the largest administrative unite in Kazakhstan) Akmola with three counties (the second largest unite): Zelinograd (C1), Atbasar (C2), and Esil (C3) and Kostanay, with two counties: Denis (C4) and Kamysty (C5) were investigated (see Fig. 7.1). Since low-resolution data were not

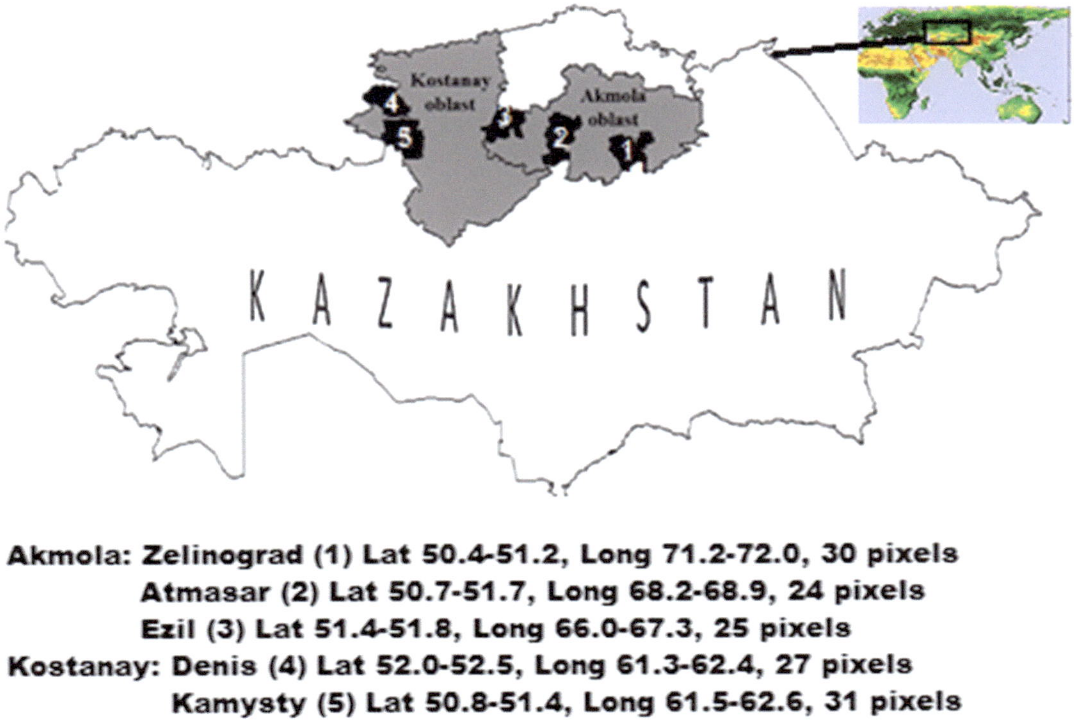

Fig. 7.1 Kazakhstan with the locations of studying counties in the northern Akmola and Kostanay oblast

aggregated for a single farm or small administrative divisions, which included several farms, the VCI and TCI were collected from all AVHRR-based 16 km^2 pixels inside each county. Mean county's values of these indices were obtained averaging all pixels inside rectangular area covering the principal part of each C1–C5 counties (Spivak et al. 2008). Since spring wheat is growing everywhere in each county and is planted in a short period of time (USDA 1994), such aggregation is warranted, since more than 90% of county area is included in the aggregation. Table 7.1 provides the area of aggregation and the number of the aggregated pixels. Spatial variation of VH indices for these pixels has not exceeded 12%, which corresponds to variation of precipitation over relatively small areas of Kazakhstan (Kogan et al. 2003).

Spring wheat is the major crop grown in Kazakhstan's Akmola and Kostanay oblast (first-order country's administration division). According to the official statistics (National Agency of Statistics of the Republic of Kazakhstan 2012), during 2003–2013, wheat occupied 81.6% and 80.4% of the sown areas, respectively. Other important crops in these regions are also grains, such as spring barley and oats. Since 1988, the prevalent crop rotation in these two grain-producing regions has been fallow-wheat-wheat-barley-wheat (Suleimenov and Akshalov 2011). Although Kazakhstan is a large country, the spring wheat calendar is quite similar for many main grain-producing regions (variations in the timing of sowing might result in a difference of 1 or 2 weeks) and from year to year. Kazakhstan's spring wheat is sown in the last 10 days of May, heading takes place at the end of June/beginning July, followed by the ear emergence phase in July, and finally ripening and harvesting in August, or at the beginning of September. In general, spring wheat phenological dates show limited variation between the main grain-producing regions in Northern Kazakhstan.

The county and farm yield data for the same time span were obtained from regional statistical offices. The wheat yield time series were tested

Table 7.1 Summary statistics of 28-year (1982–2010) mean county area-average VH indices and farm yield

Counties	Parameter	SD	Minimum	Maximum
		Farm Yield (q/ha)		
C1, Zelinograd, 12 farms	8.9	3.8	0.2	24
C2, Atbasar, 11 fams	8.8	3.8	0.8	21
C3, Esil, 7 farms	8.3	3.4	1.2	19
C4, Denis, 10 farms	10.7	5.1	0.9	26
C5, Kamysty, 7 farms	9.2	3.9	0.3	22
VCI weeks 22–36		VCI		
C1, Zelinograd	42	22	1.4	96
C2, Atbasar	42	22	0.1	98
C3, Esil	43	24	4.0	98
C4, Denis	41	24	2.1	100
C5, Kamysty	38	22	1.4	100
TCI weeks 22–36		TCI		
C1, Zelinograd	51	24	3.1	98
C2, Atbasar	51	23	1.9	97
C3, Esil	52	26	1.0	99
C4, Denis	54	24	1.3	100
C5, Kamysty	50	24	1.8	99

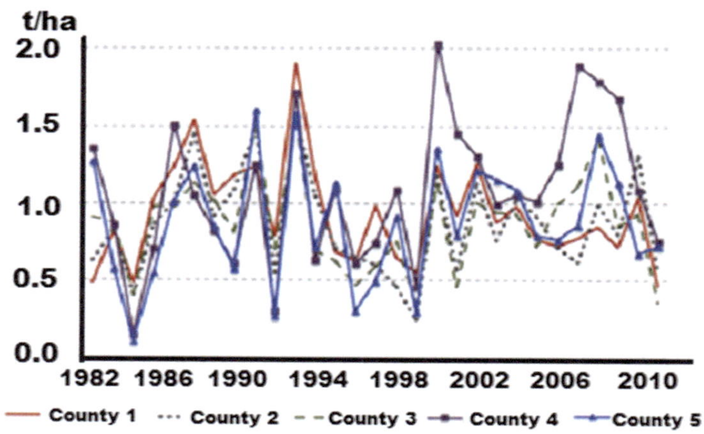

Fig. 7.2 Spring wheat yield for five counties (shown in Fig. 7.1) during 1982–2010 in northern Kazakhstan

for structural breaks and adjusted for a technological time trend using linear, and second- and third-degree polynomial functions. Figure 7.2 shows counties' average spring wheat yields. It exemplifies that severe droughts in 1984, 1989, 1991, 1995, 1997, 1998, and 2010 caused dramatic yield declines in all five of the studying counties. The marginal distributions of wheat yield and both VH indices were modeled using one of the following distribution families: Normal, Log-normal, Logistic, Log-logistic, Gamma, and Weibull distributions. The most appropriate distribution was determined, employing the Kolmogorov–Smirnov test. Summary statistics of index and farm yield data is shown in Table 7.1.

7.5 Results and Discussion

Empirical results reveal a relatively strong relationship in the joint distributions of wheat yields and each of two VH indices. The magnitudes of the Pearson's correlation coefficients (R)

Table 7.2 Copula dependence parameters estimates at county level

County	Weeks	Gaussian	Clayton	Frank	Survival Gumbel
VCI					
C1, Zelinograd	28–31	0.69	1.86	5.42	1.93
C2, Atbasar	29–32	0.78	2.10	5.95	2.05
C3, Esil	29–32	0.64	1.94	2.67	1.79
C4, Denis	28–31	0.64	1.59	4.79	1.80
C5, Kamysty	26–29	0.70	1.97	5.60	1.97
TCI					
C1, Zelinograd	31–34	0.72	1.41	4.34	1.70
C2, Atbasar	31–34	0.75	0.89	4.07	1.45
C3, Esil	29–32	0.64	1.76	2.80	1.79
C4, Denis	25–28	0.71	1.90	5.52	1.95
C5, Kamysty	26–29	0.79	1.94	5.26	1.97

between county level wheat yields and VH indices during 1982–2010 range between 0.63 and 0.77 for VCI, and 0.64 and 0.79 for TCI (Table 7.2). The estimates of dependency between VH indices and county wheat yields are very consistent across all three copula models employed in the study with critical periods coinciding across all four copula models for each county. Dependence parameters for Clayton, Frank, and survival Gumbel copulas have other measurement scales than the Gaussian copula. In the case of the Clayton copula, the dependence parameter is defined in the interval $[-1, \infty)\backslash\{0\}$, for the Frank copula the feasible interval is $[-\infty, \infty)\backslash\{0\}$, and for the Gumbel copula it takes values $[1, \infty]$. Higher values of dependence parameters indicate a stronger dependence. According to estimates, the periods with the highest dependence differ only marginally across single counties in the case of VCI. They span the period between weeks 26 (end of June) and 31 (end of July-first week in August), which correspond to the phenological phases of "ear emergence" and "grain formation," i.e., the two most influential phases in formation of spring wheat yield. However, divergent results were obtained for TCI. In Zelinograd, Atbasar, and Esil counties, TCI is the mostly influential in weeks 29 (mid-July) through 34 (mid-August), while for the other two study counties, in northwest part of Kazakhstan (Esil, Denis), vegetation surface temperature produced different results.

Using the Cramer-von-Mises criterion, the survival Gumbel copula provided the best statistical fits in farms study, modeling farm yields dependence on VCI, whereas the Frank copula was more adequate in most cases, estimating farm yields relationship to TCI. These findings suggest that stronger sensitivity of spring wheat yields to VCI can be modeled more adequately by a copula allowing for an asymmetric dependence, namely with a stronger relationship in the left tail of joint distribution, while the relationship between wheat yields and TCI seems to demonstrate a radial symmetry, suggesting the same extent of dependence in the left and right tails of the joint distribution. Considering these results, only two copula models were used. Estimates of relative risk reduction, presented in the upper part of Table 7.3, show that the study farms could limit benefit when 16 km^2 VCI- and TCI-based data are used for insurance contracts and when calculations are based on county average yields and indices. On average, VCI-based insurance could reduce the downside risk of farm wheat yields by 25% when designing insurance contracts on the survival Gumbel copula estimates. The corresponding estimate obtained employing Frank copulas is 22%. TCI-based insurance contracts seem to provide slightly higher relative risk reductions, 29 and 26% on average for survival Gumbel and Frank copulas, respectively (Table 7.3). However, the performance of both VCI- and TCI-based insurance contracts varies substantially across single counties and farms.

Table 7.3 Downside risk reduction estimates for VCI and TCI (two-index selection alternative)

County	VCI		TCI	
	Surv. Gumbel	Frank	Surv. Gumbel	Frank
County-based index selection				
C1, Zelinograd	0.28	0.25	0.34	0.32
C2, Atbasar	0.28	0.24	0.17	0.14
C3, Esil	0.12	0.10	0.15	0.12
C4, Denis	0.14	0.12	0.39	0.37
C5, Kamysty	0.44	0.40	0.28	0.34
Entire sample	0.25	0.22	0.29	0.26
Farm-based index selection				
C1, Zelinograd	0.57	0.52	0.38	0.34
C2, Atbasar	0.26	0.22	0.17	0.15
C3, Esil	0.18	0.14	0.16	0.12
C4, Denis	0.37	0.35	0.42	0.39
C5, Kamysty	0.52	0.48	0.38	0.24
Entire sample	0.39	0.35	0.30	0.27

The average risk reduction was the highest in Kamysty county in case of VCI-based insurance, while for TCI-based contracts the highest average risk reductions were found for Denis and Kamysty counties. In the next step, the study investigated if the risk-reducing effectiveness of satellite index-based insurance can be improved, selecting indices on the basis of farm yields Table 7.3. The later procedure caused practically no shift in the relevant periods for most farms of Atbasar county, but caused some shifts for majority of farms from other four counties. In the later cases, the critical periods were 1–2 weeks later for VCI, spanning overall the period between weeks 30 and 35. Regarding TCI, no significant changes in the critical periods were found for most farms from Atbasar, Esil, and Kamysty counties; however, the critical period shifted to an earlier period (weeks 27–30) for most farms from Zelinograd county and to a later period (by 1 week) for farms from Denis county. Results presented in the lower part of Table 7.3 show that the yield-based index selection procedure for farms attained substantially higher risk reductions than for farms from Zelinograd, Kamysty, and Denis counties in the case of VCI-based contracts. A small increase in the average risk reduction due to VCI-based insurance contracts is also observable for farms from Esil county. However, the average risk reduction hardly changed for the study farms from Atbasar county, since, as mentioned above, the period for measuring VCI did not change between county-based and farm-based procedures for most farms in this county. The risk reducing performance of the TCI-based contracts have not shown considerable differences between two procedures employed for index selection. For the VCI insurance, the highest risk reductions were found for farms from Zelinograd and Kamysty counties with the average estimates of 0.57 and 0.52, respectively Table 7.3. A group of farms from Denis county would also benefit from VCI-based insurance. However, most farms in this county would achieve higher risk reductions using TCI-based contracts. The average risk reduction estimates for farms in Atbasar and Esil were found to be very low for both VCI- and TCI-based insurance contracts. Furthermore, estimates for both types of contracts suggest that a higher effectiveness of insurance contracts is attained using the survival Gumbel copula. This finding might be surprising taking into account that Frank copulas showed a better fit for most farms when modeling the dependence of farm yields on TCI. It can be explained by the fact that the goodness-of-fit test evaluates copula models considering the entire interval of the distribution. Accordingly, selection of copula model that shows a good fit in both tails of the joint distribution is recommended. Given that the survival Gumbel copula shows a strong left tail dependence and a relatively week right tail dependence, it might be tested as showing a worse fit than a copula with radial symmetry in the case of joint distributions with a considerable magnitude of dependence in the right tail. However, the later would not assure that the radial copula provides the best fit in the left tail of the distribution. In the context of the empirical analysis, this implies that the Frank copula might show best fits when evaluating the entire interval of joint distributions of TCI and farm yields, but the survival Gumbel copula might be better suited to model left tail dependence between TCI and farms yields. The empirical results support the conclusion that the relationship between TCI and farm yields in the left tail can be better captured by survival Gumbel copulas.

These results show that the performance of both index-based insurance schemes varies substantially not only among study counties but also among single farms within counties (Fig. 7.3). While a number of farms could achieve considerable risk reductions (for a group of 16 farms risk reduction estimates are found to be above 50%), there is a number of farms (12 farms primarily in Atbasar and Esil counties) for which risk reduction was estimated to be below 25%. For remaining 19 farms, only moderate magnitudes of risk reduction between 25% and 50% was assessed. The later finding implies that the county-mean VH indices used in this analysis are too gross to represent vegetation conditions for a single farm. This happens because (a) low-resolution AVHRR data (16 km^2) are used and (b) VH indices were measured at the mean-county but not at farm level.

Comparing VCI- and TCI-based insurance contracts for single farms (Fig. 7.3), a small group of farms would show higher or comparable risk reductions using TCI-based contracts instead of VCI-based contracts. These farms are situated in Denis county. This finding suggests that there are some local effects that cause TCI slightly outperform VCI in this county. Finally, it is important to emphasize a high variation in risk reduction estimates within single counties. These differences might be related to both geographical location and soil specifics, and technological aspects of spring wheat technology. During the USSR period, farms used relatively similar technologies, whereas since the mid-1990s there have been considerable differences in technologies applied and intensification of input use by single farms. These later differences may explain considerable deviations in productivity and risk exposure at the farm level. Therefore, farm-based index insurance must be based on high-resolution satellite data, specifically collected at a farm level. Such data are currently available at 0.5, 1, and 4 km^2 resolution (NOAA/NESDIS 2018).

7.6 Conclusions

This analyses indicate clearly that two satellite-based VH indices, VCI, estimating moisture conditions, and TCI, estimating thermal conditions, can be used as criteria for insuring yield losses for spring wheat-producing farms in Kazakhstan. The tested methodology indicates that applying the copula approach for estimating dependence of spring wheat yields on VH indices would be effective for designing and rating insurance contract. The empirical relevance of the study relies on the use of the farm yield time series and the application of VH-based data (NOAA/NESDIS 2018) for the development of index-based insurance for grain crops. Different copula models

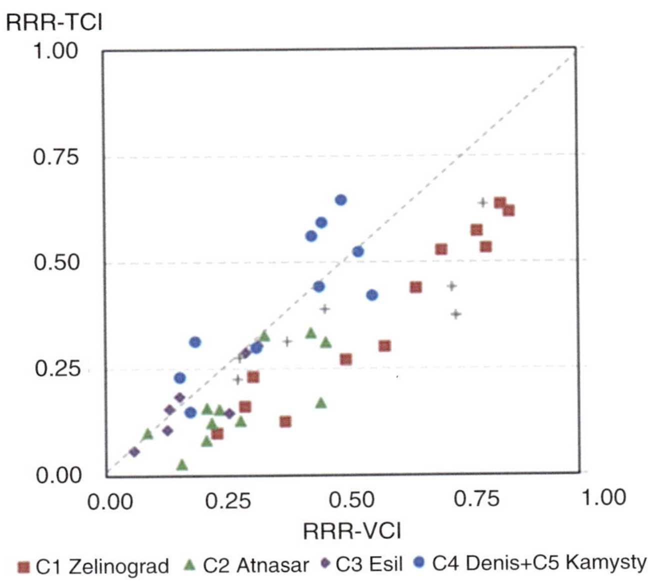

Fig. 7.3 Correlation of relative risk reduction (RRR) of moisture conditions (VCI) with RRR of temperature conditions (TCI) for five counties (shown in Fig. 7.1) in northern Kazakhstan

were employed and tested for their adequacy to represent the dependence structure in the tails of joint distributions of two VH indices and wheat yields. Our empirical results for wheat-producing farms in Kazakhstan showed that survival Gumbel copula provided a better fit when modeling joint distributions of farm wheat yields and VCI, while Frank copulas showed a better performance in describing dependence of farm yield on TCI.

Furthermore, the analysis revealed that performance of the satellite-based VH index insurance varies substantially among farms: while for one-third of farms yield risk could be reduced by 70%, for a quarter of sample farms risk reduction was estimated at 16%. These findings suggest that in general VH indices can provide a solid basis for detecting drought-related yield losses. However, their performance strongly depends on quality of yield and resolution of satellite data, as well as regional and local specifics in climate and wheat growing. The effectiveness of the satellite-based VH index insurance can be improved by employing satellite data of a higher resolution, using more sophisticated way of time aggregation of VH indices, and constructing VH indices for a single farm. Multiyear weekly indices at 0.5, 1 and 4 km^2 resolution are currently available from NOAA Vegetation Health WEB: https://www.star.nesdis.noaa.gov/smcd/emb/vci/VH/index.php (NOAA 2018). Another options to increase risk-reducing effectiveness of satellite index-based insurance would be to (a) complement it with information on crop sowing dates for more accurate selection of critical periods in plants' response to weather; (b) use VHI (combined VCI and TCI), which often correlates stronger with crop yield than individual indices (Kogan et al. 2012, 2015). Finally, differences in risk reduction estimates between two index selection procedures based on county yields and farm yield suggest that index-based insurance, designed by using aggregate index and yield data, could involve a relatively high basis risk, that might affect the demand for this kind of insurance. This puts a premium on the development of insurance products based on more precise data, i.e., farm-level yields and higher resolution satellite data.

References

Barnett, B.J., C.B. Barrett, and J.R. Skees. 2008. Poverty traps and index-based risk transfer products. *World Development* 36 (10): 2045–2102.

Bokusheva, R. 2011. Measuring dependence in joint distributions of yield and weather variables. *Agricultural Finance Review* 71 (1): 120–141.

———. 2014. Improving the Effectiveness of Weather-based Insurance: An Application of Copula Approach. Working paper, MPRA (Munich Personal RePEc Archive).

Bokusheva, R., and G. Breustedt. 2012. The effectiveness of weather-based index insurance and area-yield crop insurance: how reliable are post predictions for yield risk reduction? *Quarterly Journal of International Agriculture* 51 (2): 135–156.

Bokusheva, R., F. Kogan, I. Vitkovskaya, S. Conradt, and M. Batyrbayeva. 2016. Satellite-based vegetation health indices as a criteria for insuring against drought-related yield losses. *Journal of Agricultural and Forest Meteorology* 220: 200–206.

Breustedt, G., R. Bokusheva, and O. Heidelbach. 2008. The potential of index insurance schemes to reduce farmers' yield risk in an arid region. *Journal of Agricultural Economics* 59 (2): 312–328.

Claassen, R., and R.E. Just. 2010. Heterogeneity and distributional form of farm-level yields. *American Journal of Agricultural Economics* 93: 144–160.

Embrechts, P., A. McNeil, and D. Straumann. 2002. Correlation and dependence in risk management: properties and pitfalls. In *Risk Management: Value at Risk Beyond*, ed. M.A.H. Dempster, 176–223. Cambridge: Cambridge University Press.

Goodwin, B.K., and A. Hungerford. 2015. Copula-based models of systemic risk in U.S. agriculture: implications for crop insurance and reinsurance contracts. *American Journal of Agricultural Economics* 97 (3): 879–896.

Hazell, P., J. Anderson, N. Balzer, A. Hastrup Clemmensen, U. Hess, and F. Rispoli. 2010. Potential for Scale and Sustainability in Weather Index Insurance for Agriculture and Rural Livelihoods. International Fund for Agricultural Development and World Food Program.

Kellner, U., and O. Musshoff. 2012. Precipitation or water capacity indices? An analysis of the benefits of alternative for index insurance. *Agricultural Systems* 104: 645–653.

Kogan, F.N. 1995. Drought of the late 1980s in the United States as derived from NOAA polar-orbiting satellite data. *Bulletin of the American Meteorological Society* 76: 655–668.

———. 1998. A typical pattern of vegetation conditions in southern Africa during El-Nino years detected from AVHRR data using three-channel numerical index. *International Journal of Remote Sensing* 19 ((18): 3689–3695.

Kogan, F., A. Gitelson, E. Zakarin, L. Spivak, and L. Lebed. 2003. AVHRR-based spectral vegetation index for quantitative assessment of vegetation state and productivity: calibration and validation. *Photogrammetric Engineering and Remote Sensing* 69 (8): 899–906.

Kogan, F., L. Salazar, and L. Roytman. 2012. Estimation of crop production using satellite-based vegetation health indices in Kansas. *International Journal of Remote Sensing* 33 (9): 2798–2814.

Kogan, F., W. Guo, A. Strashnaia, A. Kleshenko, O. Chub, and O. Virchenko. 2016. Modelling and prediction of crop losses from NOAA polar-orbiting operational satellites. *Geomatics Natural Hazards and Risk* 7 (3): 886–900. https://doi.org/10.1080/19475705.2015.1009178.

Lui, W.T., and F.N. Kogan. 1996. Monitoring regional drought using the vegetation condition index. *International Journal of Remote Sensing* 17: 2761–2782.

Mainik, G., and E. Schaanning. 2012. On Dependence Consistency of CoVaR and Some Other Systemic Risk Measures. RiskLab, Department of Mathematics, ETH Zurich. http://arxiv.org/pdf/1207.3464v3.pdf.

Makaudze, E.M., and M.J. Miranda. 2010. Catastrophic drought insurance based on remotely sensed normalized difference vegetation index for smallholder farmers in Zimbabwe. *Agrekon* 49 (1): 418–432.

McNeil, A., R. Frey, and P. Embrechts. 2005. *Quantitative Risk Management*, 211. Princeton: Princeton University Press.

Miranda, M.J., and K. Farrin. 2012. A index insurance for developing countries. *Applied Economic Perspectives and Policy* 34 (3): 391–427.

National Agency of Statistics of the Republic of Kazakhstan. 2012. http://www.stat.kz/digital/selskoehoz/Pages/default.aspx. Accessed 11 Jan 2012.

Nelsen, R.B. 1999. *An Introduction to Copulas*, 357. New York: Springer.

NOAA/NESDIS. 2018. V*egetation Health indices*. https://www.star.nesdis.noaa.gov/smcd/emb/vci/VH/vh_browse.php.

Skees, J.R., J.R. Black, and B.J. Barnett. 1997. Designing and rating an area yield crop insurance contract. *American Journal of Agricultural Economics* 79: 430–438.

Sklar, A. 1959. Fonctions de répartition àn dimension et leurs marges. *Publications of the Statistical Institute of the University of Paris* 8: 229–231.

Spivak, L., I. Vitkovskaya, and M. Batyrbaeva. 2008. Analysis of intra-seasonal variations in the productivity of vegetation in Kazakhstan, using time series remote sensing. *Bulletin of the National Academy of Sciences of the Republic of Kazakhstan: Series in Physics and Mathematics* 4: 29–32.

Suleimenov, M.K., and K.A. Akshalov. 2011. Productivity and quality of spring wheat grain subject to cultivation technology, crop rotation and weather. *Bulletin of Agricultural Science of Kazakhstan* 630: 17–21.

Turvey, C.G. 2001. Weather derivatives for specific event risks in agriculture. *Review of Agricultural Economics* 23: 333–351.

Unganai, L.S., and F.N. Kogan. 1998. Drought monitoring and corn yield estimation in Southern Africa from AVHRR data. *Remote Sensing of Environment* 63 (3): 219–232.

USDA. 1994. Major World Crop Areas and Climate Profile. *USDA Agricultural Handbook,* No 664, pp. 157–170.

USDA/FAS. 2012. Kazakhstan 2011 Grain and Feed Annual Report. http://gain.fas.usda.gov/.

Vedenov, D.V. 2008. Application of copulas to estimation of joint crop yield. *Annual Meeting of the AAEA 2008*, Orlando, USA, July 27–29.

Vedenov, D.V., and B.J. Barnett. 2004. Efficiency of weather derivatives as primary crop insurance instruments. *Journal of Agricultural and Resource Economics* 29: 387–403.

8 Advanced VH-Based Long-Term Drought and Food Security Prediction

8.1 Introduction

The twenty-first century has begun with a series of widespread, long, and intensive droughts around the world. Droughts of severe-to-exceptional (SE) intensity covered 7–16% and extreme intensity—2–6% of the world land (Kogan et al. 2013). A wide range of economic and societal activities have been affected, especially agriculture, forestry, water, energy, ecosystems, transportation, and recreation in multifacet ways. These droughts had adverse consequences for societal sustainability since they reduced agricultural production, caused a misbalance between grain supply and demand, shortages of food, malnutrition, human health deterioration, poverty, regional disturbances, population migration and death. Global food security (FS) was the twenty-first century's issue 8 years out of 17 when drought reduced global grain production (the main source of food and feeds for the world's population), which was below steadily grown consumption. Additional FS concern arises from anticipated drought intensification, expansion, and penetration to the new areas due to the current and future climate changes (IPCC 2014). Drought detection and watch has successfully served the global society since the satellite-based vegetation health (VH) method has been introduced in the past three decades. The recent scientific advances and the new generation of operational satellites have provided ways for drought forecasting applying satellite-based vegetation health (VH) method and ocean–atmosphere system oscillations. One of such systems, where VH showed some success, is El Niño southern oscillation (ENSO). ENSO events served as a reliable phenomenon for up to 6-month advance drought prediction, crop losses assessments, addressing food security problem.

ENSO, which is the climate forcing phenomenon, disturbs weather worldwide, affecting land ecosystems. The total societal impact of ENSO is estimated in billions of dollars (Kovats 2000). The associated weather vagaries affect agriculture, fisheries, transportation, tourism, energy consumption, human health, environment, and water resource in many world regions (WMO 1997; Suplee 1999; Potter et al. 2003; Buermann et al. 2003; Lemonick 2013). In the twentieth century, several world disasters occurred following a very strong ENSO events such as 1902–1903, 1913–1915, 1918–1920, 1940–1941, 1972–1973, 1982–1983, 1988–1989, 1991–1992, and 1997–1998 (Kovats 2000; Kogan 1998, 2000). In the mid-2010s the next strong ENSO has occurred. Its development started during 2015, when sea surface temperature (SST) in the central tropical Pacific (TP) was much warmer than normal. However, the first signs of that El Niño appeared in September 2014 when SST in TP exceeded multiyear normal by 0.5 °C (NOAA 2015). In the first 2–3 months of 2015, westerly

F. Kogan, *Remote Sensing for Food Security*, Sustainable Development Goals Series,
https://doi.org/10.1007/978-3-319-96256-6_8

wind blew over the central TP, moving warmer surface waters eastward, increasing subsurface heat content, strengthening down welling, and triggering Kelvin waves, which created stable warmer SST in the central and eastern TP. During the summer 2015, SST in the TP continued to increase quickly, so that by the end of summer, SST anomaly in the most central part of TP, called Niño-3.4, crossed up +1.5 °C threshold (NOAA2 2015; NOAA 2016). This trend continued through the autumn and by January 2016, SST anomaly reached +2.9 °C (NOAA2 2016). This anomaly was 0.3 °C higher and occurred 1–2 months earlier than in the recent strongest 1997–1998 El Niño.

Following ENSO, some regions experience hot and dry but the others wet and cool weather. Both weather types create strong environmental and societal impacts. One of the worst disasters associated with ENSO is drought, which in some vulnerable regions might lead to food crises. Such events were recorded in 1972–1973 in southeastern Asia and northeastern China and in 1991–1992 and 1997–1998 in southern Africa and northern Brazil. Wet and cool weather associated with ENSO might stimulate a very good harvest, but in some regions such as Sub-Sahara Africa, India, and tropical South America might lead to enhancement of mosquito- and rodent-related diseases. For example, several major disease outbreaks occurred in late 1997 following ENSO-triggered heavy rain and flooding in Chad, Guinea-Bissau, Kenya, Somalia, and United Republic of Tanzania and at the end of 1999–early 2000, in India, Africa (Mozambique), the Americas, and Asia (Kovats 2000; WER 1998; Bouma and Dye 1997; Bouma and van der Kaay 1996; Hales et al. 1996; Dilley and Heyman 1995; Nicholls 1993; Anyamba et al. 2001, 2002, Buermann et al. 2003; Bastos et al. 2013; Hilker et al. 2014).

Analysis of the past 30-year ENSO's environmental impacts showed that in the affected land regions, ecosystems responded strongly to the associated changes in weather (Ropelewski and Halpert 1987; Halpert and Ropelewski 1992; Kogan 1998, 2000; Anyamba et al. 2002; Buermann et al. 2003; Huber and Fensholt 2011; Bastos et al. 2013). A combination of ENSO-based weather and ecosystems response usually impacts socioeconomic activities. The strength of the impacts depends on the intensity of ENSO, duration, and the affected area. During ENSO some of the areas have hot/dry and others have wet/cool conditions. The same conditions might produce negative and positive impacts on socioeconomic activities. For example, on a negative side, hot and dry conditions normally reduce crop production but in the regions of mosquito-borne diseases such conditions might weaken mosquitoes' activities reducing some epidemics, especially malaria.

Following ENSO research and sea surface temperature dynamics in 2015, the 2015–2016 El Niño was expected to affect weather, ecosystems, and socioeconomics in South America, Southeast Asia, Africa, Australia, and some parts of North America (Nicholls 1993; Ropelewski and Halpert 1987, 1996; Halpert and Ropelewski 1992; Glantz 1996; WMO 1997; Trenberth and Hoar 1997; Kogan 1998, 2000; Anyamba et al. 2002; Bastos et al. 2013; Hilker et al. 2014; Mathews 2014; IRI 2016). However, the 2015–2016 El Niño turned out to be stronger than the recent two strongest events in 1982–1983 and 1997–1998 since 1980 (Monastersky 2016; Mecray 2016). Therefore, the recent ENSO-VH teleconnection analysis followed the goals to investigate drought prediction with the appearance of the first signs of ENSO development. Besides, the goals included verification of (a) the affected area, ecosystem type, and the intensity of the impact; (b) similarity and differences of the 2015–2016 El Niño impacts with the other strong events; (c) VH-ENSO teleconnection for all and the strongest events; (d) transition of VH-ENSO impacts from boreal winter to spring and summer; (e) since the strongest La Niña occurred 27 year ago (1988–1989) the attempt was made to predict the development of the new one and identify the areas affected by droughts.

8.2 Data and Method

Two 36-year data sets of sea surface temperature (SST) and vegetation health (VH) derived from the radiances measured by the Advance Very High Resolution Radiometer (AVHRR) on NOAA operational afternoon polar-orbiting environmental satellites were applied. SST was used to characterize ENSO event and VH to characterize conditions of land ecosystems (Reynolds et al. 2002; Reynolds and Smith 1995; Kidwell 1997; NOAA 2015; Kogan et al. 2011; Kogan 1990, 1997, 2001). Sea surface temperature (SST) was derived from infrared (IR) AVHRR-based 4 km^2 resolution daily radiances in the 10–11 and 11–12 μm bands (channels 4 and 5, respectively). SST retrieval algorithm was designed to minimize the effects of atmospheric water vapor. The data were tuned by regression analysis against quality-controlled buoy data using the multichannel technique converting the radiances to skin SST temperature (roughly 0.5 m in depth). Finally, daily SST was aggregated to mean monthly values for each 1° latitude–longitude box of the global ocean (Reynolds et al. 2002; Reynolds and Smith 1995). Monthly SST data for every 1° box were available from January 1982 through May 2016 (NOAA2).

8.3 El Niño Southern Oscillation

El Niño Southern Oscillation (ENSO) phenomenon is ocean–atmosphere coupled climate forcing, which is the strongest natural and climate fluctuation on global timescale. ENSO's typical characteristic is the emergence of huge mass of warmer or cooler water from time to time in central and eastern TP. ENSO has two phases: warm (El Niño) with positive and cold (La Niña) with negative SST anomaly in TP. These oscillations occur irregularly every 3–7 years, affecting global weather and earth surface conditions. ENSO onset, intensity, development, and impacts are assessed from mean monthly SST anomaly (SSTa) averaged over the TP between 5° N and 5° S. Several regions are normally selected to aggregate SST over the longitudes in order to characterize ENSO onset, intensity, progression, and impacts. Among these regions, the central part (from 120° W to 170° W) called Niño 3.4 (Fig. 8.1b) is considered as one of the most important areas, where following weather data, SST anomaly characterizes Niño intensity, duration, and impacts (Ropelewski and Halpert 1987, 1996; Halpert and Ropelewski 1992; NOAA 2015). Since the beginning of NOAA operational environmental satellites era in 1981, SST was measured regularly from space for every 1 km^2 of global ocean and aggregated to 4 km^2 data set. By 2018, 37-year 4 km^2 (aggregated to 1° latitude–longitude box) monthly SST and SSTa (deviation from 35-year climatology) data are available. These data presented in Fig. 8.1a are investigated together with vegetation health (VH) indices for an advanced drought prediction. The VH indices (NOAA/NESDIS 2018), which include vegetation condition index (VCI), characterizing availability of moisture for vegetation, TCI, characterizing thermal vegetation condition, and VHI, estimating moisture–thermal conditions together (Chap. 4) were used to predict drought start/end, affected area, intensity, and impacts.

8.4 Weather and Ecosystem Patterns During ENSO

Numerous climate studies indicated that opposite precipitation (dry and wet) and temperature (hot and cool) patterns occur in some world areas during the two phases (warm and cold) of ENSO (Trenberth 1997; Trenberth and Hoar 1997; Ropelewski and Halpert 1987, 1996; Halpert and Ropelewski 1992; Suplee 1999; Buermann et al. 2003; NOAA 2016). In boreal winter, El Niño years are characterized by dry and mostly hot weather in southeastern Asia, southern Africa, and northern Brazil, but wet and cool weather is typical for north–central Argentina, southern Brazil, and southwestern USA (specifically California). It was also observed that central Canada, southeast Australia, and east–central Asia are warm. In La Niña years, the weather situation is normally opposite. Although boreal summer weather is less sensitive to ENSO events,

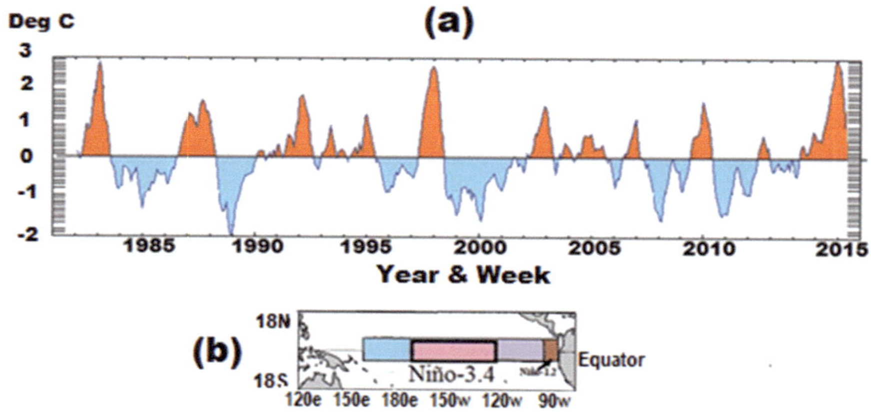

Fig. 8.1 Monthly sea surface temperature anomaly (SSTa,°C) in 3.4 Niño area of tropical Pacific (**a**) and the location of Niño area (**b**); Land east of 90° W is north-western South America and between 120 and 150° E is New Guinea

still El Niño induces dry weather is typical for southeastern Asia, eastern Australia, India, and Central America; oppositely, La Niña-type weather is normally wet in those regions.

The impacts of ENSO on ecosystems were investigated to find out which areas of global ecosystems are historically teleconnected to SSTa changes in TP. Monthly 36-year (1981–2016) VH indices (VCI, TCI, and VHI) for each 4 km^2 global land pixel were used as indicators of ecosystem condition (or health). Monthly VHI for each global pixel was correlated with monthly Niño-3.4 mean SSTa during the same period. The first results of this analysis were presented in 2000 (Kogan 2000) for a short (13-year, 1985–1997) period ENSO years. They supported climate-based regional precipitation and temperature response to SSTa changes (Ropelewski and Halpert 1987; Halpert and Ropelewski 1992; NOAA 2016). These results were later confirmed for the strong El Niño events in some regions with a combination of satellite indices and in situ data (Anyamba et al. 2001, 2002; Buermann et al. 2003; Potter et al. 2004; Huber and Fensholt 2011; Bastos et al. 2013; Hilker et al. 2014). Finally, all ENSO cases during the entire 1981–2016 period and the entire globe were reevaluated using VHI and SSTa observation, investigating reaction of ecosystem conditions to different SST anomalies.

As seen in Fig. 8.1 and Table 8.1, monthly SSTa in the Niño-3.4 is changing from ±0.6 °C to ±3.0 °C. However, a few research showed that small anomalies are not causing widespread and intensive ENSO-type impacts on the globe. Following our research (Kogan 2000), the Niño-3.4 index greater than absolute value of 1.5 °C during December–February is a very good indicator of ENSO impacts on Earth ecosystems. Since the late 1950s, El Niño events with |SSTa| > ±1.5 °C occurred during winters of 1957/1958, 1965/1966, 1972/1973, 1982/1983, 1987/1988, 1988/1989, 1991/1992, 1997/1998, 1998/1999, 1999/2000, 2007/2008, 2009/2010, 2010/2011, and 2015/2016 (NOAA 2015; WRCC 2015; SCCONC 2015; Trenberth and Hoar 1997). During the 1997–1998 strong El Niño, the impacts on ecosystems were much stronger when SSTa was greater than +2.0 °C (Huber and Fensholt 2011; Brown et al. 2010; Anymba et al. 2001, 2002; Kogan 2000; Kogan et al. 2017). Such SSTa were observed four times for strong El Niño since 1982. However, we investigated first, how world ecosystems react if all ENSO cases in the past 36 years, when 3.4 Niño's |SSTa| > 0.5 °C, are included in the correlation analysis. These results were compared with ecosystem reaction to the strongest ENSO cases (|SSTa| > 1.5 °C and > 2.0 °C).

Table 8.1 indicates that from 1981/1982 through 2015/2016 there were 16 cases of ENSO (nine El Niño and seven La Niña) with |SSTa| > 0.5 °C. Of those cases, 13 had |SSTa|

Table 8.1 Years of El Niño (E) and La Niña (L) for SSTa (°C) exceeding absolute values of some thresholds during November–February 1982/1983–2015/2016

SSTa	1982–1983	1983–1984	1984–1985	1987–1988	1988–1989	1991–1992	1994–1995	1997–1998	1998–1999	1999–2000	2002–2003	2005–2006	2007–2008	2009–2010	2010–2011	2015–2016
>0.5	E	E	L	E	L	E	E	E	L	L	E	L	L	E	L	E
>1.5	E		L	E	L	E		E	L	L	E		L	E	L	E
>2.0	E				E			E								E
Highest	+2.8	−0.9	−1.5	+1.7	−2.4	+1.7	+0.6	+2.7	−1.7	−1.8	+1.5	−0.8	−1.8	+1.6	−1.7	+3.0

>1.5 °C and only four of them had the strongest ENSO with |SSTa| ≥ 2.0 °C (1982/1983, 1988/1989, 1997/1998, and 2015/2016). Figure 8.2 shows color-coded maps of Pearson correlation coefficients (PCC × 100) between pixel-based monthly VHI and Niño-3.4 area mean monthly area-average SSTa during November–February for the (a) all 16 ENSO cases and (b) cases with |SSTa| > 1.5 °C. Figure 8.3a, b shows VHI-SSTa correlation of all ENSO cases (|SSTa| > 0.5 °C) during 1981–2016. Two cases are presented with positive (Argentina, blue areas on Fig. 8.3e map) and negative (southern Africa, white/red areas on Fig. 8.3f). In Argentina's 16 total ENSO cases (each case is represented by 4 months mean SSTa and mean VHI for area indicated in Fig. 8.3e), if SSTa < − 0.8 °C during November–February of La Niña years vegetation experiences drought-related stress (VHI < 48 for the majority of points), but favorable conditions (VHI > 55) during El Niño years with SSTa > +0.5 °C. In southern Africa (white-red color in Fig. 8.3f), vegetation conditions are opposite to Argentina: favorable during La Niña (VHI > 50 for the majority of points) and stressful during El Niño (VHI < 45). It is important to take into consideration that for some month during November–February VHI values might deviate slightly from the indicated thresholds.

Figure 8.3c, d shows distribution of monthly VHI versus monthly SSTa for only the strongest four ENSO years (Table 8.1) such as El Niño 2015/16. Following Preacher et al. (2005), it is not advisable to calculate correlation for a small data sample extracted from a big one similar to that shown in Fig. 8.3a, b. Therefore, the correlation for data shown in Fig. 8.3c, d has not been calculated. However, sampled data displayed in Fig. 8.3c, d indicate that for both strongest and all ENSO cases VHI-SSTa relationship is similar in values, slopes, and time. Since (1) the current El Niño was the strongest in the past 36 years, (2) the last strongest La Niña was 27 years ago (1988–1989), and (3) there is some potential for

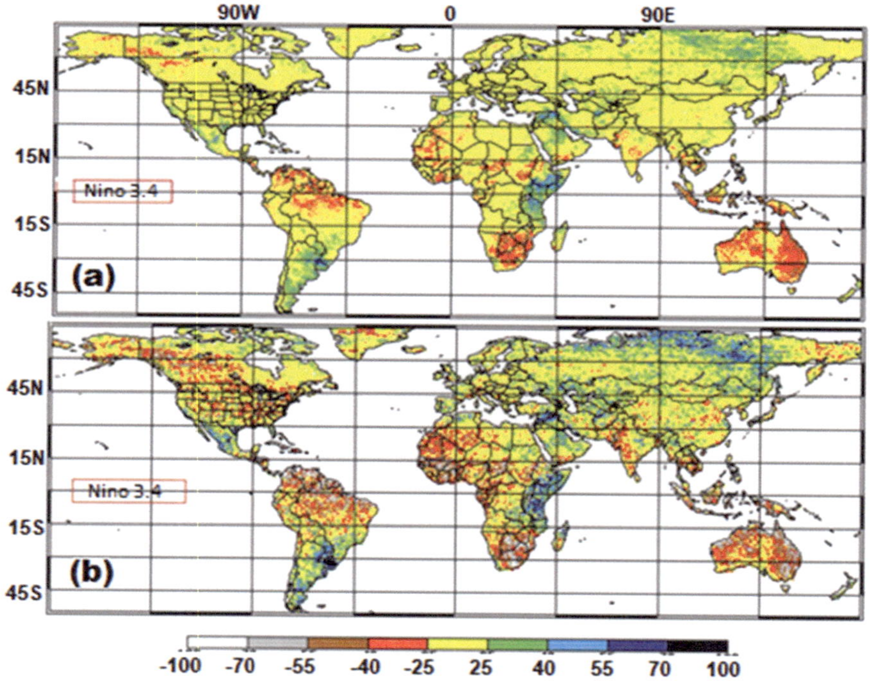

Fig. 8.2 Color-coded maps of Pearson correlation coefficients (PCC × 100) of 4 km² pixel-based monthly mean vegetation health index (VHI) with monthly and area mean SSTa over 3.4 Niño during 1981–2016 ENSO years (**a**) all 16 ENSO cases (|SSTa| > 0.5 °C) and (**b**) 13 ENSO cases (|SSTa| > 1.0 °C); Axis labels: Latitude (°), Longitude (°)

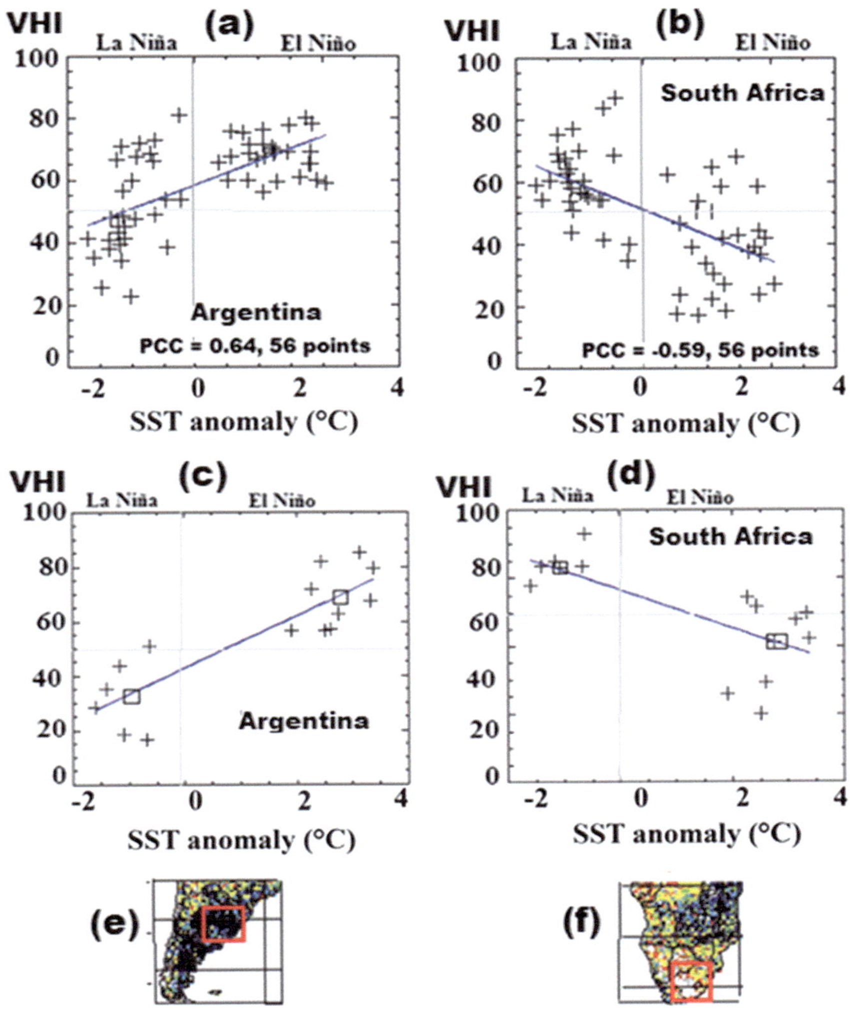

Fig. 8.3 Scattered plot of mean monthly average VHI inside the area's (shown by red square in e, f) versus Niño-3.4 area mean monthly average SSTa for November–February all 16 ENSO years (|SSTa| > 0.5 °C, **a**, **b**) and the strongest four ENSO cases (|SSTa| > 2.0 °C, **c**, **d**); regression lines are shown in (**a**–**d**)

development of another such strong event, we investigated possible vegetation health response to such SSTa. Following this goal, we applied cluster analysis advised by Preacher et al. (2005) to estimate VHI response (value, location, time) to |SSTa| > 2.0 °C if the events similar to 2015–2016 (El Niño) and 1988–1989 (La Niña) will occur in the future.

Table 8.2 compares VHI values for the strongest and all ENSO cases in the world regions sensitive to SSTa changes in the 3.4 Niño area. Two types of VHI-SSTa teleconnections are seen for all 16 cases: with positive and negative correlation (Fig. 8.3a, b). Countries and regions with positive correlation include Argentina (north), the USA's California (south and central), Mexico (west), Horn of Africa, and Saudi Arabia (central). The strength of the correlation for all 16 ENSO cases is changing from 0.29 in Saudi Arabia to 0.86 in Argentina. Ecosystems in these areas experience stress during La Niña (VHI range 35–45) and healthy condition during El Niño (VHI range 53–70). In the countries with the negative VH-ENSO correlation in Table 8.2

Table 8.2 Vegetation health during ENSO years and VH-SSTa correlation

Region	Years	VH during ENSO El Niño	La Niña	VH-SSTa correlation during November–February
Argentina	16 years	45	70	0.86
	4 strongest	37	70	
Horn of Africa	16 years	35	71	0.44
	4 strongest	27	45	
USA (California)	16 years	39	53	0.47
	4 strongest	43	47	
Mexico	16 years	45	54	0.35
	4 strongest	43	53	
Saudi Arabia	16 years	40	52	0.29
	4 strongest	38	58	
Brazil	16 years	53	61	−0.70
	4 strongest	54	41	
Rep. S. Africa	16 years	63	40	−0.59
	4 strongest	65	41	
Australia	16 years	63	45	−0.52
	4 strongest	64	35	
Canada	16 years	50	37	−0.52
	4 strongest	43	38	
Ghana	16 years	59	36	−0.53
	4 strongest	59	61	
Borneo	16 years	53	38	−0.51
	4 strongest	49	34	

(Brazil north, Republic of South Africa-central, Australia east, Canada central, Republic of Ghana south, and Borneo south) vegetation experiences healthy conditions during La Niña (VHI > 50) and stressed during El Niño (VHI < 45). The correlation ranges between −0.52 for Australia and Canada to −0.70 for Brazil.

Although the correlation of VHI versus SSTa was estimated only for all ENSO years with |SSTa| > 0.5 °C, the values of VHI for the strongest ENSO cases (|SSTa| ≥ 2.0 °C) in all affected regions (Table 8.2) is similar to all cases (|SSTa| > 0.5 °C) since 1980. For the majority of the strongest four cases the values and proportion of stress-no stress vegetation are identical to the all ENSO cases. However, it should be pointed out that for a few locations the strongest cases show more intensive stress (lower VHI for Argentina and the Horn of Africa) or better health conditions (higher VHI for the Horn of Africa and California). Following this analysis, it is important to emphasize that some statistical restrictions for modeling VHI-SSTa relationship for the limited number of the strongest ENSO events only (Preacher et al. 2005) can't limit our ability for accurate estimation of land conditions in the ENSO-affected areas. If a very strong ENSO (with |SSTa| ≥ 2.0 °C) event occurs in the future, Table 8.2 might be a good source to estimate VHI range during that event.

Summarizing, it should be emphasized that following Fig. 8.2 and Table 8.2, few areas of VHI-SSTa teleconnections are clearly identified for both all and the strongest ENSO cases during 1981–2016. For most areas (Table 8.2) the correlation is stronger than 0.45 and significant at 5% level. For nearly 50% of pixels inside each area (Fig. 8.2b), the correlation is stronger than 0.55. Among the strong VHI-SSTa teleconnections, large and stable areas with positive correlation (blue color) is typical for central and northern Argentina, southern Brazil, Horn of Africa, eastern Asia, and the USA's far southwest (mostly California). These areas experience favorable vegetation condition (VHI > 55) during El Niño and stressful (VHI < 45) conditions during La Niña. Northern Brazil, southern Africa, southeast Australia, Africa's Sahel (west), and central Canada are typically experiencing vegetation stress during El Niño and healthy vegetation condition during La Niña (Figs. 8.2 and 8.3). These results coincide generally with climate studies where regional precipitation and temperature anomalies were correlated with Niño's 3.4 SSTa (Ropelewski and Halpert 1987; Halpert and Ropelewski 1992; Trenberth and Hoar 1997; NOAA 2016) and also with satellite-based research done after 1997/1998 strongest El Niño, which estimated ENSO teleconnection to the global temperature, rainfall, greenness, phenology, net primary production, and other metrics (Lao 2016; Hilker et al. 2014; Brown et al. 2010; Anyamba et al. 2001, 2002; Kogan and Guo 2017). If ENSO years include both strong and less strong cases (Table 8.1 and Fig. 8.2a), then VHI-SSTa correlation is reduced and the area of the impacts shrinks. However, in the Southern Hemisphere the areas of teleconnection remained similar to the strong ENSO events (Argentina, Brazil, Africa, and Australia).

The results in Table 8.1 characterize Niño 3.4 SSTa impact on VHI, which is composited of two indices VCI (estimates moisture conditions) and TCI (thermal conditions). The next step was to find out how VCI and TCI correlate with the 3.4-Niño area's SSTa. Table 8.3 shows the results of correlation analysis between each of the areal-mean indices for all ENSO cases with Niño 3.4 SSTa (|SSTa| > 0.5 °C). For all regions VCI-SSTa and TCI-SSTa correlation keep the same sign as the correlation of the joint index (VHI-SSTa). This means that for a positive index-SSTa correlation, vegetation experiences moisture and thermal stress during El Niño and no stress during La Niña; for negative correlation, vegetation conditions are opposite (Table 8.3 and Fig. 8.3b for VHI). Meanwhile, in the majority of regions (Argentina, Horn of Africa, Brazil, Republic of South Africa, Australia, and Canada) both moisture and thermal indices determine vegetation conditions, although thermal index (TCI), having higher correlation, shows some advantages before VCI. The thermal index shows also considerable advantages before the moisture index in California and Asia islands. In Sub-Sahara Africa, the correlation of VCI and TCI with Niño 3.4 SSTa is not strong.

Table 8.3 Pearson correlation coefficients between area-mean (shown on Fig. 8.3) monthly VH (VHI, VCI, and TCI) indices and mean monthly SSTa in Niño-3.4 area during all ENSO events (|SSTa| > 0.5 °C) in 1981–2016

Region number	Region name	VHI	VCI	TCI
1	Argentina	0.64	0.59	0.80
2	Horn of Africa	0.44	0.58	0.71
3	Asia north-east	0.76	0.63	0.03
4	USA California	0.46	0.025	−0.59
5	Brazil	−0.70	−0.51	−0.73
6	Rep. S. Africa	−0.59	−0.60	−0.69
7	Australia	−0.52	−0.56	−0.59
8	Sub-Sahara Africa	−0.57	−0.39	−0.01
9	Canada	−0.52	−0.65	−0.75
10	Asia south-east (islands)	−0.88	−0.02	−0.80

PCCs >0.7 are significant at 1%, the rest at 5%

8.5 Vegetation Health During the Strongest 2015–2016 El Niño

As has been mentioned, for the strongest 2015/2016 El Niño (since 1980s), warmer than normal SST (nearly +0.5 °C) appeared in the TP at the end of 2014 to early 2015. However, such SSTa was small to cause widespread and intensive ENSO-type impacts on the globe. The signs of the strong 2015/2016 El Niño, characterized by unusually warm TP in the equatorial region, appeared in autumn 2015 when SST exceeded 2.0 °C, the level not seen since the strongest known El Niño in autumn 1997. By the end of 2015, El Niño intensified more when SSTa in Niño-3.4 area reached almost 3 °C exceeding the 1997 SSTa by 0.3 °C. The indicators of the mature strong El Niño included also upwelling in the eastern TP, continuation of low-level westerly wind and upper-level easterly wind, enhanced convection over the central TP and suppressed convection over Indonesia (NOAA2 2015, NOAA2 2016).

Based on vegetation health and in situ (precipitation and temperature) data for autumn 2015, it was forecasted that the strongest El Niño is going to produce considerable impact on ecosystems in the identified (from climate and ecosystems studies) land areas of South America (Argentina and Brazil), Africa (south, Horn, and eastern Sub-Sahara), Southeast Asia, and southeast Australia (Ropelewski and Halpert 1987;

Halpert and Ropelewski 1992; Kogan 2000, 2001; Mecray 2016; NOAA2 2016; NOAA 2016). El Niño is also expected to bring warmer weather in the northern regions of North America and wetter weather in the southwestern and partially southern USA bringing some relief to drought-affected southwestern states, especially California (NOAA2 2016). The VHI-based assessment of vegetation health at the end of December 2015, shown in Fig. 8.4, confirms these impacts. An intensive drought has developed in the northern Brazil, southern Africa, the islands of southeastern Asia, and most of Australia. Ecosystems in Argentina and most of the Horn of Africa were healthy at the last week of December 2015. There are also some inconsistencies. Most of the Horn of Africa shows healthy vegetation condition except for eastern part, where vegetation was under stress. This problem appears because ecosystem condition in the entire region strongly depends on SSTa in two ocean regions: Niño-3.4 and Indian Ocean (IO). In the case of cooler IO, which was observed during December 2015–January 2016, drought is normally developing in the eastern Horn only (WFP 2016). During 1997/1998 strong El Niño, drought in the Horn was not developed since, in addition to positive SST anomaly in Niño-3.4 area, IO was also warm. Oppositely, during the 1982/83 El Niño, cooler IO (similarly to the current event) triggered vegetation stress in the eastern Horn. Besides the Horn, Sub-Sahara Africa (south of 15° N) had healthy vegetation instead of expected stressed during El Niño (see Figs. 8.2 and 8.3).

Since Northern Hemisphere land (north of 30° N) is covered with snow during boreal winter, only thermal conditions (TCI) of ecosystems were evaluated at the end of December 2015. They are shown in Fig. 8.5 in comparison with TCI of the two other strong El Niño, 1997–1998 and 1982–1983. As expected (Table 8.2), the 2015–2016 strong El Niño produced intensive thermal vegetation stress in Brazil, southern Africa, eastern Australia, and the island of southeast Asia (regions 5, 6, 7, and 10, respectively). Similar thermal conditions were also observed in 1997–1998 and 1982–1983. However, they were less intensive in Brazil and especially in southern Africa. Following positive TCI-SSTa correlation (Table 8.3), healthy thermal condition developed in Argentina (region 1) during all three strong El Niño years. But in the Horn of Africa (region 2), mostly normal thermal conditions (TCI = 50–55) developed in 2015, while in two other years, vegetation was very healthy (TCI = 70–100). Central Canada (region 9) is characterized by

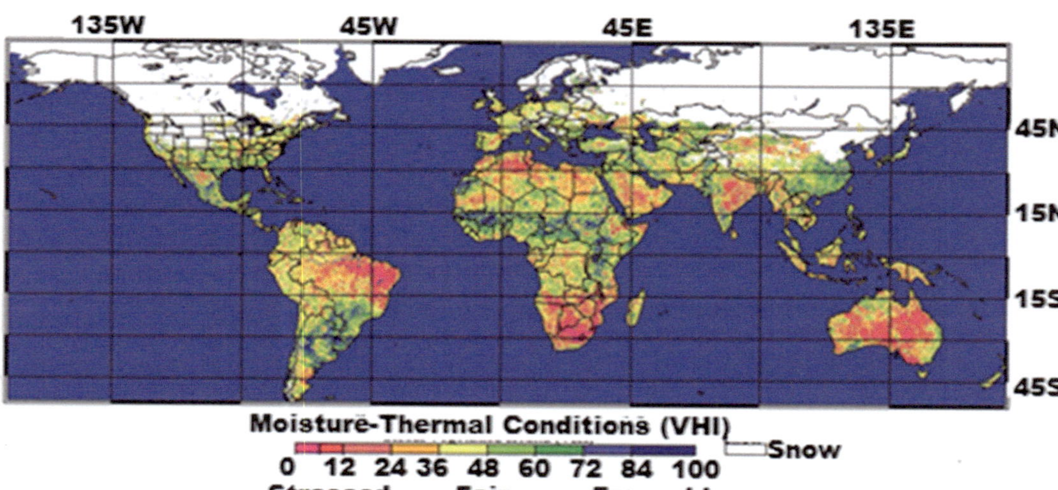

Fig. 8.4 Color-coded map of global vegetation health (VHI) for each 4 km^2 pixel at the end of December 2015; Axis labels: Latitude (°), Longitude (°)

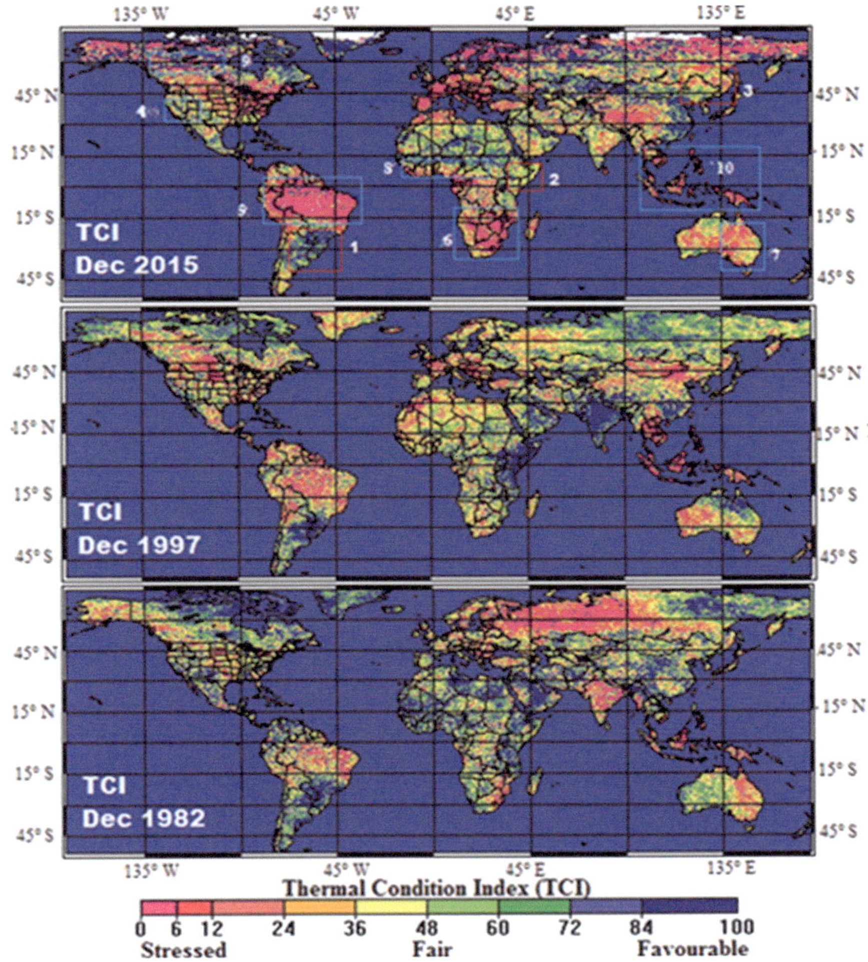

Fig. 8.5 Color-coded map of global thermal condition (TCI) for each 4 km^2 pixel at the end of December 2015 for the three strongest El Niño years (2015–2016, 1997–1998, and 1982–1983); axis labels: Latitude (°), Longitude (°)

TCI < 40, indicating thermal drought-type conditions, which is in agreement with the available forecasts (NOAA2 2016; AccuWeather 2016). Finally, southwestern USA, including California, is normally wet and cool during December–February of El Niño years resulting in healthy thermal condition. This is supported by high TCI, which was larger than 60 in December 1997 and 1982. December 2015 thermal health in California was mixed, with some areas under thermal stress due to a slow-going recovery of this region from a very intensive multiyear drought.

8.6 Prediction of 2015–2016 El Niño Impacts on Ecosystems During Spring and Summer 2016

Since ENSO appearance and impacts are season-specific, further investigation of VHI-SSTa correlation was focused on vegetation response during spring and summer 2016. The question was if the impacts of the strong ENSO are transitioned from boreal winter to boreal spring and summer, what areas are affected and an intensity

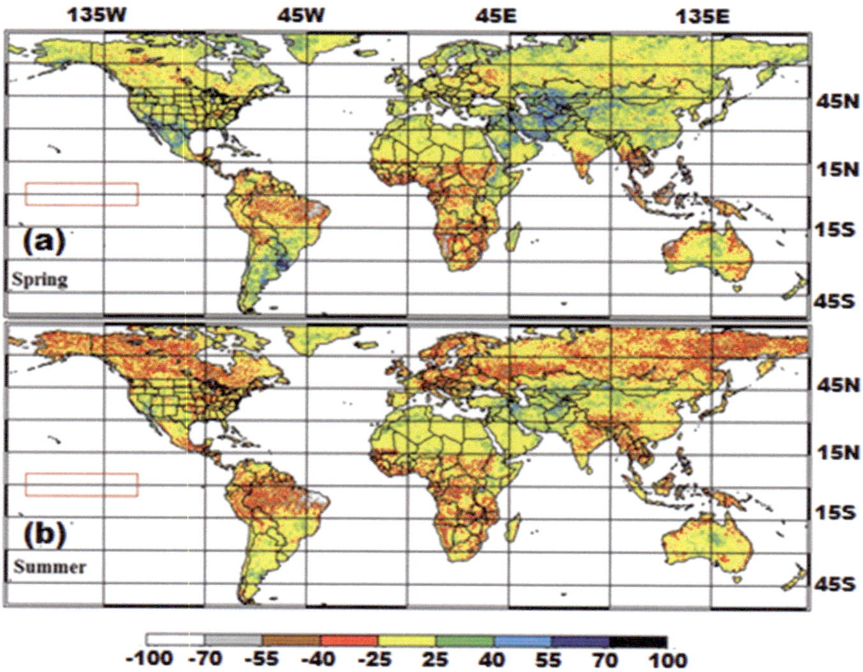

Fig. 8.6 Color-coded maps of Pearson correlation coefficients (PCC × 100) between 4 km² pixel-based mean monthly VHI for (**a**) spring (March–May) and (**b**) summer (June–August) with 3.4-Niño area mean monthly area-average SSTa for November–February for all 16 ENSO years (|SSTa| > 0.5 °C); Axis labels: Latitude (°), Longitude (°)

of ecosystems' response. Figure 8.6 shows PCC's color-coded maps of spring (March–May) and summer (June–August) monthly VHI with mean monthly (November–February) Niño-3.4 SSTa for all ENSO cases (|SSTa| > 0.5 °C). As seen, in North and South America, Africa, Australia, and Southeast Asia VHI-based enhanced sensitivity of land ecosystems toward ENSO in winter (Fig. 8.2) proceeds into spring (|PCC| > 0.55). It is expected that in northern Brazil, southern Africa, central Canada, eastern Australia, and southern parts of Borneo and Sumatra islands vegetation will be under stress, specifically thermal-based. Healthy vegetation conditions are expected in Argentina, Horn of Africa, and southwestern USA. However, as has been mentioned, vegetation health in the Horn of Africa is also affected by the Indian Ocean temperature. Currently, IO is cool triggering drier condition in the eastern part of the Horn. In summer (June–August), VHI-ENSO teleconnection in most areas becomes weaker, except for two regions: northern Brazil, where vegetation will continue to be under stress, and southwestern USA, where vegetation is expected to be healthy but not everywhere due to the impacts of the 10-year drought. In the other areas the signal of VHI-SSTa teleconnection was either weak or was shown on a very smaller area (southern Ethiopia, northern Namibia, and Botswana).

8.7 Conclusion

This chapter provides satellite-based numerical evaluation of global land ecosystem response toward ENSO during the last 36 years. Intensity and areas of ENSO and conditions of land ecosystems were evaluated with satellite-derived SSTa in the Niño-3.4 TP regions and vegetation health indices (VHI, VCI, and TCI) for each 4 km² land pixels. The VH indices were used to estimate land area's moisture, thermal, and the total health conditions. Teleconnection was derived by correlating pixel-based monthly VH indices in boreal winter, spring, and summer with monthly mean SST anomaly during November–February for all and the strongest ENSO events.

The results indicate that there are areas on all continents where vegetation is sensitive to ENSO during boreal winter. These areas and intensity of the impacts are in agreement with previous studies of teleconnection between ENSO versus weather parameters (precipitation and temperature). However, the advantages of this study are in derivation of vegetation response to moisture, thermal, and combined conditions during the two ENSO phases. ENSO impact was evaluated for different 3.4 Niño TP's SST anomalies ($|SSTa| > 0.5$, ≥ 1.5, and ≥ 2.0 °C). The results of ENSO impact on vegetation are stable for all thresholds. In ENSO years, vegetation of northern Brazil, southern Africa, islands of Southeast Asia, eastern Australia, western part of Africa' Sahel, and central Canada experiences moderate-to-strong stress during El Niño and healthy condition during La Niña. The opposite situation (healthy conditions in El Niño and unhealthy in La Niña) occurred in central and northern Argentina, southern Brazil, Horn of Africa, eastern Asia, and some parts of southwestern USA (especially California). Since the 2015–2016 strong El Niño has triggered drought in Brazil, southern Africa, southeastern Asia islands, and eastern Australia during December–February, agricultural crops and pastures were predicted to experience some losses of production, deteriorating food situation. On an opposite side, dry condition in the areas of mosquito-borne diseases will reduce the number of affected people in those regions. Such areas as Argentina, southwestern USA, and western Horn of Africa were predicted to experience favorable conditions for crops and pastures. Eastern part of the Horn experienced dry conditions in February 2016 because IO was warm (temperature was 1.5 °C above normal). Since IO temperature became cooler than normal in March 2016, the eastern Horn of Africa (including Ethiopia and Somalia) vegetation conditions have improved predicting plentiful crop production during the minor agricultural season in March–May, which is supported by Shemm (2016).

In some world regions affected by the strongest El Niño (SSTa >2.5 °C), vegetation conditions will be transitioned from boreal winter to spring. Thus, northern Brazil, southern Africa, central Canada, and southeastern Asia would be affected by drought, while Argentina, southwest USA, and the Horn (except eastern part) would have wet and normal vegetation condition. Summarizing, it is important to emphasize that ENSO is a very good indicator of 2–4 months advanced prediction of drought and non-drought conditions in the affected land areas (Figs. 8.2, 8.5 and 8.6). Farther research comparing VH indices with other ocean-atmospheric forcing (especially, due to mostly oceanic component, which determines the oscillation period), such as Indian ocean temperature, the quasi-periodic variations of the coupled ocean-atmosphere system in the North Atlantic (Atlantic Multidecadal Oscillation), and others would considerably improve drought prediction area, start time, intensity, duration, etc., helping to improve global and regional advanced food security assessments.

References

AccuWeather. 2016. Canada: Spring of 2016 May Rank in Top 10 Warmest on Record. http://view.s6.exacttarget.com/?j=fed112727164067d&m=fe8b12727d65027471&ls=fe8915737c610d7a71&l=ff5f12717c&s=fecf12717564067a&jb=ffcf14&ju=fe5011747c6d02787612&r=0.

Anyamba, A., C.J. Tucker, and J.R. Eastman. 2001. NDVI anomaly patterns over Africa during the 1997/98 ENSO warm event. *International Journal of Remote Sensing* 22 (10): 1847–1859. https://doi.org/10.1080/01431160010029156.

Anyamba, A., C.J. Tucker, and R. Mahoney. 2002. From El Niño to La Niña: Vegetation response patterns over East and Southern Africa during the 1997–2000 period. *Journal of Climate* 15: 3096–3103. https://doi.org/10.1175/1520-0442(2002)015<3096:FENOTL>2.0.CO;2.

Bastos, A., S.W. Running, C. Gouveia, and R.M. Trigo. 2013. The global NPP dependence on ENSO: La Niña and the extraordinary year of 2011. *Journal of Geophysical Research: Biogeosciences* 118: 1247–1255. https://doi.org/10.1002/jgrg.20100.

Bouma, M.J., and C. Dye. 1997. Cycles of malaria associated with El Niño in Venezuela. *Journal of the American Medical Association* 278: 1772–1774. https://doi.org/10.1001/jama.1997.03550210070041.

Bouma, M.J., and H.J. van der Kaay. 1996. The El Niño Southern oscillation and the historic malaria epidemics on the Indian subcontinent and Sri Lanka: An early warning system for future epidemics? *Tropical Medicine & International Health* 1: 86–96. https://doi.org/10.1046/j.1365-3156.1996.d01-7.x.

Brown, M.E., K. de Beurs, and A. Vrieling. 2010. The response of African land surface phenology to large scale climate oscillations. *Remote Sensing of Environment* 114: 2286–2296. https://doi.org/10.1016/j.rse.2010.05.005.

Buermann, W., B. Anderson, C.J. Tucker, R.E. Dickinson, W. Lucht, C.S. Potter, and R.B. Myneni. 2003. Interannual covariability in northern hemisphere air temperatures and greenness associated with El Niño-Southern oscillation and the Arctic oscillation. *Journal of Geophysical Research: Atmospheres* 108 (D13): 4396–4418. https://doi.org/10.1029/2002JD002630.

Dilley, M., and B. Heyman. 1995. ENSO and disaster: Droughts, floods, and El Niño/Southern oscillation warm events. *Disasters* 19: 181–193. https://doi.org/10.1111/disa.1995.19.issue-3.

Glantz, M.H. 1996. *El Nino Impacts on Climate and Society*, 345. New York: Cambridge University Press.

Hales, S., P. Weinstein, and A. Woodward. 1996. Dengue fever epidemics in the South Pacific: Driven by El Nino southern oscillation? *The Lancet* 348: 1664–1665. https://doi.org/10.1016/S0140-6736(05)65737-6.

Halpert, M.S., and C.F. Ropelewski. 1992. Surface temperature patterns associated with the Southern oscillation. *Journal of Climate* 5: 577–593. https://doi.org/10.1175/1520-0442(1992)005<0577:STPAWT>2.0.CO;2.

Hilker, T., A.I. Lyapustin, C.J. Tucker, F.G. Hall, R.B. Myneni, Y. Wang, J. Bi, Y. Mendes de Moura, and P.J. Sellers. 2014. Vegetation dynamics and rainfall sensitivity of the Amazon. *Proceedings of the National Academy of Sciences U. S. A.* 111 (45): 16041–16046. https://doi.org/10.1073/pnas.1404870111.

Huber, S., and R. Fensholt. 2011. Analysis of teleconnections between AVHRR-Based sea surface temperature and vegetation productivity in the semi-arid Sahel. *Remote Sensing of Environment* 115 (12): 3276–3285. https://doi.org/10.1016/j.rse.2011.07.011.

IPCC. 2014. Climate change 2014. *Synthesis Report*. 5th Assessment, eds. R. K. Pachauri and L. Meyer. Geneva, Switzerland, 151 pp. https://www.ipcc.ch/pdf/assessment-report/ar5/syr/SYR_AR5_FINAL_full_wcover.pdf.

IRI (International Research Institute). 2016. Why Do We Care about El Niño and La Niña? http://iri.columbia.edu/our-expertise/climate/enso/why-do-we-care-about-el-nino-and-la-nina/.

Kidwell, K.B. 1997. *Global Vegetation Index Users Guide*, 67. Washington, DC: Department of Commerce, NOAA/NESDIS, National Climate Data Center.

Kogan, F., and W. Guo. 2017. Strong 2015–2016 El Niño and implications to global ecosystems from space. *International Journal of Remote Sensing* 38 (1): 161–178. https://doi.org/10.1080/01431161.2016.1259679.

Kogan, F., T. Adamenko, and W. Guo. 2013. Global and regional drought dynamics in the climate warming era. *Remote Sensing Letters* 4: 364–372.

Kogan, F., W. Guo, and A. Jelenak. 2011. Global vegetation health: Long-term data records. In *Use of Satellite and In-Situ Data to Improve Sustainability.*, ed. F. Kogan, A. Powell, and O. Fedorov, 247–255. Dordrecht: Springer.

Kogan F., Z. Popova and P. Alexandrov 2016. Early forecasting corn yield using field experiment dataset and Vegetation health indices in Pleven region, north Bulgaria. *Ecologia i Industria (Ecology and Industry)* 9 (1): 76–80.

Kogan, F.N. 2001. Operational space Technology for Global Vegetation Assessment. *Bulletin of the American Meteorological Society* 82: 1949–1964. https://doi.org/10.1175/1520-0477(2001)082<1949:OSTFGV>2.3.CO;2.

———. 2000. Satellite-observed sensitivity of world land ecosystems to El Niño/La Niña. *Remote Sensing of Environment* 74: 445–462. https://doi.org/10.1016/S0034-4257(00)00137-1.

———. 1998. A typical pattern of vegetation conditions in southern Africa during El Nino years detected from AVHRR data using three-channel numerical index. *International Journal of Remote Sensing* 19 (18): 3689–3695.

———. 1997. Global drought watch from space. *Bulletin of theAmericanMeteorologicalSociety* 78:621–636.https://doi.org/10.1175/1520-0477(1997)078<0621:GDWFS>2.0.CO;2.

———. 1990. Remote sensing of weather impacts on vegetation in non-homogeneous areas. *International Journal of Remote Sensing* 11 (8): 1405–1419. https://doi.org/10.1080/01431169008955102.

Kovats, R.S. 2000. El Nino and human health. *Bulletin of the World Health Organization* 78 (9): 1127–1135.

Lao, N.C. 2016. Model diagnosis of El Niño teleconnections to the global atmosphere-ocean system. *Bulletin of the American Meteorological Society* 97 (6): 981–988.

Lemonick, M.D. 2013. Global warming: El Niño link stronger but still not proven. http://www.climatecentral.org/news/global-warming-el-nino-link-stronger-but-still-not-proven-15427.

Mathews, K. 2014. How California can survive a prolonged drought. http://www.care2.com/causes/how-california-can-survive-a-prolonged-drought.html.

Mecray, E. 2016. NOAA El Niño resources and coordination. http://www.nrcc.cornell.edu/services/special/reports/NEelnino_jan2016.pdf.

Monastersky, R. 2016. Monster El Niño probed by meteorologists. *Nature* 529: 267–268. https://doi.org/10.1038/529267a.

Nicholls, N. 1993. El Nino–Southern oscillation and vector-borne disease. *The Lancet* 342: 1284–1285. https://doi.org/10.1016/0140-6736(93)92368-4.

NOAA. 2015. ENSO discussion: El Niño is here. https://www.climate.gov/news-features/blogs/enso/march-2015-enso-discussion-el-ni%C3%B1o-here.

NOAA2. 2015. MonthlySST data. ftp://ftp.emc.ncep.noaa.gov/cmb/sst/oimonth_v2/YEARLY_FILES/.

NOAA. 2016. Teleconnection El Niño. https://images.search.yahoo.com/yhs/search;_ylt=A0LEVvxuFLJW3WsAmQYnnIlQ;_ylu=X3oDMTByMDgyYjJiBGNvbG8DYmYxBHBvcwMyBHZ0aWQDBHNlYwNzYw–?p=Teleconnection+El+Nino&fr=yhs-mozilla-003&hspart=mozilla&hsimp=yhs-003.

NOAA2. 2016. El Niño Advisory. https://www.climate.gov/enso.

NOAA/NESDIS 2018. Vegetation Health indices and products. http://www.star.nesdis.noaa.gov/smcd/emb/vci/VH/index.php

Potter, C., S. Klooster, M. Steinbach, P. Tan, V. Kumar, S. Shekhar, R. Nemani, and R. Myneni. 2003. Global teleconnections of climate to terrestrial carbon flux. *Journal of Geophysical Research* 108 (D17): 4556–4563. https://doi.org/10.1029/2002JD002979.

Potter, C., S. Klooster, M. Steinbach, P.N. Tan, V. Kumar, S. Shekhar, and C.R. De Carvalho. 2004. Understanding global teleconnections of climate to regional model estimates of Amazon ecosystem carbon fluxes. *Global Change Biology* 10 (5): 693–703. https://doi.org/10.1111/j.1529-8817.2003.00752.x.

Preacher, K.J., D.D. Rucker, R.C. Maccallum, and W.A. Nicewander. 2005. Use of the extreme groups approach: A critical reexamination and new recommendations. *Psychological Methods* 10 (2): 178–192. https://doi.org/10.1037/1082-989X.10.2.178.

Reynolds, R.W., N.A. Rayner, T.M. Smith, D.C. Stokes, and W. Wang. 2002. An improved in situ and satellite SST analysis for climate. *Journal of Climate* V15: 1609–1625. https://doi.org/10.1175/1520-0442(2002)015<1609:AIISAS>2.0.CO;2.

Reynolds, R.W., and T.M. Smith. 1995. A high-resolution global sea surface temperature climatology. *Journal of Climate* V8: 1571–1583. https://doi.org/10.1175/1520-442(1995)008<1571:AHRGSS>2.0.CO;2.

Ropelewski, C.F., and M.S. Halpert. 1987. Global and regional scale precipitation patterns associated with the El Niño/southern oscillation. *Monthly Weather Review* 115: 1606–1626. https://doi.org/10.1175/1520-0493(1987)115<1606:GARSPP>2.0.CO;2.

———. 1996. Quantifying southern oscillation-precipitation relationships. *Journal of Climate* 9: 1043–1059. https://doi.org/10.1175/1520-0442(1996)009<1043:QSOPR>2.0.CO;2.

SCCONC (State Climate Office of North Carolina). 2015. Global Patterns—El Niño-Southern Oscillation (ENSO). https://climate.ncsu.edu/climate/patterns/ENSO.html.

Shemm. 2016. Trying to hold off disaster. *The Washington Post*, February 23, 1 & 11.

Suplee, C. 1999. El Niño/La Niña. *National Geographic* 195: 73–95.

Trenberth, K.E. 1997. Short-term climate variations: Recent accomplishments and issues for future progress.

Trenberth, K.E., and T.J. Hoar. 1997. El Niño and climate change. *Geophysical Research Letters* 24: 3057–3060. https://doi.org/10.1029/97GL03092.

WER. 1998. Cholera in 1997. *Weekly Epidemiological Record* 73 (27): 201–208.

WFP (World Food Program). 2016. El Niño: Implications and Scenarios for 2015. http://documents.wfp.org/stellent/groups/public/documents/ena/wfp276236.pdf.

WMO (World Meteorological Organization). 1997. The 1997–1998 El Niño: A Scientific and Technical Retrospective. *Bulletin of the World Meteorological Organization,* Geneva (WMO No 905).

WRCC (Western Regional Climate Center). 2015. Will El Niño Make a Difference? http://www.water.ca.gov/waterconditions/docs/Drought_ENSO,Jul2015_handout.pdf.

Climate Change and Food Security Current and Future

9

9.1 Introduction

Since the beginning of the nineteenth century, the Earth's climate has changed significantly. Earth's warm up resulted in never before experienced environmental events and strong disasters. Some countries during that period suffered from anomalous heat, enormous shortages of rains and harsh and/or snowy winters, atypical for those places. As many scientific and social publications have described, climate changes triggered many unusual environmental, economic, and social events, from a number of extreme droughts followed by hunger in Africa, to withering heat in South Asia resulting in human death, to multiyear forest fires in North America to devastating hurricanes and land floods in 2017. One of the biggest climate-warming concerns is how these changes have affected agricultural production and food security (FS) and what to expect in the near and distant future. Will FS intensify, leading to more people suffering from a lack of food and hunger, especially considering that Earth's population is increasing faster than food production grow (see Chap. 2). Experts from the United Nations and some countries are warning that climate warming will negatively affect crop yield, especially underdeveloped countries of Africa, Asia, and Latin America, which will lead to more problems with food security. According to some forecasts, yields of corn, rice, and wheat will drop nearly 25% by 2050 (Alexandratos and Bruinsma 2012). One principal question is how the multiyear climate warming has changed the environment and land cover and how these changes have affected the development of agro-industrial complex, tendencies (deterioration/improvement) of events, what has occurred with food security and what to expect in the future.

It is currently expected that climate change will affect both developing and developed countries following an increased risk from drought, floods, hurricanes, and other natural disasters (IPCC5 2014). What kind of changes we can expect in the near future? For example, in the last 100 years, the USA's crop area has been reduced to lowest levels (WB 2017). Is this a result of an economic situation or climate warming impacts and how will this affect the rating of the world agricultural producers? Another example: Ukrainian farmers expect that warm winters might increase the yield of winter wheat. Is there any hope that in the near future, Ukraine, which is now the number eight world producer of grain (WB 2017), will be able to rise to the world's top wheat producers? How might climate warming affect agricultural production and will it affect the cost of growing crops, taking into account the adaptation of seeds and plant protection agents to the new climatic conditions? Will the climate influence the appearance of new plant diseases

F. Kogan, *Remote Sensing for Food Security*, Sustainable Development Goals Series,
https://doi.org/10.1007/978-3-319-96256-6_9

and pests? Finally, the most important question, what changes can be expected with land cover change, natural disasters, especially droughts and food security?

It is difficult to answer all of these questions since they relate not only to climate change but also to economics, policy, science, and even traditions. However, food security in relation to climate change issues, especially such as FS's culprit as drought, will be discussed with the application of 38-year satellite-based vegetation health (VH) data and products for analysis of relationship between changing climate and biological and environmental parameters, such as land cover change (greenness and temperature), drought (start/end, intensity, area, and duration) dynamics, vegetation health conditions, changes in moisture and thermal vegetation stress, tendencies in land surface changes, and others (Kogan et al. 2013, 2015a, 2016, 2017,). The 38-year weekly data used in this analysis have 1, 4, and 16 km^2 pixel resolution and can be presented by a pixel, areal mean for latitude–longitude coordinate box, administrative region, country, continent, hemisphere, and the entire world. These multiyear data and products are ready to be used without additional processing, which has been done, for assessment of the existing and near-future dynamics of changes. In addition to the first-order changes (drought, greenness, crop stress and production losses, etc.), the discussion will touch upon soil saturation with water (waterlog), fire risk, and other events and activities. These VH data can be used also for long-term predictions in agriculture, forestry, climate change and forcing, health, invasive species, diseases, land cover change, etc. Following these assessments, a very important conclusion would be drawn about current and future global and regional environmental safety, food security, and sustainability.

9.2 Earth Climate Change and Consequences

Global temperature measurements showed that in the past 100 years, Earth's climate has been warming. The average global temperature over the past 100 years (from 1906) increased 0.74 °C (IPCC4 2007). Following IPCC Fourth and Fifth assessments (IPCC4 2007 and IPCC5 2014, respectively), in the past 30 years, following a climate warming, few very important global environmental events have occurred: ice in the North Polar region was melting, the ice area was shrinking, and the sea level was rising. The recent National Academy of Science report indicated that melting ice sheets in Greenland and Antarctica are speeding up the already fast paces of sea level rise. Following this rate, the world's ocean will be at least 2 ft higher by the end of the current century. In the past 20 years, environmental observations also showed global changes in snow and ice areas, the sea level, biological systems (plants, birds, etc.), and others (IPCC5 2014). On the land, it has been reported that climate change is causing spring to begin earlier, prompting insects to move to Texas (USA) sooner and giving the bats something to eat. Therefore, bats are migrating to Texas roughly 2 weeks earlier (mid-March instead of late March) than they were moved 22 years ago (Chandler 2018). Besides, the same publication emphasizes an increase in corn and soybean production in the USA's Midwest due to a trend toward cooler and wetter summers. One very important issue of climate warming consequences is drought intensification and expansion, water scarcity and the gradual deterioration of agricultural system, which is expected to affect five billion people by 2050 (UNESCO 2018; Watts 2018).

Regarding Earth vegetation, some research has shown an early greening, especially in the northern latitudes (Lucht et al. 2002; Myneni 1997; Nemani et al. 2003; Forzieri et al. 2017). These results were obtained from the analysis of 15–17 years of the normalized difference vegetation index (NDVI), calculated from the Advanced Very High Resolution Radiometer (AVHRR) measurements on board NOAA operational polar-orbiting satellites. Currently, almost 20-year NDVI data have been added to the AVHRR-based NDVI time series, including the new S-NPP and J-1 (NOAA-20) satellite data from more advanced VIIRS sensor, improving the entire NDVI time series and adding also land cover temperature records. Processing almost 4-decade data records improved considerably through comprehensive noise correction, new results in data analysis, and, what is the most important, satellite data has been validated comprehensively against in situ biological and weather observations. Moreover, following biophysical and ecosystem laws the new theory of vegetation health (VH) algorithm was introduced, permitting to develop VH-based products and new applications for agriculture, forestry, human health, climate forcing, natural disasters, and others. All of these innovations permitted to develop the new 38-year global vegetation health (VH) dataset and products and used them for analysis of tendencies in drought, vegetation greenness, moisture and thermal-based crop stress and yield prediction (Kogan 2001; Kogan et al. 2013, 2015a, b, 2016, 2017, 2018).

9.3 Causes of Climate Change

Following IPCC4 (2007) and IPCC5 (2014) reports, continuous global warming is the result of fuel burning (coal, wood, etc.), cement production, etc., resulted in greenhouse gas emission, specifically CO_2, which intercept infrared (IR) solar radiation and warm up land surface and atmosphere. Therefore, in the past 20 years, there has been a strong focus on the need to reduce CO_2 emission in order to mitigate climate warming and develop some measures for adaptation. The Kyoto Protocol and the subsequent Adaptation Fund were the first steps to encourage the international community begin working on these goals (UNFCCC 2014). The Kyoto Protocol was issued in 1997 and in 8 years was ratified by the participants, obliging industrial countries to cut greenhouse gas emission by 5% (compared to 1990 level) by 2008–2012. An intensive global campaign began from the publication of the book "Inconvenient Truth" (Gore 2006) and a film with the same title. Those sources show a diagram with global temperature increase time series and the matching CO_2 increase and other displays as a prove that global warming is the result of CO_2 release in the atmosphere and the trapping of heat. Some of the conclusions were quite scary: "the world at the edge of climatic catastrophe" … "if not stop emitting CO_2, we come to the point of no return" (Gore 2006). In addition to CO_2 some other greenhouse gases with smaller concentration are CH_4, N_2O, and SF_6. However, according to IPCC5 (2014) "CO_2 is the largest single contributor to radiative forcing over 1750–2011 and its trend since 1970." Therefore, human activity in releasing greenhouse gases, specifically CO_2, is considered to be a cause of the current global warming (IPCC 2014). Following an intensive greenhouse gases-based United Nations' actions, in 2015, 195 countries agreed with the so-called "Paris Agreement" to reduce emission of greenhouse gases into the atmosphere (Gray 2016). According to the Agreement, countries have to cut emission of CO_2 (coming from a burn facile fuel) to keep the global temperature under 1.5 °C (and if possible below 2.0 °C).

The largest world CO_2 emitters are currently China, the USA, the European Union, India, Russia, and Japan. These countries contribute 68% of the total global CO_2 emission, having only 51% of global population (Berwyn 2016). From 195 countries, 179 signed the Agreement (UN 2016). Regarding the emission amount, it is known that in the twenty-first century, China increased coal consumption by 19% (Bastasch 2017; Ren et al. 2017; Miller 2017). More than one half of Russia's territory (east of Ural Mountains) is burning wood and coal to warm houses during very harsh winters. I worked as an agricultural meteorologist in East Siberia for 4 years and know that there are little chances to change this trend. Following the signed international documents, if CO_2 emission exceeds the level, assigned by the Agreement, countries have to pay a fine in the fund distributed to developing nations. Economists calculate that fulfilling the Agreements of CO_2 limitation goal and keeping global temperature inside the indicated limits would cost the world $10 trillion (Ward 2016a). The United States of America agreed to cut emission from 28 to 26% by 2025, otherwise the USA has to pay a fine. The US Heritage Foundation calculated that if the USA pays its share assigned by the Agreement, then by 2035, losses of Gross Domestic Product (GDP) would account for $2.5 trillion (Groves 2016).

During Earth's history, climate changed many times. As Dr. Ward (2016a, b) presented, "The Earth climate has never been settled." We know that millions of years ago (before appearance of human beings on the Earth), near-polar regions were heavily forested. For example, the Norwegian Spitsbergen (known as west Spitsbergen in the archipelago) is well known as a coal-mining place since the nineteenth century (Harland et al. 1976). Availability of coal means that millions of years ago archipelago was quite forested due to warm climate and, following severe climate cooling and ice formation, the remnants of the forest were turned into coal being inside the soil without oxygen over millions of years (Harland et al. 1976). It is also known that between 100,000 and 10,000 years ago, the Earth warmed up approximately every 3600 years from the ice age to 10–16 °C; 25 times, these changes from warm to ice age and back has happened (Ward 2016a). Around 12,000 years ago, more than a half of the Earth was covered with ice, which melted very quickly with climate warming (Schlossberg 2016). In the tenth century, the Vikings discovered Greenland, which was covered with heavy vegetation due to warm climate but currently is under a thick ice. Little ice age was during 1340–1700 (Haldon et al. 2018), which was changed by intensive warming up at the beginning of the seventeenth century (Alley et al. 2010). That means that Earth has always been affected by strong climate changes which resulted in some period with complete changes in Earth cover. During these ancient times, forest fires and volcanoes were practically the only source of CO_2 release in the atmosphere. Therefore, if no human-activities triggered CO_2 release million and thousands of years ago, but the Earth's climate fluctuated strongly from warm to cold and back with the corresponding changes of Earth cover, from heavy vegetation to ice sheets and back, then a logical question to ask is what other factors (in addition to volcanoes, fires, and currently human activity) are responsible for climate changes.

The IPCC Report (IPCC5 2014), discussing the "Past and recent drivers of climate change," correctly focused on the atmosphere, oceans, cryosphere, natural and anthropogenic radiative forcing, and human activities. However, it is well known that except for the indicated in the report factors, the climate of the Earth is strongly controlled by solar activity, the distance between the Earth and the Sun, which is cyclical such as changes in the angle of Earth's rotation axes (to the perpendicular toward the plate of Earth rotation around the Sun), atmospheric pressure, ocean-atmosphere thermal balance, and volcanoes (Ward 2016a). Some other less-frequent factors such as cosmic events (interconnection between Earth with Moon and Sun) and internal Earth forces should also be considered, in case of their events. In addition, other climate-affecting factors such as the quasi-periodic oscillations of the meridional heat transport (MHT) in the North Atlantic are usually regarded as the main mechanism for the formation of low-frequency variations in SST and heat fluxes on the ocean–atmosphere boundary in the North Atlantic (i.e., of the AMO) and others. The importance of the AMO mechanism in changing the climate is confirmed by the fact that approximately 16-year phase shift is observed between the low-frequency variations of absolute humidity in the surface layer of the atmosphere (leading to the variations of latent heat fluxes through the sea–air boundary) and SST in the northwest part of the North Atlantic (Polonskii and Voskresenskaya 2004). Also, such a large-scale atmospheric circulation pattern as the North Atlantic Oscillation (NAO), Pacific Decadal Oscillation (PDO), Atlantic Multidecadal Oscillation (AMO), Arctic Oscillation (AO), Atlantic Meridional Overturning Circulation (AMOC), and others affect earth climate (Serreze 2018; Weisberger 2018; Rice 2018; Ward 2016a). Moreover, some research showed that quasi-periodical inter-decadal warming and cooling of the North Atlantic is of the same order or even exceeds human-induced warming (Kerr 2005; Raa et al. 2004). The most recent *Science* article by Samset (2018) focuses on human activities' emission of aerosols particles, which produced an overall cooling effect on earth, blocking sunlight penetration. Moreover, climate models indicate that the cooling effect might be as large as 0.5 °C and should be taken into consideration calculating the balance of CO_2 (temperature warming) and aerosol's (temperature cooling) contribution to Earth warming.

9.4 Global Temperature and CO_2 Trend

Following Sect. 9.3 discussion, climate change is driven by multidimensional factors. However, the explanation of the current climate warming is considered to be related to greenhouse gasses, especially by human activities of CO_2 release (from fossil fuel burning) in the atmosphere (IPCC5 2014; IPCC4 2007; UNFCCC 2014; Gore 2006). Therefore, before coming to the assessment of the current climate warming consequences for land cover changes, agricultural production dynamics, drought trend, and, finally, food security, it is important to analyze the global mean temperature anomaly (TA) and CO_2 dynamics (from a well-established source) in detail for a more precise understanding of their roles in food security situation. Figure 9.1 presents TA and CO_2 trends from IPCC5 (2014) together with average North and South Hemisphere's temperature anomaly trends (Hansen et al. 2000, 2010). Very many publications, especially the UN-based (UNESCO 2018, UNFCCC 2014, 2015; IPCC4 2007; IPCC5 2014), indicate a matching *general* upcoming trend in global CO_2 (Fig. 9.1b) with both global (Fig. 9.1a) and hemispheric (Fig. 9.1c) temperature anomaly warming up. Sometimes it is hard to compare the TA data, presented in publications, since they estimate global earth temperature relative to different basic climatic periods (1950–1990 or 1961–2000 or 1986–2005 or others). Our analysis is based on the recent IPCC report, which uses 1986–2005 base climatic period (IPCC5 2014). Following IPCC5 (2014), during 162 years (1850–2012), global Earth mean temperature anomaly increased almost 100% (from −0.7 to +0.3 °C) and CO_2 has increased from 2 to 40 $GtCO_2$/Yr (190%). Hemispheric temperature anomaly increased since 1980 from −0.3 to +0.7 °C for the Northern Hemisphere and to nearly +0.5 °C for the Southern Hemisphere. The strongest both CO_2 and temperature increase occurred after 1980. Moreover, as IPCC5 (2014) indicates "… the period from 1983 to 2012 was likely the warmest 30-year period of the last 1400 years in the Northern Hemisphere," when "the global average combined land and ocean temperature data calculated as linear trend, show a warming of 0.85 °C (0.65–1.06 °C) between 1980 and 2012."

However, a more precise analysis of global temperature anomaly and CO_2 trends (Fig. 9.1a, b) indicate that at the background of the *stably increasing general 150-year trend in CO_2 up to 2015, temperature anomaly (both global and hemispheric) has experienced three opposite type short-term (17–35 year) trends.* They are shown with red lines compared to a gray line of the general TA trend from 1850 to 2012 in Fig. 9.1a. This mismatch was first indicated by Dr. Ward (2016). Our analysis is focused on short-term trends in global temperature anomaly, on the reasons of this mismatch, and how these problems will affect the food security situation in the currently changing climate. First, during the period of *slow CO_2 increase* in the past 100 years (1850–1950) global temperature anomaly experienced three short-term trends: (a) 1850 through 1880s, (b) 1880s through 1910, and (c) 1910 through mid-1940s. The analysis indicates that, at the general background of 2.5 times CO_2 increase (from 2 to 7 $GtCO_2$/Yr, Fig. 9.1b) during 1850–1880 (31-year, case (a)), global temperature anomaly trend was *flat* (no increase, TA around −0.3 °C for 31-year), *decreasing* TA trend (intensification of negative anomaly from −0.3 °C to −0.5 °C for 25–30 years in case (b, 1880s–1910) and a strong *increasing* TA (changed from −0.5 °C to 0 °C) between 1910 and mid-1940s), which was continuing for 30–35 years. Second, following the strongest, *4.7 times CO_2 increase* (from 7 to 40 $GtCO_2$/Yr, Fig. 9.1b) from the approximately mid-1940s, there was *declining* trend in TA during 30-year, up to mid-1970s (TA dropped nearly 0.2 °C from −0.3 to −0.5 °C), *strongly increasing TA (upward trend)* during

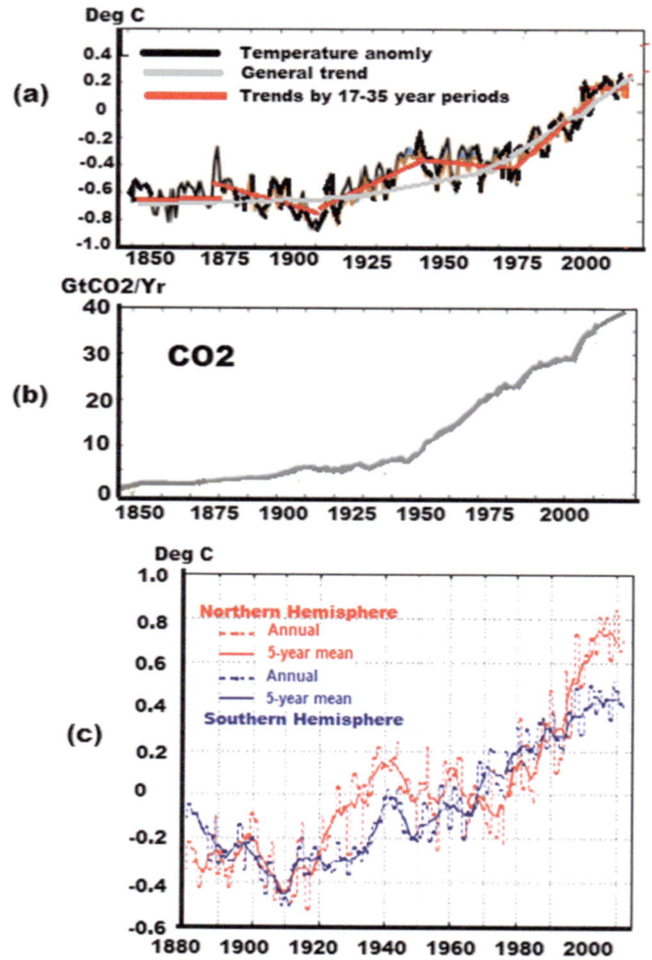

Fig. 9.1 Global (**a**) mean land and ocean temperature anomaly (TA, relative to the 1986–2005 mean (IPCC5 2014)), Lines: Black—TA, Gray—general trend (1850–2012), Red—17–35 years trend; (**b**) atmospheric CO_2 emission from burning fossil fuel, cement production and flaring, as well as from forestry and other land use (1850–2012), (**c**) North and South Hemisphere temperature anomaly (1880–2009, Hansen et al. 2010)

25-year from mid-1970s to late 1990s (TA increase from −0.5 to +0.15 °C), and, finally, the most important, *flat trend in TA* (around +0.2 °C) during a 17 year period, between 1998 and 2014 (Ward 2016a), while CO_2 increased strongly (nearly 50% from 28 to 40 $GtCO_2$/Yr in the same 17-year). The 17 year (1998–2014) period of flat trend is called by climatologists as a "*hiatus*" (Karl et al. 2015). It is interesting that short-term trends in TA for both Hemispheres (Hansen et al. 2010) are similar to the described global short-term trend, although Northern Hemisphere's TA is showing more precise match with the globe.

Unfortunately, the most important, United Nations-approved recent climate warming publication (IPCC5 2014) has not provided well-determined explanations to mismatches between strong CO_2 increase trend and the up/down and hiatus trends in global temperature anomaly. Some discussions have been about the 30-year (mid-1940s to mid-1970s) decline in TA trend at the background of a very intensive CO_2 increase, considered as a cause of climate warming (IPCC4 2007; IPCC5 2014; Kennel 2014; Brahic 2007; Kerr 2005; Berardelli 2010; NOAA/NCDC 2017). As some analysis suggest, the 1945–1975 period of climate cooling, followed by elevated industrial and volcanic aerosols in the atmosphere during the post-World War II period of return to a peaceful life (Kennel 2014; Brahic 2007; Kerr 2005). Following the indicated climate cooling, by the mid-1970s, the world was very concerned about a 30-year global temperature reduction, specifically, was it possible for this reduction to lead to an ice age and what impact it may have on human activities?

I remember that time very well since I worked as a senior agricultural meteorologist at the USSR Hydrometeorological Center, the main forecasting weather, ocean, hydrology, and agrometeorology organization. I was responsible for weather-based modeling and forecasting of USSR's annual grain production. Grain in the USSR was the most important agricultural product, used for estimation of food security. The grain production forecast has been issued regularly for the central government only in May and June. The USSR had been generally very concerned about the amount of grain produced by Soviet agriculture annually, since that amount was much below what was needed for food and feed. According to nonofficial estimates, Soviet agriculture had to produce every year one million tons of grain per each person. Between 1946 and 1980 (the year of my emigration from the USSR), that goal had never been achieved. The maximum grain collected during that period was 208 million metric tons (MMT) in 1978 (Kogan 1983) for the total USSR population of about 260 million people (Anderson and Silver 1990). In 1972, the USSR was affected by the strongest drought (since 1946), when only 160 MMT of grain was collected (my prediction issued in early May 1972 was 162 MMT). Following a lack of grain in general (compared to USSR population), extremely low grain production following the 1972 drought and the cooling global temperature trend, the USSR Government asked the Hydrometcenter to provide a long-term forecast, what changes in Soviet grain production might be expected due to climate cooling. We had performed a lot of modeling, calculations and estimated changes in the amount of Soviet grain production due to the reduction of grain crops area in the northern European USSR, potential improvement in grain production in Kazakhstan and others following continuation of global cooling. Fortunately, no measures had been taken by the USSR Government, since in mid-1970s, global warming has started again.

In summary, the recent years' United Nations (WMO, UNEP) climate change actions, following the IPCC reports and other scientific publications, have strongly emphasized that the CO_2 increase has very likely been triggering global warming and has resulted in negative consequences for the environment and society (UNESCO 2018; UNFCCC 2014, 2015; IPCC4 2007; IPCC5 2014). Available data (Fig. 9.1) confirm a matching *stably* upcoming trend in global CO_2 (Fig. 9.1b) and *general trend* (gray line) in global and hemispheric TA increase. However, during the entire 1850–2014 period of *continuing CO_2 increase*, global temperature anomaly experienced six 17–35-year periods with opposite trends: *flat* from 1850 to 1880s, *decreasing* from 1880s to 1910, *increasing* from 1910 to mid-1940s, *decreasing* from mid-1940s to mid-1970s, *strongly increasing* from mid-1970s to late 1990s, and *flat or hiatus time trend* from 1998 to 2014. A reasonable question to ask is why, at the background of continuing initially slow and during industrial time strong CO_2 increase (a) the global mean TA changed its trend six times? and (b) during 1998-2014, global warming stopped at the background of intensive CO_2 growth?

9.5 New Ideas About the Causes of Global Warming

In the recent two decades, there have been many different ideas about global warming in addition to CO_2-triggered TA upward-going (general) trend. They included changes in solar activity, El Nino Southern Oscillation (ENSO), large-scale atmospheric circulation patterns changes (NAO, AO, etc.), and others (Kennel 2014; Brahic 2007; Kerr 2005; Berardelli 2010; NOAA/NCDC 2017; Freedman 2017; Polonskii and Voskresenskaya 2004; Hansen et al. 2000, 2010; Lucht et al. 2002; Raa et al. 2004; Ren et al. 2017; Chandler 2018). One of the new ideas came from the Book of Dr. Ward (2016a). The Book (Ward 2016a) not only discusses the new cause for global warming but also investigates the deeply physical principle of greenhouse gases-based global warming, its history, development, duration, and impacts.

Following Dr. Ward's (2016a) investigation, the greenhouse warming idea has appeared in 1859 from physical experiments by J. Tyndall, showing that greenhouse gases (GHG) absorb some infrared (IR) radiation. Later, spectral physicists have learned in laboratory experiments that IR radiation is absorbed into oscillations of the bonds, holding each molecule of GHG together. However, it has been unknown how the oscillatory bond energy is transferred to kinetic energy, which is heating the atmosphere and land. It was assumed that energy transfer is going through numerous collisions of GHG molecules, resulting in broken chemical bonds holding molecules together. In 1900, Angsröm, a well-known physicist, showed in physical experiments that the IR warming effect is minimal. Meanwhile, the recent 20–40 years, climate models, built on GHG theory, has shown reasonable results of the Earth's warming. But until 1998, during the 17-year period, between 1998 and 2014, a flat global TA (around 0.2 °C) trend was not matching with CO_2 strong upward trend (increase from 30 to 40 $GtCO_2$/Yr, or 33% (IPCC5 2014)). This was also confirmed by Dr. Christy in his testimony before the U.S. House Committee on Science, Space and Technology in March 2017. He demonstrated that the rates of warming Tropical Mid-Tropospheric temperature depicted by the 102 climate models for the period 1979–2016 was significantly different from the observations (Christy 2017). Other scientific publication blamed declining solar irradiance during the two last 11-year cycles and ENSO (Coddington et al. 2016; Kennel 2014; Berardelli 2010) for TA hiatus during 1998–2014.

Dr. Ward (2016a) explained that the upward global temperature trend from the early 1970 was due to chlorine-induced ozone depletion in the atmosphere and a flat temperature trend during 1998–2014, due to ozone restoration. The principle of this process has the following explanation. Ozone (O_3) is a colorless gas, in the stratosphere is a shield absorbing ultraviolet type B (UV-B) radiation protecting living things on the Earth. The process of ozone depletion began in the 1960s following human activities in using widely chlorofluorocarbon gases (CFC) as refrigerants, paints, perfumes, lubricants, cooking oil, and others. By 1974, scientific research has shown that when CFC gases rise to the lower stratosphere, they are broken down by UV-B radiation, resulting in a release of chlorine, which destroys the ozone. It was estimated that one chlorine atom might destroy up to 100,000 molecule of ozone through a catalytic reaction (Ward 2016a, b). The process of ozone destruction was very extreme in polar regions, resulting in development of the so-called ozone hole over Antarctic in 1985, where ozone was depleted in half. It was slightly less depleted in the Arctic, up to 15% in mid-latitudes and very little in the tropics (Bromwich et al. 2013). Ultraviolet radiations are high energy electromagnetic waves emitted by the Sun which, if enters the Earth's atmosphere, can lead to a number of health-related issues for all living organisms and also various environmental issues including global warming (Ankit 2015). Following the chlorine-based destruction of the ozone shield, Earth began to warm up. It was clear that the global society must stop releasing CFC gases into the atmosphere. Following this goal, the United Nations Montreal Protocol limited manufacturing of CFC gases. As the result, by 1993, chlorine increase in the atmosphere stopped, by 1995, ozone depletion stopped, and by 1998, the global temperature increase stopped as well (Ward 2016a).

Following this idea, a few research and measurements (Ward 2016a, b; Staehelin et al. 1998; Solomon 1999; NOAA/NCDC 2016; Levitus et al. 2012; NOAA 2016, 2017; NASA 2018; Farman et al. 1985; Bromwich et al. 2013) a diagram was developed showing this process. Figure 9.2 shows 1945–2014 trends in global temperature anomaly and CO_2 in the atmosphere, O_3 and chlorine accumulation in the stratosphere. Ozone trend clearly indicates its depletion along with accumulation of chlorine in the stratosphere. This diagram supports the previous paragraph's discussion on the physical principal of ozone depletion. It shows that ozone depletion has initiated as soon as chlorine began to increase in the early 1970s. At the same time, TA began to increase strongly. CFC release in the atmosphere reduced considerably in 1993, following intensive

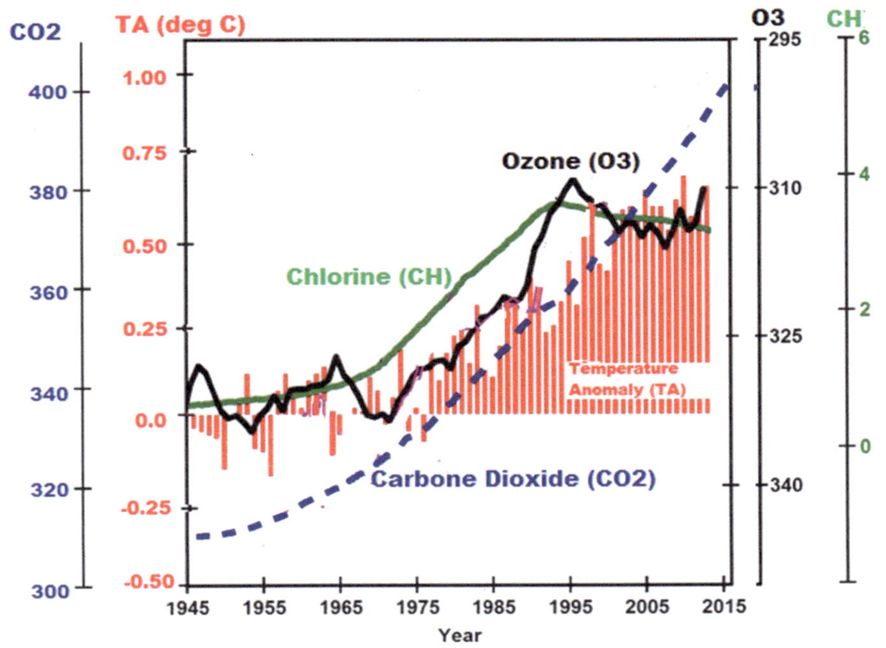

Fig. 9.2 1945–2014 global mean temperature anomaly (TA, Degree C), carbon dioxide (CO_2, in Parts per million), ozone (O_3, in Dobson units), and chlorine (in Parts per billion) from CFC gases

international efforts to limit using CFC gases. From that time, the amount of chlorine started to reduce as well and the most important that ozone depletion practically stopped in the next 17 years (between 1998 and 2014).

Trend analysis in Fig. 9.2 indicates that during 1975–1997, the strongest upward trend was observed for three parameters: increase in TA and CO_2 and depletion in O_3. During that 23-year period, the temperature anomaly increased from 0 to +0.5 °C with matching CO_2 growth from 330 to 350 parts per million (PPM) and ozone decreased from approximately 332 to 313 Dobson Units (DU). Since both O_3 depletion and CO_2 increase had approximately the same rate (around 6%) during 1945–1995, it is hard to understand which of these parameters triggered Earth's warming (strong TA increase). The answer came from an analysis of these three parameters dynamics after 1997. The most interesting differences in matching O_3 and CO_2 trends with TA changes began in 1998 and continued through 2014. The amount of CO_2 in the atmosphere during those 17 years continued to increase intensively (from 375 to 400 PPM, or 7%), while TA did not change (flat trend), remaining at the level of +0.6 °C during hiatus time (Karl et al. 2015). The CO_2 and O_3 trends' mismatch indicated that carbon dioxide increase has not stimulated rising global temperature during 1998–2014. Here where ozone comes as the potential source of global temperature stabilization during the 17-year period, since ozone restoration in the lower stratosphere recreated the natural screen, protecting the Earth from dangerous UV-b radiation's penetration to the earth surface. It is known that ultraviolet radiations are high energy electromagnetic waves emitted by the Sun, which if it enters the Earth's atmosphere, can lead to a number of health-related issues for all living organisms and also intensifies global warming (UCS 2017; Ankit 2015). The diagram in Fig. 9.2 clearly shows a very good match between TA trend and O_3 depletion trend from the mid-1970s through 2014. Two cycles in both data are perfectly seen: strong increase in TA and O_3 during 1975–1997 (23-year) and, what is the most important, matching almost flat trend in both parameters during 17-year from 1998 through 2014. Therefore, following Fig. 9.2 and physics of ozone concentration as Earth protective shield (Ward 2016a, b; Ankit 2015), strong depletion of ozone layer was accepted as additional to CO_2 cause of climate warming during 1975–1997 and the only cause of no warming during hiatus time from 1998 through 2014.

The only point of concern is the causes of considerable global warming in the 3-year (2015, 2016, and 2017) after-hiatus time (NOAA 2016; NOAA/NCDC 2016, 2017; NOAA/NCEI 2017). Following NOAA reports, the global land and ocean mean temperature was above the twentieth century average by 0.90 °C in 2015, 0.94 °C in 2016, and 0.84 °C in 2017 (NOAA/NCEI 2017). Figure 9.3 demonstrates IPCC5 (2014) based mean land temperature anomaly (annual lines) and trends (general and short-term) with added TA (bars, adjusted to the IPCC5 2014 climatology) during 2013–2017 from the NOAA reports (NOAA 2016; NOAA/NCDC 2016, 2017; NOAA/NCEI 2017). The 2013–2014 TA does not indicate any increase compared to the 1998–2012 trend. However, during 2015–2017, global TA was the highest since 1880 (NOAA/NCDC 2017; NOAA/NCEI 2017). Some scientific publications blame strong El Nino of 2015–2016 for such a warm up, others blame solar activity, ocean-atmosphere interaction, etc. (ClimateBet 2018; Haldon et al. 2018; GWCh 2018; UCS 2017; Samset 2018; NOAA/NCEI 2017; NOAA/NCDC 2017; Kogan et al. 2013; Kogan and Guo 2014, 2015, 2017). The most interesting fact is that the highest global TA between 2015 and 2017 period occurred after 17 years (1998–2014) of stable (no increase) global TA during climate hiatus time. Another interesting fact is that NOAA reports indicate only the values of global TA and its distribution over the Earth, but do not indicate the causes of such warming, which is very important for understanding if TA increase is going to continue. If it is, then a new multiyear global warming trend should be expected after a hiatus time of flat global TA trend. If global temperature is going to reduce after 2017, then hiatus-type stable global TA trend would continue.

Early 2018 analysis of publications related to this matter indicates that the first 3 months of (January–March) 2018 show much cooler global temperatures. Following ClimateBet (2018) report, the global TA (deviation from 1981 to 2010) mean for January–March was 0.71 °C in 2016, 0.35 °C in 2017, and 0.23 °C in 2018. So, in the first 3-month 2018, global TA cooled off considerably compared to the previous 2 years. It is simply a fact, which does not indicate what TA will be for the entire 2018 and it is not known currently, what to expect. Summarizing this climate discussion for our food security goals, it is important to emphasize that (a) during 1981–1997, there was a match in the trend of TA and CO_2 increase and O_3 depletion; and (b) during 1998–2014, global TA mismatched with CO_2 increase and matched with O_3 depletion trends;

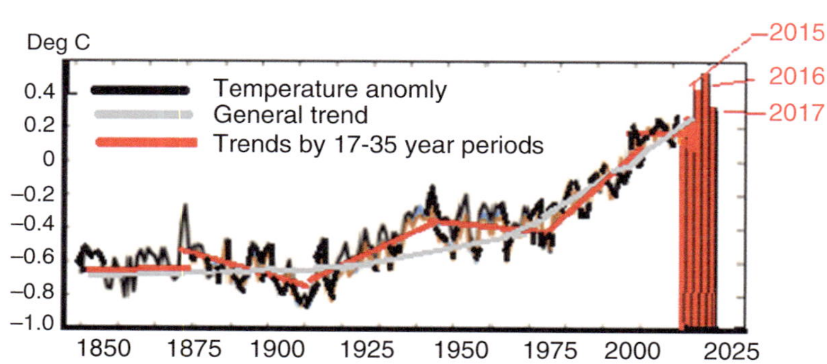

Fig. 9.3 Global mean land temperature anomaly (relative to the 1986–2005 mean) during 1850–2012 (IPCC5 2014), Trends: Gray—general, Red—short-term (17–35 years) and temperature anomaly (adjusted to the IPCC5 2014 climatology) during 2013–2017 (red bars)

(c) the world was very warm during 2015–2017, without naming the definite causes (except strong El Nino in 2015–2016 and possible solar activity); (d) global January-March mean TA considerable decrease in 2018 compared to previous 2 years. These conclusions do not explain what TA changes are expected in the future. Therefore, further discussion related to analysis of the food security situation will cover investigation of land cover changes, specifically, greenness, radiative temperature of vegetation, droughts, potential crop losses during the two periods of strong global warming (1981–1997), and flat TA trend during hiatus time from 1998 through 2014. The tendencies of the indicated parameters will be analyzed in relation to food security potential.

9.6 Global Land Cover Changes During Climate Warming and Hiatus Time

The 38-year vegetation health data (NOAA/NESDIS 2018) has been used for this analysis. The greenness and radiative temperature were investigated by the no noise normalized difference vegetation index (NDVI), called smoothed NDVI (SMN) and brightness temperature (BT), called smoothed temperature (SMT). Moisture and thermal condition of vegetation were investigated with the vegetation health indices, VCI (moisture), TCI (thermal), and VHI (combined moisture–thermal). Crop losses were assessed by VH-based drought area, intensity, and produced impacts. Almost four-decade trends all of these parameters were investigated during the periods of intensive global warming (TA increase in 1981–1997) and flat TA trend during hiatus time from 1998 through 2014. The interest to these two periods of global climate change were stimulated by coincidence of an intensive increase in TA and CO_2 trends and O_3 depletion trend in 1981–1997 and *mismatch between TA and* CO_2 *trends versus match between TA and* O_3 *depletion trends in 1998–2014*. The analysis covers the entire world, continents, countries, and major grain areas. Land cover greenness, temperature, moisture, and thermal conditions trends were estimated by a linear regression of weekly index value (k) against weeks and years. An intensity of the trend was estimated by statistical analysis of linear trend through (a) slope (S) and (b) trends' relative differences (RD). The slope (y) was estimated by linear equation's trend (9.1). The RD was calculated for the corresponding parameters' values taking the difference between the trend end (t_j) end trend beginning (t_i) relative to the t_i (Eq. 9.2).

$$\text{Trend}_k = x_k + y_k \times \text{week and year}_k \quad (9.1)$$

$$\text{RD}_k = 100 \times \left(t_{kj} - t_{ki}\right) / t_{ki}, \quad (9.2)$$

where k is the investigated index (SMN, SNT, VCI, TCI, VHI), x is intercept, y is slope, RD—relative difference, t_{kj}—values at the trend end, and t_{ki}—values at the trend beginning.

9.6.1 World

Table 9.1 presents the changes in the global mean vegetation greenness (SMN) and surface temperature (SMT) during a 37-year by the cumulative periods 1981–1990, 1981–2000, 1981–2010, and 1981–2017. Following the table, it is seen that since the NOAA operational afternoon polar-orbiting satellite began measurements in 1981,

Table 9.1 Trend intensity in global mean vegetation greenness (SMN) and temperature (SMT) by cumulative intervals during 1981–2017

Parameter	Measured intensity	1981–1990 (10-year)	1981–2000 (20-year)	1981–2010 (30-year)	1981–2017 (37-year)
SMN from NDVI	%	7.63	7.13	11.91	11.61
	Weekly slope ($\times 10^{-3}$)	1.4	0.7	0.7	0.6
SMT from BT	%	20.45	6.58	10.24	20.90
	Weekly slope ($\times 10^{-3}$)	240	42	43	69

global warming has led to a greening up of global land surface with intensity of slightly over 7% in the first 10 (1981–1990) and 20 (1981–2000) years and nearly 12% thereafter. The strongest weekly SMN increase occurred in the first 10-years (weekly slope 1.4), with the stable smaller increase (0.6–0.7) thereafter. Mean global SMT (temperature) of land surface shows a 20% increase (quite strong) in the first 10 years, coinciding with the strongest global warming (TA increase), considerable SMT reduction to 7% during 1981–2000 when the hiatus period began in 1998 (but CO_2 has still continued its strong upward trend), and further SMT increased to 21% at the end of 2017, when the last three the warmest years are included in the trend analysis. Since scientific literature does not clearly explain the causes of strong TA increase during 2015–2017, the next analysis includes two periods only: strong TA increase after mid-1970s (VH data from 1981 to 1997 or 17-year) and hiatus time of stable TA trend during 17-year between 1998 and 2014. NDVI-based SMN and BT-based SMT for these two periods are presented in Fig. 9.4.

Following Fig. 9.4a, it is clearly seen that during an intensive global warming from 1981 through 1997, the entire world became 8.8% greener by 1997 (compare to 1981), which approximately coincides with the 20-year results in Table 9.1. Meanwhile, during the hiatus time (1998–2014, with a flat global TA trend) greenness reduced considerably to 3.3%, which is opposite to the 30-year trend intensity in Table 9.1. So, stable global TA during the hiatus time reduced intensity of land cover greenness by almost three times. Brightness temperature (SMT, Fig. 9.4b) during these two periods (global TA increase and hiatus-based no increase) enhanced 16–18%, but no big changes occurred between the two periods (15.7% versus 18.2%, correspondingly). These changes in vegetation greenness (SMN) and vegetation cover brightness temperature (SMT) during hiatus time correspond to TA stability in 1998–2014 and do not support the concept of CO_2 stimulating global warming. Since global ecosystems are very different over the entire global land, the next step was to investigate changes in SMN and SMT over latitudes and longitudes during the same two periods of TA trends: intensive increase and no increase, or stable during the hiatus time.

Global changes in land cover greenness (SMN) and brightness temperature (SMT) were investigated over each 16 km^2 latitude lines (Fig. 9.5). As seen in the figure, SMN and SMT changes are not uniform over latitudes. In the first period (1981–1997) of the strongest global warming (Fig. 9.1), vegetation became much greener (SMN changed between 10 and 25%) in the Northern Hemisphere but no changes in greenness are seen (0–2% SMN increase) in Southern Hemisphere (diagram on the left of the world image of Fig. 9.4a). Latitudes 40–70° N (Canada, northern Russia, and Europe) showed

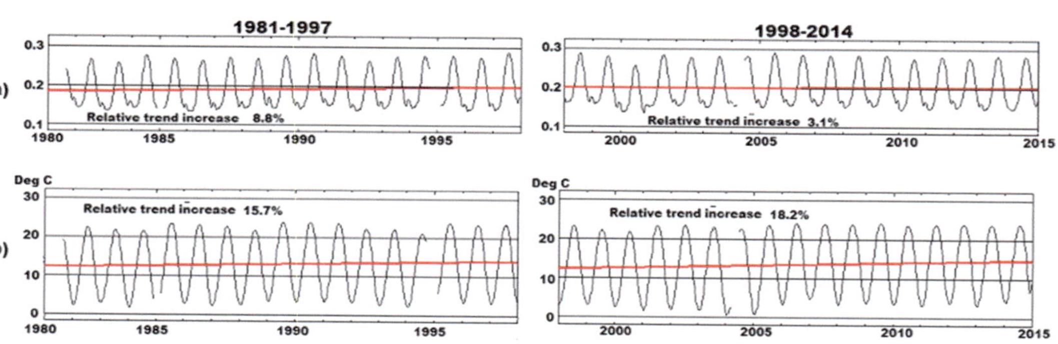

Fig. 9.4 Global area-mean annual (**a**) SMN and (**b**) SMT weekly time series and linear trend (red line) during 1981–1997 (the period of the strongest global warming or TA increase) and 1998–2014 (hiatus period of the stable TA or flat trend)

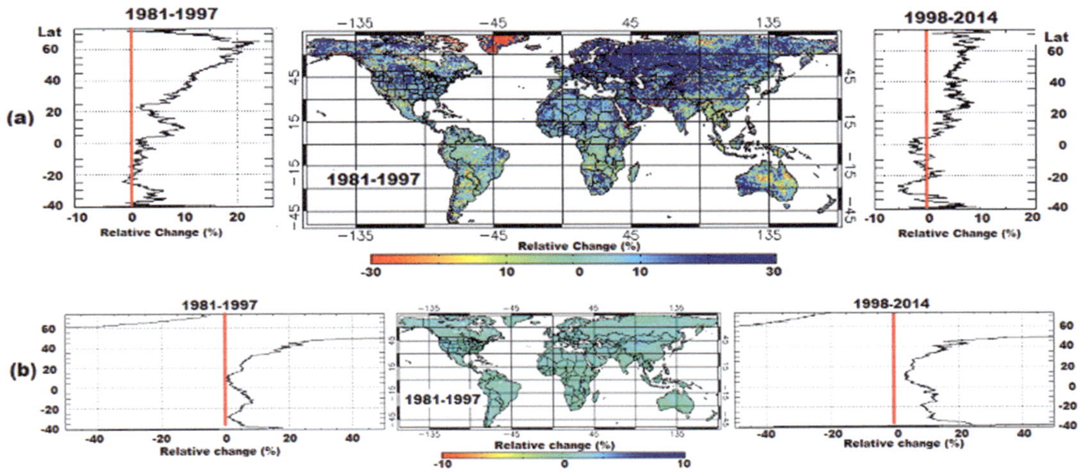

Fig. 9.5 Global changes in vegetation greenness (SMN) and brightness temperature (SMT) for each 16 km^2 lines (between −180° and +180°) from 75° N to 45° S for two periods strong global warming during 1981–1997 and hiatus-time flat global TA trend from 1998 through 2014

the strongest greening up, although for the central Sahara (20° N), SMN trend is near zero, which is natural for desert. Comparison of greenness during the hiatus-time (1998–2014, stable global TA) with the strongest global warming during 1981–1997 clearly indicates considerable greenness reduction to 7–9% in the Northern Hemisphere and even slight reduction in the Southern Hemisphere, which matches with no increase (flat trend) in TA. The most important conclusion from 16 km^2 global lines of SMN trends indicates increase in vegetation greenness during 1981–1997, no greenness increases in the Southern Hemisphere, and considerable greenness reduction during hiatus time of flat trend in global TA (1998–2014).

Land cover brightness temperature (SMT) analysis by 16 km^2 lines indicates more complicated than SMN lat-long changes (Fig. 9.5b). At the background of strong global warming in 1981–1997, the very northern latitudes of the globe (north of 60° N) experienced strong cooling, up to 40% relative decrease in SMT (left diagram from the image in Fig. 9.5b); area between 60 and 35° N experienced strong warming (SMT increase up to 50%) and the area below 35° N warmed up slightly (up to 10% SMT increase). The Southern Hemisphere shows only minor warming (up to 8% SMT increase) during the period of strong global TA growth. It is interesting that during the hiatus time (right diagram from the image in Fig. 9.5b) SMT is showing similar to the 1981–1997s 17-year trends, with up to 15% relative warming (SMT increase) in the Southern Hemisphere.

Summarizing latitudinal temperature (SMT) analysis, it is important to emphasize that, at the background of periods with strong global warming in earlier years (1981–1997) and no uprising TA trend during hiatus years (after 1997), (a) northern latitudes have been cooling, (b) Southern Hemisphere experienced much smaller warming, and (c) two investigated periods showed very similar relative temperature changes over latitude lines. Longitudinal global SMT trend (diagrams over and under the image) is also similar for the two periods of increasing (before 1998) and flat global TA trends (from 1998): 20–60% relative cooling (SMT decline) over Alaska (125–160° W), 10–50% warming (SMT increase) over the North (except Alaska) and South America (120–40° W), stable temperature (no SMT change) over Europe and Africa (0–50° E), and 20–50% warming up (SMT increase) over Asia (except Russian's Kamchatka) and Australia (50–150° E).

9.6.2 Continents

Since trends in global land surface greenness and radiative temperature follows the mean global TA trends in the two 17-year (1981–1997 of strong warming and 1998–2014 no warming) and there are strong differences moving from north to south (based on 16 km^2 latitude lines), the next discussion provides an analysis of continental land surface SMN and SMT trends for the indicated two periods. Table 9.2 presents relative changes in SMN-based vegetation greenness and SMT-based surface temperature for the continents. As has been mentioned in global analysis, following an intensive global warming before 1998, global earth surface became nearly 9% greener, but that greenness was reduced to 3% during hiatus-based stability of global TA (Table 9.2). Similarly, all continents showed mostly an increase in land surface greenness. However, the greenness rate is quite different. Europe shows the strongest greening up (22.5% almost three times more than the world rate) and Asia follow Europe with a 12% greenness increase during an intensive global warming. North America greenness increase (8.4%) is at the level of global rate. However, the most important fact is that all continents in the Southern Hemisphere (South America, Africa, and Australia) showed a very small rate of vegetation greening up (2–5%) compared to the intensity of global warming. The next very important fact is that land surface greenness increased for all continents during the strong global warming (1981–1997) but reduced in half during the hiatus period of stable global TA. Moreover, continents of the Southern Hemisphere showed even negative greenness (slight reduction, −1% for South America to −5% in Australia). Summarizing, we should emphasize that (a) continental land cover greenness is matching well with global warming trend in the two periods (1981–1997 and 1998–2014), (b) intensity of greenness is changing considerably between the continents, (c) the continents in Southern Hemisphere are showing the lowest greenness rate, and (d) considerable greenness decreases during the hiatus time.

Table 9.2 Relative changes (%, formula (a) and (b)) in greenness (SMN) and brightness temperature (SMT) during the two 17-year periods of global strong warming in 1981–1997 and no warming in 1998–2014 (hiatus time)

Continents	Greenness (SMN) 1981–1997	Greenness (SMN) 1998–2014	Brightness temperature (SMT) 1981–1997	Brightness temperature (SMT) 1998–2014
World	8.8	3.1	14.7	21.6
North America	8.4	3.0	62.1	50.0
Europe	22.5	12.1	33.6	49.6
Asia	12.1	6.6	45.5	78.4
South America	2.2	-0.7	8.0	12.3
Africa	5.1	−2.1	4.2	6.6
Australia	2.9	−5.2	1.7	9.2

Similar to trends in vegetation greening (SMN), radiative temperature (SMN) of continental lands is warming up for both periods. However, most continents (except North America) are showing a continuation of a warm up during the hiatus time (1998–2014) with an increased rate compared to the previous period. Unfortunately, we did not find any reasonable explanation of this event why mean radiative temperature of vegetation cover has continued to increase during the 17-year period of stable global temperature. Following Fig. 9.5, the SMT rate changes over latitudes considerably, even including changes in sign (from negative values in the far north and west to positive values in the south and east). This fact would require further analysis of the SMT rate change in agriculturally important countries, presented below. Another interesting fact is that at the background of strong global warming during 1981–1997, the continental vegetation of Southern Hemisphere warmed up slightly (2–8%), especially compared to the Northern Hemisphere (15–62%).

9.6.3 Principle Grain-Producing Countries

For the purposes of the current food security assessment and future prediction, it is important to analyze the connection between the global warming situation and changes in land cover. This has been done for the world and continents (see above). This part covers the main grain-producing countries since they have a very strong contribution to global FS estimation and prediction. The countries were selected from several hundred world countries based on their main contribution to global grain production in 2014, following WB (2017) data. Table 9.3 shows the principal grain producers on each continent and explains why the countries were selected. The three main contributors to the global grain production, which was 2818 MMT in 2014, were China (19.8% contribution), the USA (15.7%), and India (10.5%). It is interesting that area-wise China (number one grain contributor) and India (number 3) had practically the same grain area (98.5 and 96.4 million hectares (MH)), while the USA, being the number two contributor to the global grain production, collected grain crops from the area (58.4 MH) of 70% less than other two countries. This is an indicator of strong agricultural technology applied to growing grain crops in the USA. The fourth, fifth, and sixth contributors to the global grain production in 2014 were Russia, Indonesia, and Brazil (3.2–3.7% contribution). Russia collected grain from an area two times larger than Indonesia and Brazil, indicating that agricultural technology in Russia is much poorer. From other important world countries with less than 3% contribution to the 2014 global grain production, we should indicate France, Ukraine, Germany, Canada, and Argentina (1.8–2.5%). Most of these countries were investigated in terms of correspondence of their land cover changes to climate warming. Although Australia contributed only 1.3% to the world grain production, it was selected for analysis as well, since it is one of the principal contributors (4–8%) to the world grain market (Kogan et al. 2018). From the African continent, Ethiopia was selected for analysis, being one of the largest Africa grain producer, although it contributes only 0.8% to the world's grain production in 2014.

The further discussion covers the three principal grain producers of China, the USA, and India. Since these countries have a large area, the investigation covers not only changes in greenness and radiation temperature of the entire countries but also area changes (by 16 km^2 lines) in both directions latitudes (north-south) and longitudes (west-east). Table 9.4 shows relative changes in greenness (SMN) and radiation temperature (SMT) during the same two periods: intensive global warming during 1981–1997 and flat global TA trend during 1998–2014. The principal grain-producing countries show some similarities and differences in changes of vegetation greenness

Table 9.3 Grain production (million metric tons (MMT)), area (million hectares (MH), WB 2017), and their (%) from the world total

Continent	Country	Grain production in 2014		Grain area in 2014	
		Amount (MMT)	% from the world	Amount (MH)	% from the world
N. America	USA	442.8	15.7	58.4	8.1
	Canada	51.3	1.8	14.1	2.0
	Mexico	36.5	1.3	10.3	1.4
Europe	France	73.3	2.5	9.6	1.3
	Germany	52.0	1.8	6.3	0.9
	Ukraine	63.4	2.2	14.0	2.0
Asia	China	557.4	19.8	96.4	13.4
	India	296.4	10.5	98.5	13.7
	Russia	103.1	3.7	44.4	6.2
	Indonesia	89.8	3.2	18.1	2.8
S. America	Brazil	101.4	3.6	20.1	2.8
	Argentina	51.0	1.8	13.2	1.8
Africa	Rep. S. Africa	16.6	0.6	2.7	0.4
	Ethiopia	23.6	0.8	10.2	1.4
Australia	Australia	38.4	1.3	17.0	2.4
World		2818.5	100.0	718.1	100.0

Table 9.4 Relative changes (%) in greenness (SMN) and radiation temperature (SMT) in the principal grain-producing countries (contribute 10–20% to total world grain production) during periods of intensive global warming (1981–1997) and no warming (flat global TA trend) during hiatus time (1998–2014)

Countries	Greenness (SMN)		Radiation temperature (SMT)	
	1981–1997	1998–2014	1981–1997	1998–2014
China	4.8	11.9	17.8	18.3
USA	8.0	1.4	11.6	10.6
India	8.4	8.0	2.8	6.9

and radiation temperature. During the intensive global warming (1981–1997) all countries indicated greenness intensification (5–8.4%) by 1997. However, the transition to a hiatus period is quite different. The USA, located farther to the north, shows a considerable decrease in greenness rate (1.4%), which is in agreement with transition of the world from an intensive warming to no warming (flat global TA trend) in 1998–2014. China doubled the greenness rate to 11.9% and India's greenness rate remained the same (8.0%). In the rate of radiation temperature, China and the USA showed considerable vegetation temperature increase (12–18%) by 1997, but India showed only a slight increase (2.8%). Moreover, no changes between the periods of global warming and no warming (flat trend). Summarizing, the principal grain producers (China, the USA, and India) showed country-specific features in vegetation greenness and surface temperature changes not completely related (except US greenness) to the global temperature trends. Since the principal grain-producing countries have a large area, the following analysis covers greenness-temperature changes by 16 km^2 latitude–longitude lines.

Figure 9.6 displays area (along latitudes and longitudes) dynamics of vegetation greenness (SMN) and brightness temperature (SMT) in the three major grain-producing countries, China, the USA, and India. Although all of China shows a small greenness increase (4.8% in Table 9.4) during the intensive global warming (1981–1997) two-thirds of its territory (in north-south direction) indicates a much stronger greenness increase (6–15%, especially at the latitudes 34–40° N, left diagram from the image) by the end of 1997 (Fig. 9.6Aa). Most interesting is that southern China (south of 26° N) shows greenness reduction (up to −5%). Transition to hiatus time (right diagram from the image) of stable global TA intensified greenness up to 20% (including positive relative greenness up to 10% in the southern latitudes, which had negative greenness in the previous period). Moreover, northern China (latitudes 43–46° N) in 1998–2014 experienced greenness reduction (up to −6% relative values). Similar to latitude trends, China's greenness along longitudes did not changing uniformly: far west and most of China's east showed an elevated greenness (up to 5%) by the end of 1997 (diagram over the image (Fig. 9.6Aa). Greenness intensified strongly (up to 20%) during the hiatus time (diagram under the image (Fig. 9.6Aa). Temperature of vegetation cover (Fig. 9.6Ab) showed more changes (compared to greenness) over latitudes and longitudes. During an intensive global warming (1981–1997), all of China's latitudes experienced warming as well, but more intensive (20–35%) in the northern half. During hiatus time (1998–2014), latitudinal temperature continued the warming trend; however, in the middle latitudes (28–36° N) relative changes increased up to 40%, while in the northern and southern latitudes relative greenness shrank to 5%. Relative longitudinal changes in vegetation temperature for the two discussed periods (1981–1997 and 1998–2014) in China were approximately similar (upper and lower diagrams around the image in Fig. 9.6Ab).

Among the three principal grain-producing countries, the US vegetation greenness is following global temperature tendencies. During the strongest global warming (1981–1997), the USA experienced relative greenness increase (up to 10% by 1998) in all latitudes and longitudes (left and upper diagrams from the US image on Fig. 9.6Ba). Following a stable global temperature during the hiatus time (1998–2014), the US vegetation relative greenness reduced to 0–5% by 2015 over all latitudes and longitudes (right and lower diagrams from the US image on Fig. 9.6Ba). The USA's latitude–longitude trends of vegetation surface temperature (relative changes 8–25%) is matching with strong global warming during 1981–1997 (left and

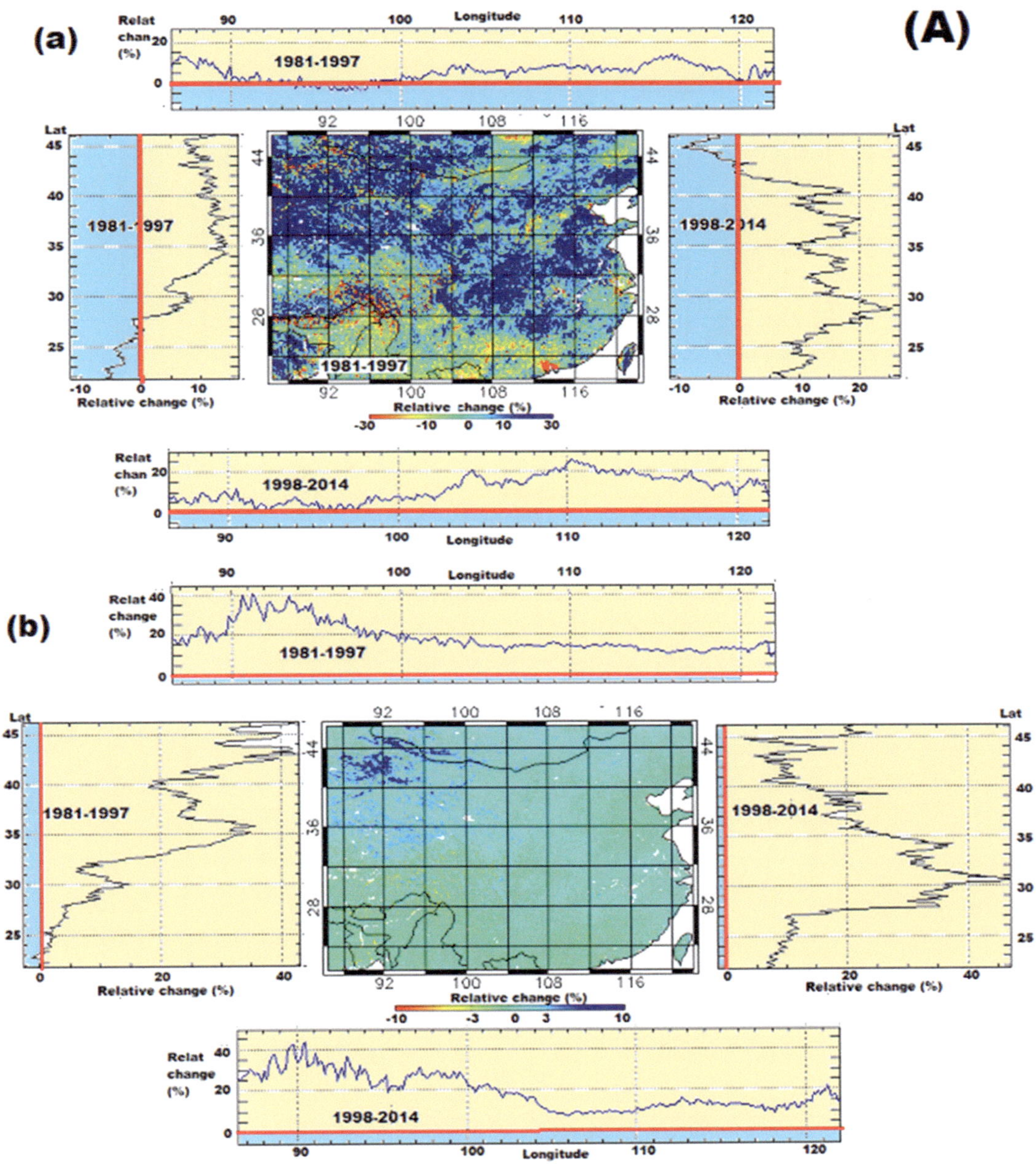

Fig. 9.6 Changes in vegetation greenness (SMN) and brightness temperature (SMT) for each 16 km^2 line area along latitudes (left and right diagrams from the images) and longitudes (upper and lower diagrams from the images) in the three principal grain-producing countries: (A) China, (B) the USA, and (C) India (in each country, (**a**) is greenness and (**b**) is radiation temperature) during the two periods of strong global warming (1981–1997) and hiatus-time (no warming or flat global TA trend) between 1998 and 2014

upper diagrams from the US image in Fig. 9.6Bb). During the stable global TA (1998–2014), the USA's north-south and west-east line trends (right and lower diagrams from the US image) continue, although with slightly smaller relative changes (5–20% for north-south and 2–40% for west-east).

Being closer to the equator, India shows its own specific features in vegetation greenness and surface temperature distribution over latitudes

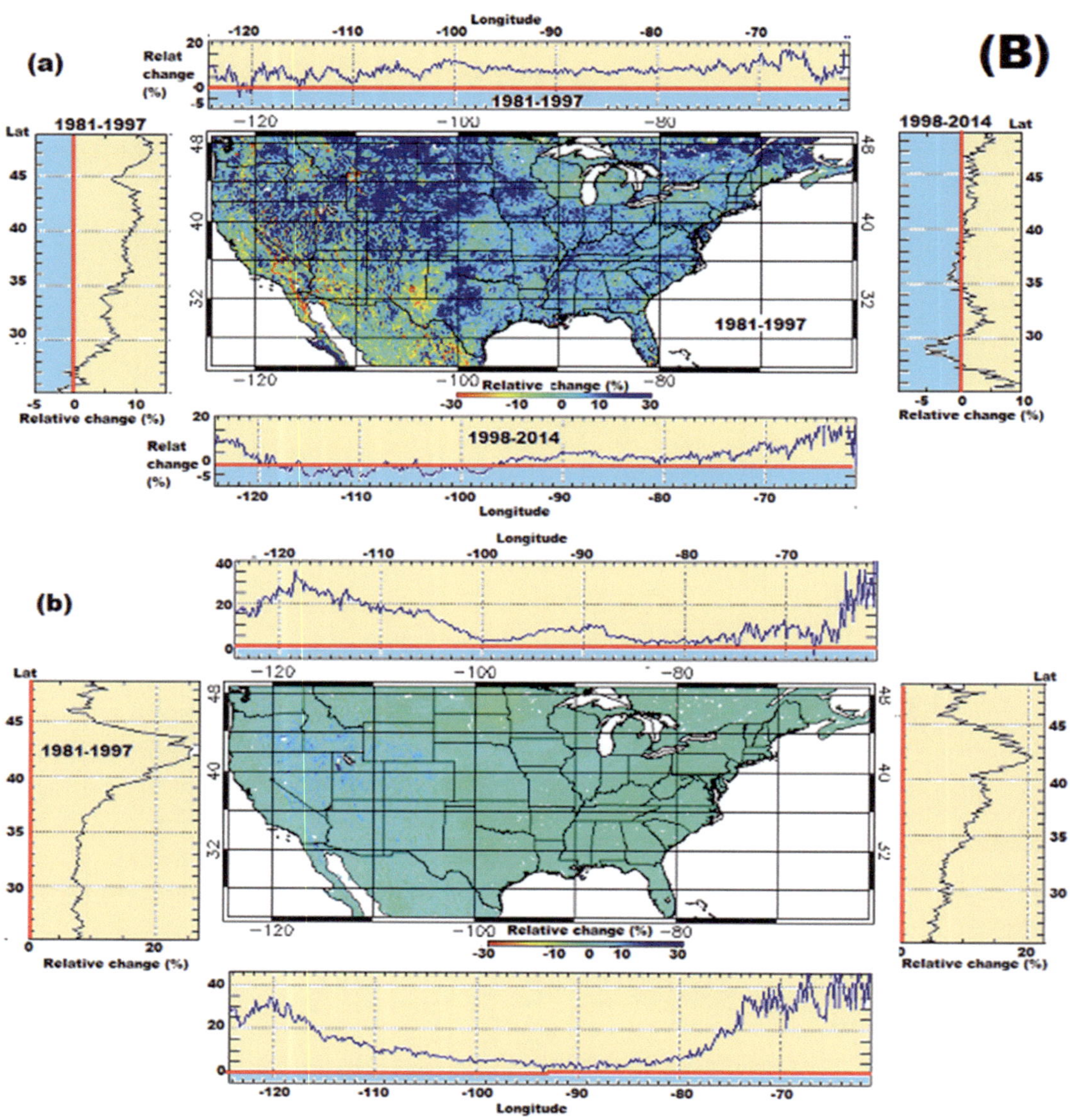

Fig. 9.6 (continued)

and longitudes. During the period of intensive global warming, all 16 km^2 latitudes' lines greened up with relative intensity of 5–15% (diagram on the left side of the image on Fig. 9.6Ca), although the greening changes frequently from line to line compared to China and the USA. In the period of stable global TA during hiatus time, greening is continuing (diagram on the right side of the image on Fig. 9.6Ca), although with slightly reduced intensity (up to 10%). Moreover, in the southern tip of India, the greening up stopped (the sign of relative change turned negative, up to −10%, from a positive in the previous period). Greening is still changing frequently along latitudes. Longitudinal greenness over India is changing similarly in the two investigated periods (intensive global warming and no warming in hiatus time): western half of India is greening up intensively, especially in the far west (68–73° E), where relative greenness increased 20–40%. Vegetation cover temperature in most of India has not changed much except for far north

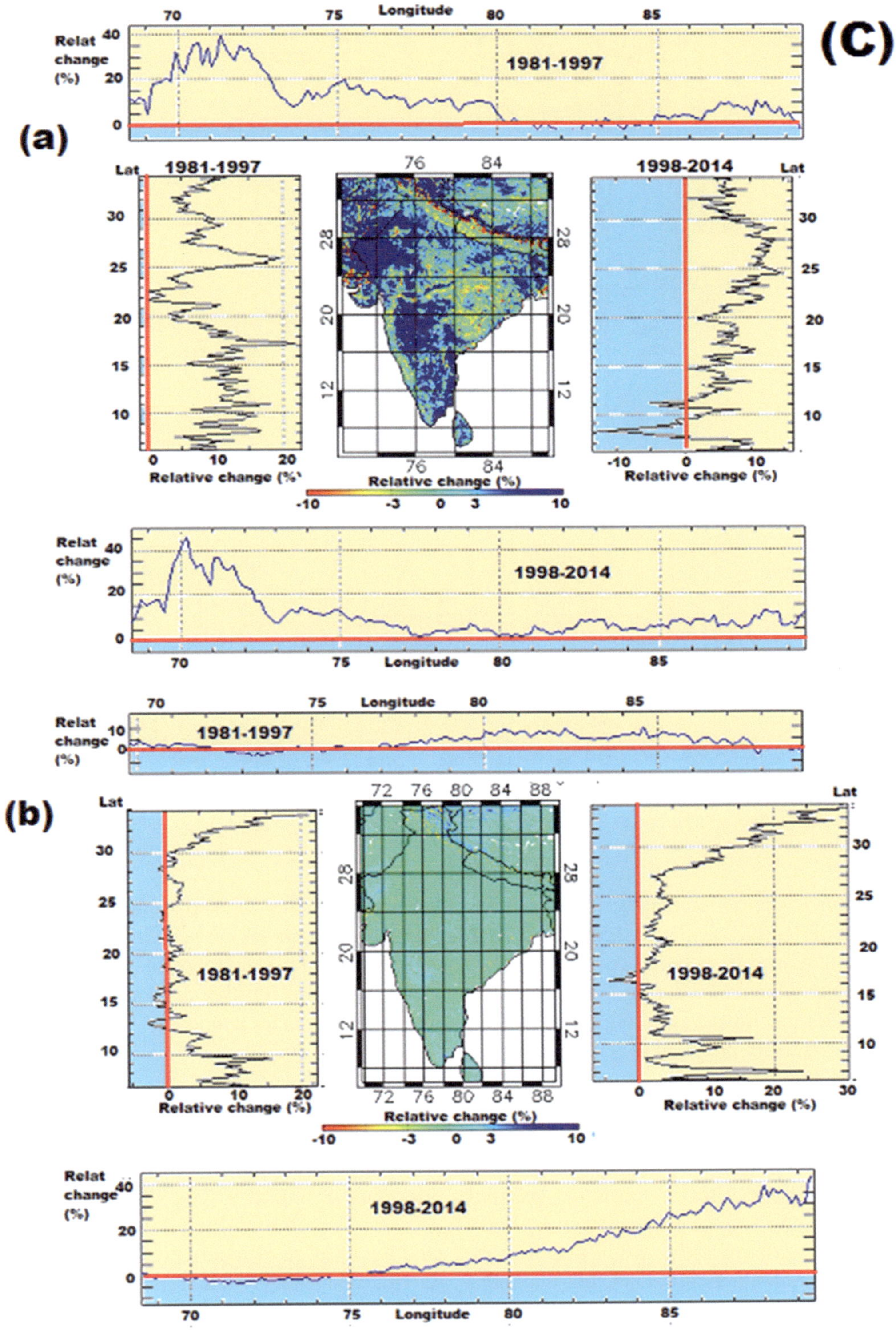

Fig. 9.6 (continued)

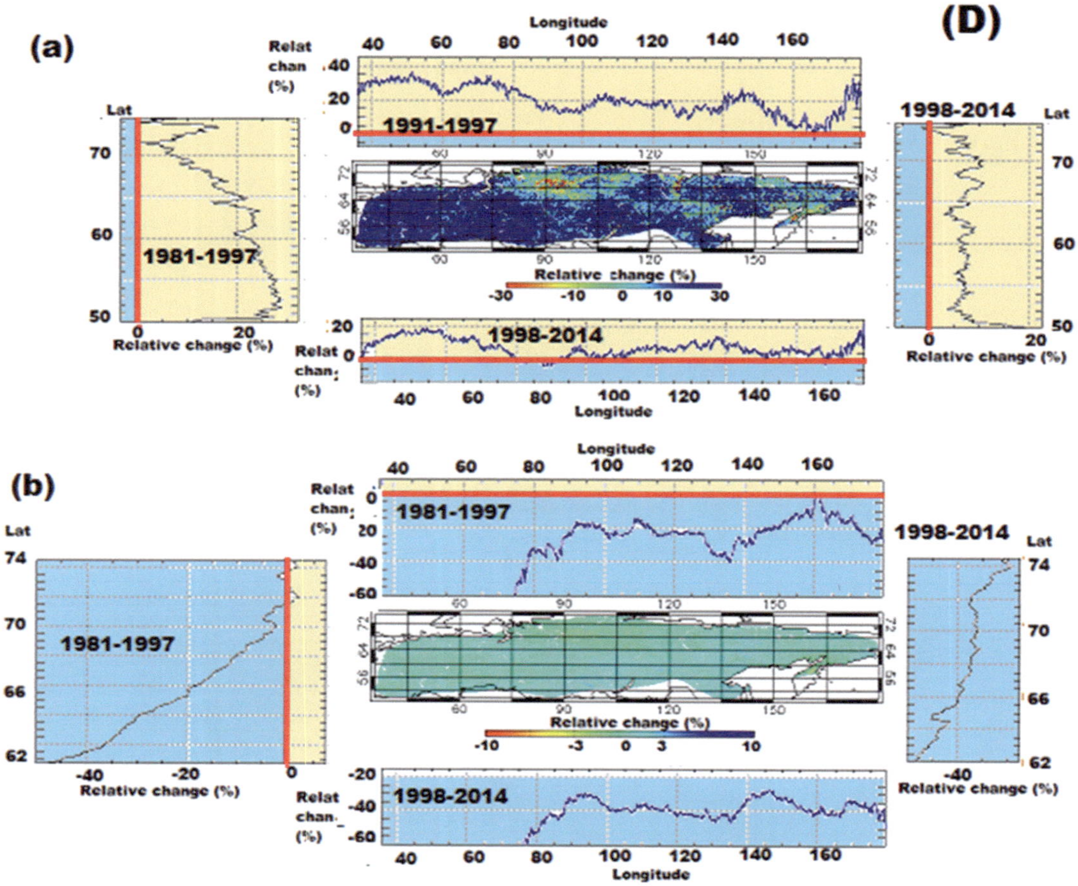

Fig. 9.6 (continued)

and south, where temperature warmed up 10–15% by 1997 and 10–25% by 2014. Vegetation temperature changes over India's longitudes lines were minor during 1981–1997 (diagram over the image on Fig. 9.6Ca), but considerable warming of up to 35% is observed in eastern India (84–89° E, diagram under the image in Fig. 9.6Cb).

In summary, it is important to emphasize that the major grain producers showed mostly (a) country-specific features in vegetation greening up and warming surface temperature; (b) in the USA, changes in greenness are in line with the global temperature trends, but in China and India, they are not, specifying country environmental and economic individuality; (c) temperature changes are even less in line with the global temperature trends for all three countries, especially during the hiatus time; (d) changes along latitudes and longitudes are showing more countries specific features (environmental, economic, and even traditions). Further analysis includes main grain crop areas for countries.

9.6.4 Main Grain Crop Area

The major grain crops area normally occupies less than a country's total area (much less in Russia). Therefore, having 4 km^2 VH pixel data and the location of main grain crops (GOOGLE 2018), the mean VH-based indices (SMN, SMT, VCI, TCI, and VHI) and products (drought area, intensity, impacts, etc.) values were calculated for each main grain crops area, which is shown in NDVI map (Fig. 9.7).

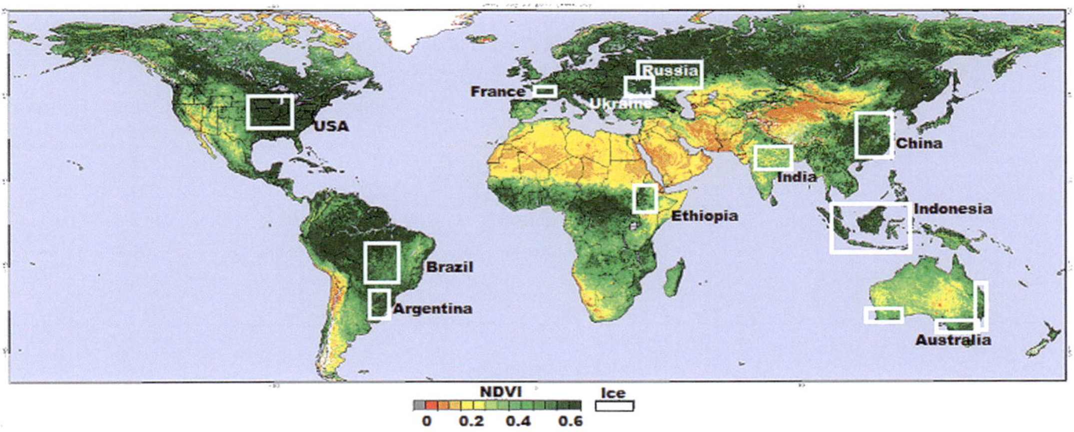

Fig. 9.7 Major grain crop areas (GOOGLE 2018), for which area-mean VH indices were calculated from all 4 km^2 pixels inside the area

Greenness, temperature, moisture, and thermal conditions of the principal grain crop area were calculated as area-mean SMN, SMT, VCI, TCI, and VHI from all 4 km^2 pixels inside the white rectangular (approximately major grain crop area). Their time series were used in the analysis of not only VH-based vegetation greenness and temperature but also moisture and thermal conditions, drought characteristics, and their tendencies during the entire period of VH data (NOAA/NESDIS 2018). Specific emphasis was also placed on the two climate warming periods, 1981–1997 (intensive global warming up) and 1998–2014 (flat global TA trend during the hiatus time). Since grain is an important component of production-consumption tendency and food security prediction, the investigation covered all countries producing more than 2% of total world grain production in 2014 (WB 2017), plus Australia (Southern Hemisphere's continent) and Ethiopia, one of the largest grain producer in Sub-Sahara Africa. These data, for all countries, are presented in Table 9.5. An example of weekly VH-based time series, estimated linear trend (red line), and analysis of relative changes for the two periods of global temperature changes is shown for the USA in Fig. 9.8. The following discussion is focused on specific features of US trends, relative changes and if these changes are matching or not matching with global temperature trends during 1981–1997 (intensive global warming) and 1998–2014 (flat global TA trend).

The VH-based parameters were aggregated from the area inside the white rectangle on the US map at the bottom of Fig. 9.8. That is the major US grain crops area. Following Fig. 9.8, during the period of intensive global warming (1981–1997), the major US grain crops' greenness (SMN) increased 11% by 1998. Meanwhile, moisture conditions (VCI), derived from SMN (related to climatology) improved much stronger (*RD* increased 30% by 1998). The trend in area-mean temperature over the US grain area during that period also increased nearly 7%, matching with global warming trend. However, there are some mismatches. The US trend of thermal vegetation conditions (TCI, derived from SMT relative to its climatology) has not changed (RD = −3%) during the period of intensive global warming. Moreover, between 1981 and 1997, TCI experienced two trends: declining by 1992–1993 and recovery thereafter. Combined moisture–thermal conditions (VHI) during the same period improved nearly 12% by 1998, reflecting equal VCI and TCI contribution to the VHI index. Transition from intensive global warming (1981–1997) to a flat trend in global TA during the hiatus time (1998–2014) indicates more changes in vegetation conditions. First, greenness reduced, experiencing flat trend (RD = 0%), which is matching with hiatus-time flat global TA trend. Brightness temperature trend has not changed compared to the 1981–1997 period (RD = 7% versus 6%). However, the most interesting fact is that moisture

Table 9.5 Relative changes (%, Eqs. 9.1 and 9.2) in vegetation greenness (SMN), brightness temperature (SMT), moisture (VCI), and thermal (TCI) conditions in the main grain area of each country during the 17-year of 1981–1997 intensive global warming and 1998–2014 hiatus time flat global TA trend

Country	Contribution to global grain (%)	Parameter	Change (%) from 1981 to 1997	Change (%) from 1998 to 2014
China	19.8	Greenness	9.8	22.3
		Temperature	12.5	12.0
		Moisture condition	19.9	46.5
		Thermal condition	−29.6	−28.4
USA	15.7	Greenness	11.1	−0.0
		Temperature	6.8	6.0
		Moisture condition	30.2	−6.9
		Thermal condition	−3.1	−16.5
India	10.5	Greenness	7.0	6.7
		Temperature	1.0	2.7
		Moisture condition	43.0	15.4
		Thermal condition	−9.4	−19.2
Russia	3.7	Greenness	29.7	11.2
		Temperature	−11.0	−38.2
		Moisture condition	57.8	18.9
		Thermal condition	−11.3	−30.3
Brazil	3.6	Greenness	2.7	4.4
		Temperature	7.2	8.0
		Moisture condition	19.9	20.4
		Thermal condition	−27.6	−35.2
Indonesia	3.2	Greenness	15.3	27.2
		Temperature	33.6	88.4
		Moisture condition	28.6	58.2
		Thermal condition	−32.8	−59.6
France	2.5	Greenness	10.6	6.5
		Temperature	19.3	18.2
		Moisture condition	29.2	18.1
		Thermal condition	−28.0	−36.9
Ukraine	2.2	Greenness	31.5	7.9
		Temperature	27.9	45.2
		Moisture condition	69.4	11.5
		Thermal condition	−6.4	−45.0
Australia	1.3	Greenness	2.2	−0.1
		Temperature	1.9	10.3
		Moisture condition	13.5	−7.1
		Thermal condition	−18.3	−44.7
Ethiopia	0.8	Greenness	−4.1	−5.2
		Temperature	1.7	9.6
		Moisture condition	−9.8	−24.4
		Thermal condition	−11.0	−45.4

and thermal conditions during 1998–2014 deteriorated (RD changes −7 through −16%) reflecting no increase in global TA. So, descaling the area of assessments (greenness, temperature, moisture–thermal conditions) from continents and countries to the major grain crop areas indicates a lot of specific trend features in grain crops conditions. Some of these trends have matched with global temperature dynamics, especially during a hiatus period (1998–2014), when moisture and thermal conditions of grain crops have deteriorated (declining trends) following a flat trend in global

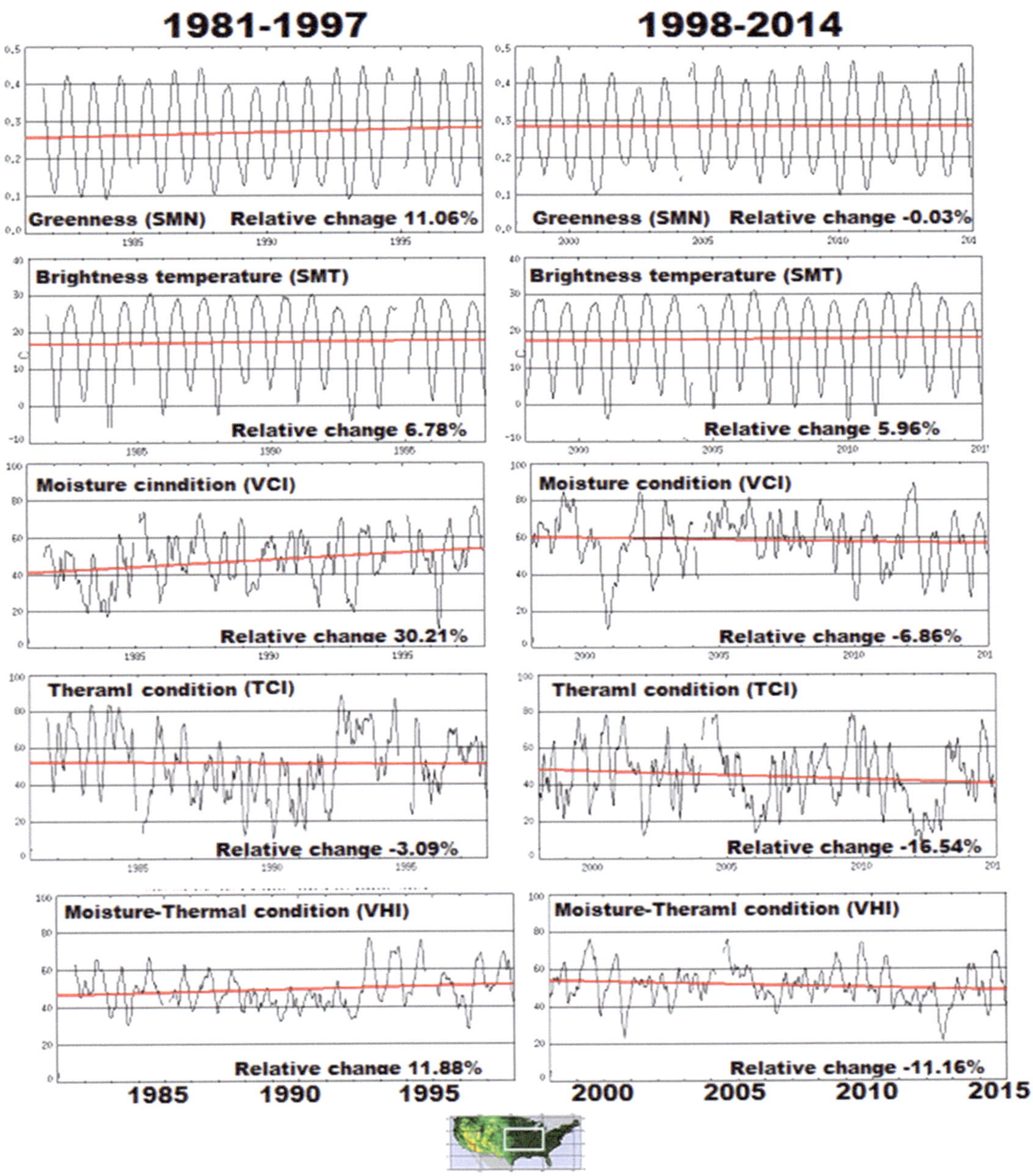

Fig. 9.8 Relative changes in area-mean vegetation greenness (SMN), surface temperature (SMT), moisture (VCI), thermal (TCI), and moisture–thermal (VHI) conditions inside US major grain crops area (white rectangular at USA map) during intensive climate warming (1981–1997) and no warming during the hiatus time (1998–2014)

TA. Another emphasis: land surface moisture (VCI) and thermal (TCI) indices provide better than greenness and brightness temperature tools for assessing crop conditions. Meanwhile, the moisture and thermal trends in grain area show also specificity of local environment. Figure 9.9 shows these differences between western and eastern parts of US major grain areas. When the entire period (1981–2015) of moisture and thermal trends matched in Illinois (eastern USA) and

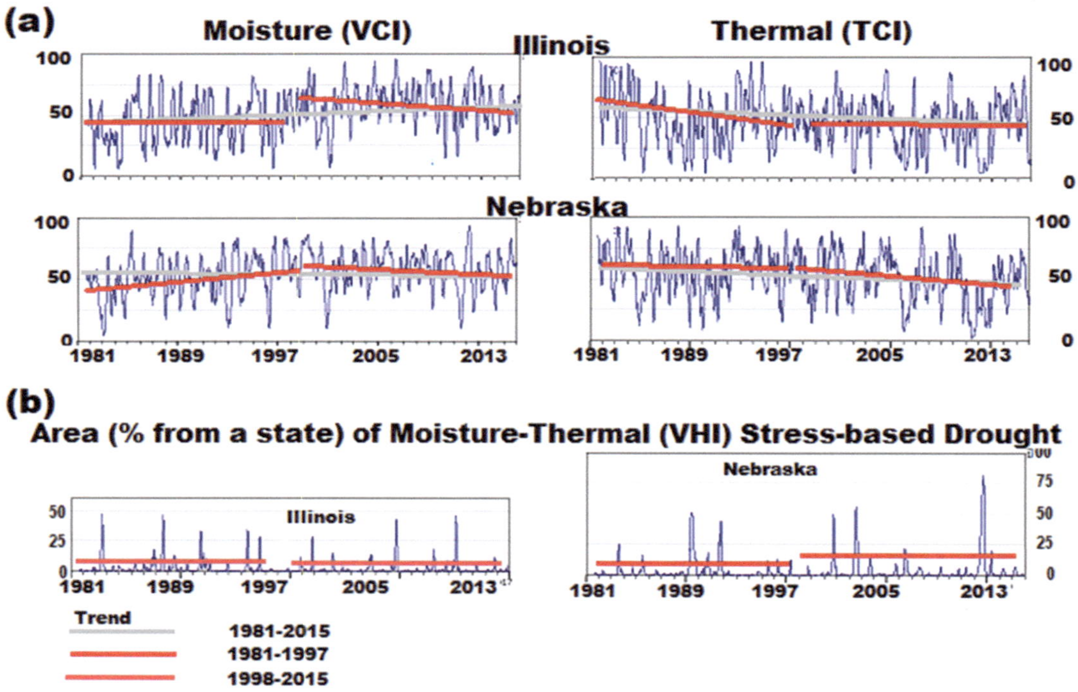

Fig. 9.9 Multiyear weekly values and trends (1981–2015, 1981–1997, and 1998–2014) in (**a**) moisture- (VCI) and thermal-based (TCI) conditions and (**b**) area (% from a state) of extreme-to-exceptional moisture–thermal (VHI) drought in two major grain-producing US states, Illinois (east) and Nebraska (west)

Nebraska (western USA), the 17-year trends mismatched during 1981–1997 in moisture conditions (flat VCI trend in Illinois and increasing trend in Nebraska) and thermal conditions (decreasing TCI trend in Illinois and flat trend in Nebraska). Also, some thermal mismatches during 1998–2014 (flat TCI trend in Illinois and declining trend in Nebraska). However, the two states matched in their moisture conditions (VCI declined by 1998) during the strong global warming period in 1981–1997.

Figure 9.8 showed only one of the countries selected for assessment of VH conditions. Table 9.5 presents similar to Fig. 9.8 relative changes in the indicated parameters for the major grain areas in the selected major grain-producing countries. Comparing relative changes in the presented parameters between the two periods (an intensive global warming during 1981–1997 and flat TA trend during the hiatus time in 1998–2014) indicates some matches and mismatches between countries crop condition trends with global warming trends for the identified two periods. Main grain areas of the USA, Russia, France, and Ukraine (all north of 30° N) showed a matching greenness increase during the intensive global warming (1981–1997), which was the strongest in Ukraine and Russia, plus considerable greenness reduction during the hiatus time (1998–2014). China and Indonesia (south of 40° N) showed opposite greenness trends (double increase in RD to 22–27%) compared to the previous countries and world's stable TA during the hiatus time. The remaining countries (India, Brazil, Australia, and Ethiopia) showed small greenness increase (RD < 5%) for both periods, mismatching with the global climate TA trends during hiatus time.

Figure 9.9 Trends in VH-based area-mean weekly moisture and thermal conditions and area of extreme-to-exceptional drought in the USA's eastern (Illinois state) and western (Nebraska state) parts of major grain area.

Regarding vegetation surface temperature there are more than in greenness mismatches with the global tendencies. China, the USA, India, France, and Brazil do not show much difference in temperature of vegetation surface between the two periods, although elevated temperature trends (RD > 5%) are still observed. Russia, which is growing major grain crops in the southern half of European territory, indicates (surprisingly) cooling off (RD = −11%), compared to intensive global warming. This Russian trend intensified considerably (RD = −33%) during hiatus time at the general background of global flat TA trend. In Indonesia and Ukraine, vegetation surface temperature warmed up considerably (around 28–34%) in the first period and even more (surprisingly) in the second, hiatus time period (45–88%). Much smaller vegetation surface temperature warm up (2–10%) is seen in Australia and Ethiopia and no temperature changes between the two periods are in France and Brazil.

From the presented four parameters in Table 9.5, the most important for monitoring grain production and food security prediction are indices for estimation moisture (VCI) and thermal (TCI) conditions. The analysis of these conditions and their impacts on grain crops is complicated by their joint impacts, especially if one of the indices is showing extreme stress. However, since the extreme cases are not frequent, further discussion covers their individual performance. During the period of intensive global warming (1981–1997), all grain-producing countries (except Ethiopia) indicated improvement (growing trend) in moisture conditions (VCI), which has RD between 14 and 43% (increase by 1998), and considerable deterioration (declining trend) in thermal conditions (TCI), when RD reduced to −10 through −30%). The lowest deterioration was in the USA and Ukraine (almost negligible RD, −3% and −5%, respectively) and the highest was in China, Brazil, Indonesia, and France (RD between −28 and −33%). In transitioning to the hiatus period, thermal conditions deteriorated even more (RD between −35 and −60%), except for China, which preserved the same conditions (RD around −30%). Ethiopia is the only grain producer, where moisture and thermal conditions deteriorated (trend reduction) in both periods. In the first one, RD reduced to −10 and −11% by 1998, and in the second one, RD reduced to −24 through −45% by 2015.

Drought area and intensity are very important components of global warming impacts on crops and pastures and food security. Many scientific and media publications emphasize drought intensification during the period of climate warming and further deterioration of this situation (area expansion, duration increase, etc.) in the future (Seager 2018; Chandler 2018; UNESCO 2018; Watts 2018; AG 2016, 2017; Forzieri et al. 2017; Freedman 2017; ClimateBet 2018; IPCC4 2007; IPCC5 2014; NOAA/NCDC 2016, 2017; Schlossberg 2016; Alexandratos and Bruinsma 2012). Therefore, next is the analysis of drought tendencies during the entire period of operational satellites' observations, focusing on the two periods of global warming. Figure 9.10 shows weekly time series of drought area and intensity and their tendencies during 1981–2015 and the two investigated periods of global warming (1981–1997 and 1998–2014).

The analysis of weekly data in Fig. 9.10 during 1981–2015 indicate that (1) major grain-producing areas in all countries are experiencing very frequent droughts, (2) in several countries the frequencies are relatively uniform (example, Ethiopia) and in others it is variable (Ukraine), (3) in some countries (France, Ukraine) up to 100% of major grain area have been affected by drought in others (Brazil) less than 80%. Following the entire 35-year period (1981–2015) of drought dynamics, a few features can be singled out. Moisture-based droughts are gradually reducing area and intensity (thin red and yellow lines) by 2015 in China, the USA, India, Russia, Brazil, Indonesia, France, Ukraine, and Argentina. These countries contribute 63% to global grain production. Australia and Ethiopia do not show much changes in these two types of drought during 35 years. Trends in thermal-based droughts between 1981 and 2015 showed two types of tendencies. Most countries, specifically China, the USA, Russia, Brazil, Indonesia,

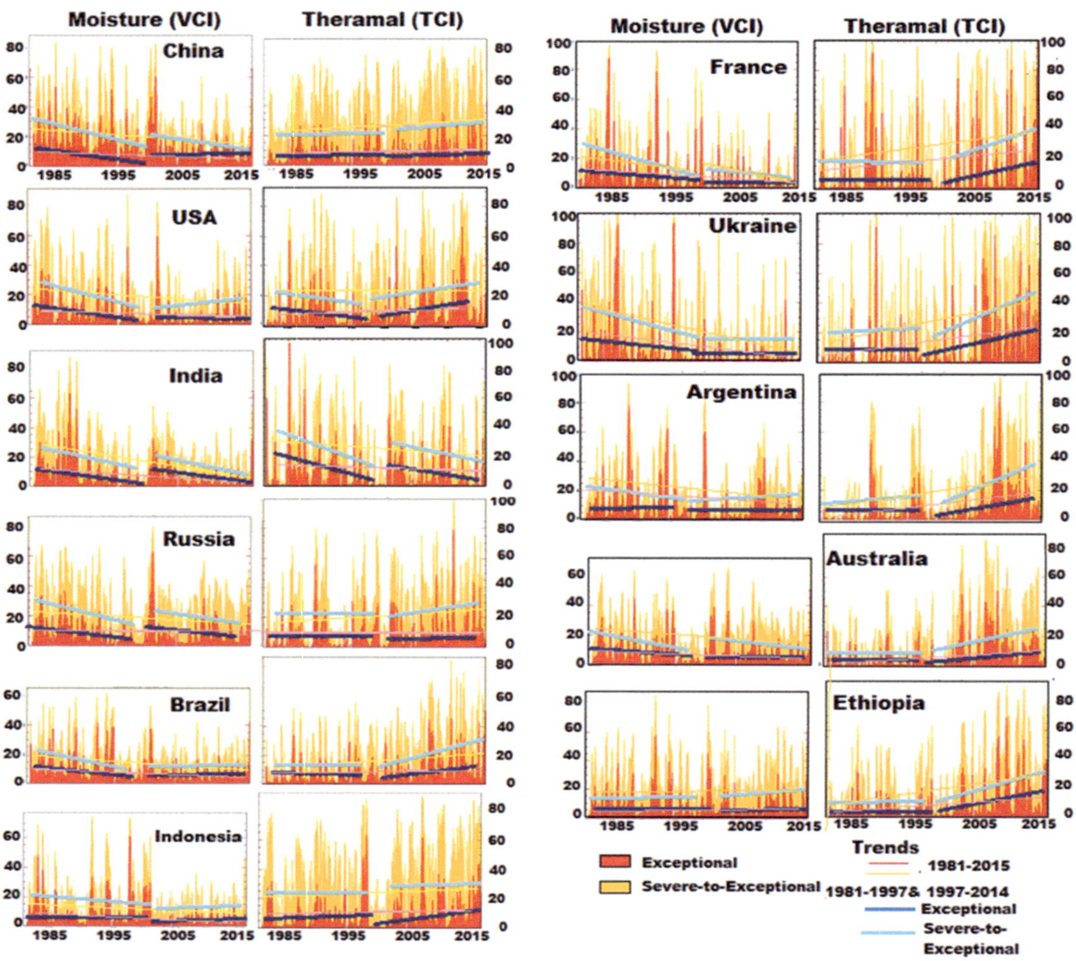

Fig. 9.10 Multiyear weekly drought area (% from a major grain area) and intensity (severe-to-exceptional and exceptional intensity) and multiyear trends (1981–2015, 1981–1997 (strong global warming) and 1998–2014 (flat global TA trend)) from moisture (VCI) and thermal (TCI) vegetation stress

France, Ukraine, Argentina, Australia, and Ethiopia (contribute 53% to global grain production) in Fig. 9.10 indicate intensification of thermal-stress (larger area in both intensities (10–20%) in the exceptional, the strongest category and up to 40% in severe-to-exceptional were affected by 2015). Only India showed some reduction in drought area for both intensities by 2015 (percent area change is from 13 to 9% for exceptional and from 25 to 18% in the severe-to-exceptional).

The most interesting analysis of drought trends would be for the two periods of intensive global warming in 1981–1997 and no warming intensification during the 1998–2014 hiatus time. Following the indicated above publications, it was logical to assume that global warming in the past 35 years (since 1981), especially during the first 17 years, was supposed to stimulate an intensification and areal expansion of regional droughts, including major grain crop areas. However, looking at the Fig. 9.10 (thick blue and light blue lines), different drought tendencies are identified. In spite of intensive global warming during 1981–1997 (IPCC5 2014; WMO 2014, 2016), intensity and area of thermal-based droughts (TCI-type) has remained stable (flat trend, between 20 and 40% of the total grain country area for droughts severe-through-exceptional and 5–15% for exceptional intensity) in the major grain areas of China, the USA, Russia, Brazil, Indonesia, France, Ukraine, Argentina, Australia, and Ethiopia, which together contributed 52% to global grain production in 2014 (WB 2017). In India, which contributes 10.5% to global grain production, thermal-based drought area and intensity (both sever-trough-exceptional and exceptional) reduced gradually from 35 to 20% (for the first category) and from 15 to 9% (for the strongest type), respectively, by 1998, in spite of intensive global warming. Thermal-based drought trend during 1998–2014 in the major grain areas is mismatched with the global TA flat trend, since most of the countries in Fig. 9.10 (China, the USA, Russia, Brazil, France, Ukraine, Argentina, Australia, and Ethiopia) showed intensification of thermal-based (TCI) vegetation stress, indicating drought area increase (for exceptional drought—from 3–5% in 1998 to 8–20% in 2014 and for severe-to-exceptional drought from 15–25% in 1998 to 40–45% by 2015).

Moisture-type drought (VCI-based) dynamics in the major grain areas (Fig. 9.10) indicates more unusual features. During the intensive global warming in 1981–1997, drought area surprisingly reduced from 30–45% in 1981 to 15–20% by 2015 for severe-through-exceptional drought category and from 10 to 3% for the strongest, exceptional drought intensity in China, the USA, India, Russia, Brazil, Indonesia, France, Ukraine, Argentina, and Australia. These countries contribute 64% to global grain production. The only Ethiopia shows a flat drought-area trends in both intensities. During the period of flat global TA trend in the hiatus time, drought area in the major grain crops either continue to decrease or does not change between 1998 and 2014. Therefore, drought trends (both moisture- and thermal-based) in the size of area and intensity for the major grain crops area is principally mismatching with global warming trends: instead of drought intensification expected due to global warming, drought area is reducing in both 17-year periods (1981–1997 and 1998–2014) and in a few smaller countries does not change. These results contradict to some publications indicating drought intensification and area expansion following global warming (Seager 2018; Chandler 2018; UNESCO 2018; Watts 2018; AG 2017; Forzieri et al. 2017; Freedman 2017; ClimateBet 2018; IPCC4 2007; IPCC5 2014; NOAA/NCDC 2016, 2017; Schlossberg 2016; Alexandratos and Bruinsma 2012). Specifically, IPCC5 (2014) report indicated nearly 80 mm precipitation reduction during 1951–2010 in the wheat-growing areas of Australia (Fig. 1.1e from IPSS5 2014 (page 41)). Unfortunately, these IPCC5 results are wrong, since they contradict to precipitation data from Australia's Bureau of Meteorology (Murph and Timbal 2008; AWXB 2018; AG 2016, 2017), which showed no changes in annual precipitation over wheat area (nearly 500 mm during 1987–2016). Moreover, a stable 30-year precipitation (P) trend is matching with 28-year (1985–2015) moisture condition (VCI) trend in wheat area shown in Fig. 9.11. Following the figure, between mid-1980s and 2015, $P = 500$ mm and VCI = 47 in Australia's wheat area although that area was warmed up by 0.3 °C since 1981 (AWXB 2018; AG 2017; Kogan et al. 2018).

9.6.5 World Droughts

Since tendencies in drought area and intensity in the principal grain-producing countries and their major grain crop areas strongly mismatched with global warming trends, there was a concern that such fact is related to a specificity of regional environment (climate, ecosystem, hydrology, etc.) in selected countries and regions. Therefore, drought area and intensity was calculated during 1981–2015 for the entire world and hemispheres. Figure 9.12 shows weekly drought (area and intensity) time series and trends (1981–2015, and for two 17-year periods). Three drought areas and intensities (exceptional (E), extreme-to-exceptional (E-to-E, and severe-to-exceptional (S-to-E)) were investigated in terms of drought dynamics and their matching with global warming. As seen in Fig. 9.12, during 1981–2015, weekly drought area changed frequently

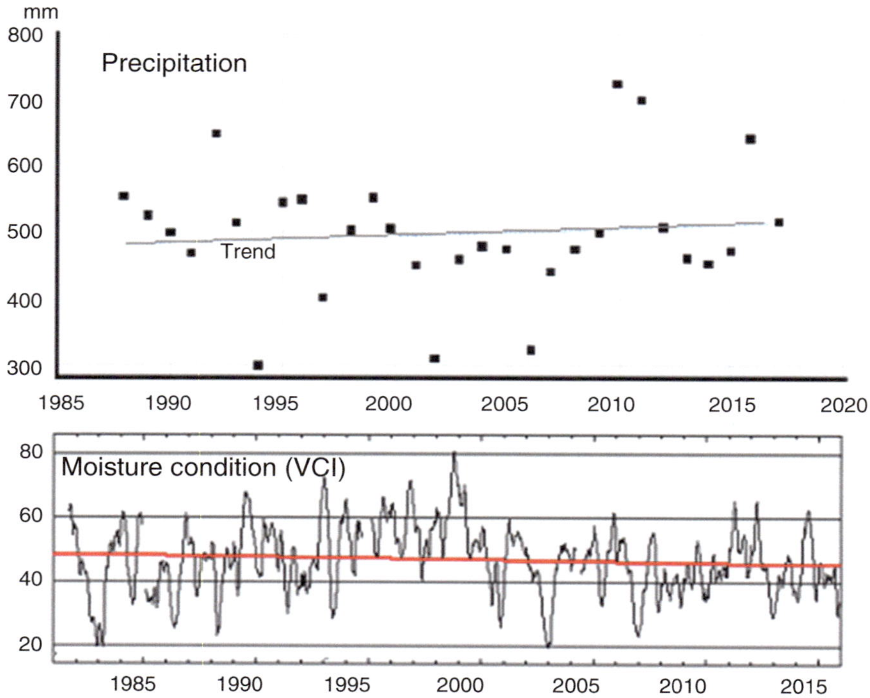

Fig. 9.11 Annual precipitation (1987–2016) and multiyear weekly VH-based moisture condition (VCI) during 1981–2016 over wheat are of Australia (shown in Fig. 9.9)

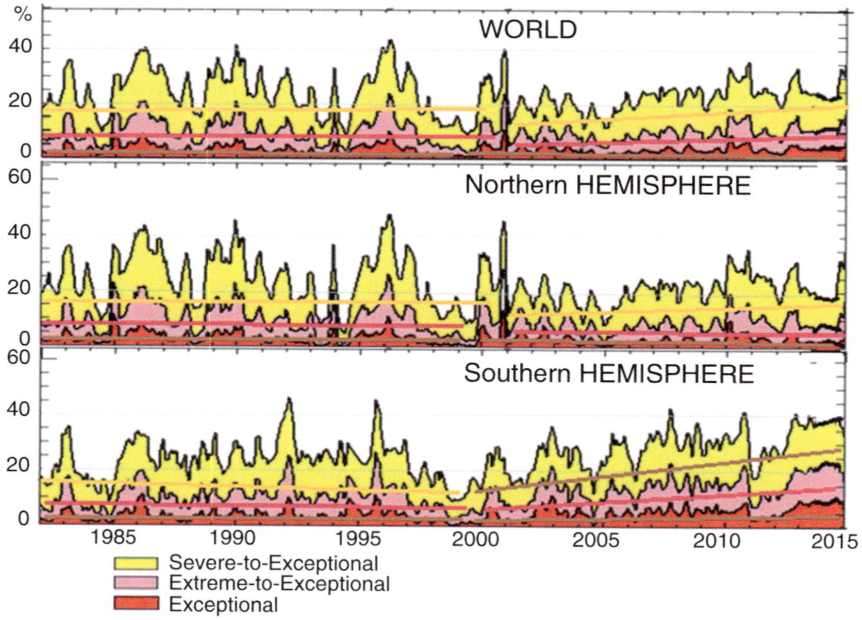

Fig. 9.12 World and Northern and Southern Hemispheres' drought area and intensity (in three category: severe-to-exceptional, extreme-to-exceptional, and exceptional)

between 15–40%, 5–20%, and 3–12% (of the total world area) for S-to-E, E-to-E, and E intensities, respectively. World drought trends during the same period show slight reduction of drought area (less than 8% by 2015) for the two intensities S-to-E and E-to-E. For the strongest, E intensity (weekly areal variations are between 5 and 10%) no change in drought area was recorded during 1981–2015. Northern Hemisphere drought frequency, area, intensity, and trends are quite similar to the world's drought tendencies. Southern Hemisphere also shows frequent droughts with slightly lower area (except the strongest, E intensity). However, opposite to the world and Northern Hemisphere, the Southern Hemisphere drought area trends are slightly increasing by 2015.

More interesting results in trend analysis are shown for the two periods of intensive global warming during 1981–1997 and flat global TA trend during the hiatus time (1998–2014, Fig. 9.12). During the intensive global warming (1981–1997) the total world area of droughts surprisingly have not changed (for all intensities) for the entire world and both hemispheres. The area of S-to-E drought covered nearly 20% of the world and Northern Hemisphere (NH) and 15% of Southern Hemisphere (SH). The area of E-to-E and E droughts was 15% and 5%, respectively, for the world and both hemispheres. Transitioning to hiatus time (1998–2014) of stable global TA indicates (1) 5–7% reduction of drought area in late 1990s to 10–16%, 5–7%, and 2–3% for S-to-E, E-to-E, and E intensities, respectively, for the world and Northern Hemisphere; and no change for Southern Hemisphere drought area. However, by 2015, the S-to-E and E-to-E drought area increased slightly (2–7%) for the world and Northern hemisphere, but no change for the extreme droughts. In the Southern Hemisphere, stronger drought increase is observed for all intensities. By 2015, drought area in Southern Hemisphere reached 28, 15, and 5% for S-to-E, E-to-E, and E, respectively. So, global drought analysis during 1981–2015 indicates (1) global and hemispheric droughts have not changed their frequency; (2) on the average (from trend) global and hemispheric droughts in S-to-E category covers nearly 15–20% of the entire area, 7–8 and 2–3% in E-to-E and E category; (3) in extreme years, drought area of three intensity types might jump up to 45, 25, and 10%, respectively; (4) during the period of intensive global warming (1981–1997), drought areas in all categories have surprisingly not increased and not intensified, which was claimed by many publications; (5) in the hiatus time (1998–2014) of a stable global TA, drought area mismatched with global temperature trend, increasing for two categories (S-to-E and E-to-E) for the globe and both hemispheres, especially for SH; drought trends have not changed for the extreme category.

As a result of this analysis, it is important to emphasize again that (a) two 17-year periods are identified in global warming: strong TA increase (0.5 °C) during 1981–1997 and flat TA trend (around 0.2 °C) during the hiatus time from 1998 through 2014; (b) vegetation greenness and surface temperature followed these two tendencies: greenness and temperature increased in the first period (9 and 15%, respectively) following greenness reduction and no temperature change (3 and 18%, respectively) in the second; (c) along with greenness increase, moisture conditions improved (trend increased) also during 1981–1997 in the majority of grain-producing countries but the improvement was 2–4 times stronger (RD from trend increased 20–40%); (d) some deterioration of these conditions was in the second period (compared to the first one); (e) surprisingly (to the publications' opinion, claiming drought intensification due to global warming) drought area and intensity have not changed neither globally, nor in the major grain areas in 1981–1997 but slightly increased during the hiatus time (1998–2014) of stable global TA. Following this summary, no major climate-related changes in grain production are expected and also long-term environmental aspects of food security would continue the past 30-year tendencies.

9.7 Conclusions

The goal of this chapter was to investigate how global warming in the past four decades (since 1981) changed weather and climate impacts globally and in the major grain-producing regions (countries, grain crop area, etc.) and if these changes worsened the environmental conditions, leading to more frequent and larger grain crop losses, following deterioration of food security. This goal was also dictated by claims from many scientific and media publications (UNESCO 2018; Rice 2018; Watts 2018; AG 2017; Forzieri et al. 2017; Freedman 2017; ClimateBet 2018; IPCC4 2007; IPCC5 2014; NOAA/NCDC 2016, 2017; Schlossberg 2016, etc.) that a warmer world has already stimulated a reduction of moisture for crops, intensified thermal stress, enlarged drought area and increased drought duration, intensity, and impacts, and resulted in a reduction of grain production. In such circumstances, food security is supposed to deteriorate without any chance for improvement.

Having a 38-year well-validated satellite-based vegetation health data, characterizing vegetation greenness and surface temperature, moisture/thermal conditions, drought area and intensity, vegetation stress, yield losses and others, the goal was to investigate if these parameters follow a global warming trend and what to expect in the near future with food security. First, the investigation covered analysis of scientific literature on an intensity of global warming, its causes and potential impacts on grain crops. Principally, many publications and international actions (IPCC5 2014; IPCC4 2007; UNESCO 2018; UNFCCC 2014, 2015; WEF 2016; WMO 2016, etc.) agreed that the world is warming up and the cause of this event is CO_2 increase. Since CO_2 is trapping infrared radiation in the atmosphere, and its 160-year trend is generally matched with very general global temperature anomaly (TA) trend (IPCC5 2014), CO_2 was considered as the major cause of the warmer Earth (Fig. 9.1). Our detail analysis of global temperature anomaly presented in IPCC5 (2014) report indicated that at the background of *generally* uprising trend in global TA since 1850, there are three different types of 17–35-year trends: two declining, two flat, and two increasing (Fig. 9.2). A reasonable question to ask is why, at the background of continuing initially slow and during industrial time strong CO_2 increase the global mean TA changed its trend six times? Moreover, why? after the period of intensive global warming from early 1970s through 1997, during 17-year (1998–2014) period after 1997 (named “hiatus” time (Karl et al. 2015; Kennel 2014)), global TA has not changed (had flat trend), while CO_2 continued strong increase. Summarizing the later statement in a view of food security assessment, it is important to emphasize again that (a) during 1981–1997, there was a match in the trends of global TA, CO_2, and O_3; and (b) during 1998–2014, global TA matched with O_3 trend but mismatched with CO_2 trend.

A few other interesting facts in the global warming. First, following a stable global TA from 1998 through 2014, during 2015–2017 global TA was the highest in the past 160 years (Fig. 9.2). Some scientific publications blamed a very strong El Nino of 2015–2016 for the warming, other solar activities, etc (Weisberger 2018; Codington et al. 2016). However, there is no definite explanation as to why. Second, in the first quarter of 2018, a considerable reduction of mean global TA was recorded (ClimateBet 2018). However, it is not known if the opposite to the highest 3-year temperature situation has begun. Third, some scientific publications blamed ozone (O_3) depletion as a cause of 17-year period of no uprising trend in mean global TA (Ward 2016a). Following these contradicting facts, 38-year of VH data were used to verify if global warming tendencies match with tendencies in vegetation greenness, its surface temperature, moisture–thermal conditions, drought, crop stress, and others and how they might affect grain production and food security dynamics.

In general, if ambient temperature is in optimal range of crop requirement, vegetation becomes greener, vigorous, heavier with biomass, etc. with a temperature increase (Lucht et al. 2002). This is confirmed by the analysis of 35-year global vegetation greenness dynamics (Fig. 9.4). During 1981–1997, global mean vegetation greenness increased nearly 9% by 1997 (compared to 1981 greenness calculated from trend) matching with an intensive global warming trend during the same period, when TA increase 0.5 °C (IPCC5 2014). Global vegetation cover temperature also increased nearly 16% during that period, matching with a global temperature rising trend. When global TA has not changed (flat trend) during hiatus time (1998–2014), the mean world greenness reduced three times (relative greenness 3.1%) between 1998 and 2014. However, vegetation surface temperature continued to increase up to 18% by 2014, mismatching with global TA. Similar results were obtained for each 16 km^2 latitude lines for the annual cycle and for winter and summer, specifically increased world and Northern Hemisphere global greenness, no greenness increase was observed in Southern Hemisphere and no greenness reduction is recorded during the hiatus time. One of the interesting conclusions is, at the background of strong global warming from 1981 through 1997, northern world latitudes (between 75° and 60° N) have been cooling, Southern Hemisphere experienced much smaller warming and two investigated periods showed very similar relative temperature changes over latitude (from north to south) and longitude (from west to east) lines.

Following matching and mismatching of global vegetation greenness and surface temperature with global TA trends and assuming that climate and ecosystems might contribute to environmental impacts on grain crops, the global area was descaled to major grain-producing countries and regions of major grain crops inside these countries (Table 9.5 and Figs. 9.6, 9.7, 9.8, 9.9 and 9.10). Especially important was assessment of grain crops' moisture (VCI) and thermal (TCI) conditions in descaled areas. Trends in moisture and thermal conditions indicate a lot of specific features in these parameters. In spite of intensive global warming during 1981–1997 (IPCC5 2014; WMO 2014, 2016), intensity and area of thermal-based droughts (TCI) has remained stable (flat trend, between 20 and 40% of the total grain country area for droughts severe-through-exceptional and 5–15% for exceptional intensity) in the major grain areas of China, the USA, Russia, Brazil, Indonesia, France, Ukraine, Argentina, Australia, and Ethiopia, which together contributed 52% to global grain production in 2014. During hiatus time (1998–2014), moisture and thermal conditions of grain crops have deteriorated following stable global warming or flat trend in global TA.

One of the most important environmental problems for grain production and food security prediction is if droughts are intensifying and their areas are expanding. In principal, many scientific publications are emphasizing deterioration of these drought features following an intensive global warming (UNESCO 2018; NOAA/NCEI 2017; Gray 2016; WMO 2016; NOAA/NCDC 2016; IPCC5 2014; Alexandratos and Bruinsma 2012; Nemani et al. 2003). However, VH data showed that 35-year (1981–2015) drought trends are principally mismatching with global warming trends: instead of drought intensification and area expansion following global warming, drought area is reducing in both 17-year periods (1981–1997 and 1998–2014) and in a few countries' drought impacts on grain crops does not change. Moreover, even global mean drought area for the three strongest intensities (severe-to-exceptional (S-to-E), extreme-to-exceptional (E-to-E), and exceptional (E)) has not changed since 1981. Meanwhile, it is important to emphasize that during 1981–2015, (a) global and hemispheric droughts are frequent as has been before; (b) on the average, global and hemispheric droughts in S-to-E category covers nearly 15–20% (calculated from trend in Fig. 9.12) of the entire area, 7–8 and 2–3% in E-to-E and E category, respectively; (c) in the extreme years, drought area in the indicated three intensities might jump up to 45, 25, and 10%, respectively; (d) during the period of intensive global warming (1981–1997), drought areas in all categories have surprisingly not increased and not intensified, which was claimed by many publications; (e) during the hiatus time (1998–2014) of a stable global TA, drought area mismatched with global temperature trend, increasing for two categories (S-to-E and E-to-E) for the globe and both hemispheres (especially southern) and drought trends have not changed for the extreme category. Finally, general climatic aspects of food security supposed to continue the past 30-year tendencies since no major changes in long-term environmental impacts on grain production is expected.

References

AG (Australia Government). 2017. http://www.agriculture.gov.au/ag-farm-food/crops/wheat.

———. 2016. Australia's Reference Climate Station Network. http://www.bom.gov.au/climate/change/reference.shtml.

AWXB. 2018. Australian Wheat. http://www.awb.com.au/customers/australianwheat/.

Alexandratos, N., and J. Bruinsma. 2012. World Agriculture towards 2030/2050: The 2012 Revision. ESA Working Paper No 12-03, June, FAO, Viale delle Terme di Caracalla 00153, Rome, Italy, pp 147. http://www.fao.org/docrep/016/ap106e/ap106e.pdf.

Anderson, B.A., and B.D. Silver. 1990. Growth and diversity of the population of the Soviet Union. *Annals of the American Academy of Political and Social Science* 510 (July): 155–177 https://deepblue.lib.umich.edu/bitstream/handle/2027.42/67141/10.1177_0002716290510001112.pdf?sequence=2&isAllowed=y.

Ankit, J. 2015. Ozone layer depletion: Causes, effects and solutions. *Continents*, June 18. http://www.7continents5oceans.com/ozone-layer-depletion-causes-effects-and-solutions.

Alley, R.B., J.T. Andrews, J. Brigham-Grette, G.K.C. Clarke, K.M. Cuffey, J.J. Fitzpatrick, S. Funder, S.J. Marshall, G.H. Miller, J.X. Mitrovica, D.R. Muhs, B.L. Otto-Bliesner, L. Polyak, and J.W.C. White. 2010. History of the Greenland ice sheet: *Paleoclimatic insights. Quarterly Science Review* 7: 1–29.

Bastasch, M. 2017. China is sticking to the Paris agreement in name only, its plans show a much different Intent. *The Daily Signal*, June 2. https://www.dailysignal.com/2017/06/02/china-is-sticking-to-the-paris-agreement-in-name-only-its-plans-show-a-much-different-intent/.

Berardelli, P. 2010. Say goodbye to sunspots? *Science, Physics, Space,* September 14, http://www.sciencemag.org/news/2010/09/say-goodbye-sunspots.

Berwyn, B. 2016. Global CO_2 emissions from fossil fuels held steady in 2016. https://psmag.com/environment/global-co2-emission-from-fossil-fuels-held-steady-in-2016.

Brahic, C. 2007. Climate myths: The cooling after 1940 show CO_2 does not cause warming. *New Scientist*, May 16. https://www.newscientist.com/article/dn11639-climate-myths-the-cooling-after-1940-shows-co2-does-not-cause-warming/.

Bromwich, D.H., J.P. Nicolas, A.J. Monaghan, M.A. Lazzara, L.M. Keller, G.A. Weidner, and A.B. Wilson. 2013. Central West Antarctica among the most rapidly warming regions on earth. *Nature Geoscience* 6: 139–145.

Chandler, D.L. 2018. *MIT News Office*, February 13. http://news.mit.edu/2018/intensive-agriculture-influences-us-regional-summer-climate-0213.

Christy, J.R. 2017. *Testimony to U.S. House Committee on Science, Space & Technology*, March 29. https://science.house.gov/sites/republicans.science.house.gov/files/documents/HHRG-115-SY-WState-JChristy-20170329.pdf.

ClimateBet. 2018. Global warming challenge. http://www.theclimatebet.com/june-2018-temperature-falls-near-middle-of-2007-base-year-range/.

Codington, O., J.L. Lean, P. Pilewskie, M. Snow, and D. Lindholm. 2016. A solar Irradiance climate data records. *Bulletin of the American Meteorological Society* 97 (7): 1265–1294.

Forzieri, G., R. Alkama, D.G. Miralles, and A. Cescatti. 2017. Satellites reveal contrasting responses of regional climate to the widespread greening of Earth. *Science* 356 (6343): 1180–1184. https://doi.org/10.1126/science.aal1727.

Farman, J.C., B.G. Gardiner, and J.D. Shanklin. 1985. Large losses of total O_3 in atmosphere reveal seasonal ClOx/NOx interaction. *Nature* 315: 207–210.

Freedman A. 2017. 2017 will rank among Earth's top 5 warmest years. *Mashable*, December 19. https://mashable.com/2017/12/19/top-5-hottest-years-global-warming-nasa-noaa/#pGwn0cadzmqy.

GOOGLE. 2018. Grain areas. https://www.google.com/search?q=grain+area+location+in+Ethiopia&rlz=1C1GGRV_enUS752US752&biw=1480&bih=831&tbm=isch&source=iu&ictx=1&fir=XQFzNC4VEf8Q3M%253A%252CiBvrZwVKfkcZcM%252C_&usg=__vfzTtw-Z1OKApdvWowEQzx9qLiE%3D&sa=X&ved=0ahUKEwjir4297LLaAhUBzVMKHX-DDqMQ9QEIdzAI#imgrc=XQFzNC4VEf8Q3M.

Gore, A. 2006. *An Inconvenient Truth*, 265. Emmaus, PA: Rodale Press.

Gray A. 2016. What is Paris Agreement on climate change? *World Economic Forum*, September 7. https://www.weforum.org/agenda/2016/09/what-is-the-paris-agreement-on-climate-change/.

Groves S. 2016. Paris climate promise: A bad deal for America. Testimony before the Committee on Science, Space, & Technology, United States House of Representatives on February 2, 2016. *The Heritage Foundation*, February 2. https://www.heritage.org/testimony/paris-climate-promise-bad-deal-america.

Haldon, J., L. Mordechai, T.P. Newfield, A.L. Chase, A. Izdebski, P. Guzowski, I. Labuhn, and N. Roberts. 2018. History meets palaeoscience: Consilience and collaboration in studying past societal responses to environmental change. *Proceedings of the National Academy of Sciences*. https://doi.org/10.1073/pnas.1716912115.

Harland, W.B., C.A.G. Picton, N.J.R. Wright, C.A. Croxton, D.G. Smith, J.L. Cutbill, and W.G. Henderson. 1976. Some coal-bearing strata in Svalbard. *Norsk Polarinstitutt Skrifter* 164: 1–89.

Hansen, J.E., R. Ruedy, M. Sato, and K. Lo. 2010. Global surface temperature change. *Reviews of Geophysics* 48: RG4004.

Hansen, J., M. Sato, R. Ruedy, A. Lacis, and V. Oinas. 2000. Global warming in the twenty-first century: an alternative scenario. *Proceedings of the National Academy of Sciences* 97: 9875–9880.

IPCC5. 2014. Climate Change 2014: Synthesis Report. 5th Assessment, eds R. K. Pachauri and L. Meyer. Geneva, Switzerland, pp 151. https://www.ipcc.ch/pdf/assessment-report/ar5/syr/SYR_AR5_FINAL_full_wcover.pdf.

IPCC4. 2007. Climate Change 2007: Synthesis Report. Geneva, Switzerland. pp 104. https://www.ipcc.ch/publications_and_data/publications_ipcc_fourth_assessment_report_synthesis_report.htm.

Karl, T.R., A. Arguez, B. Huang, J.H. Lawrimore, J.R. McMahon, M.J. Menne, T.C. Peterson, R.S. Vose, and H.-M. Zhang. 2015. Possible artifacts of data biases in the recent global surface warming hiatus. *Science* 348 (6242): 1469–1472. https://doi.org/10.1126/science.aaa5632.

Kennel, C.F. 2014. Hiatus in Global Warming. Scripps Institute of Oceanography. https://olli.ucsd.edu/display/documents/ClimateChange.pdf.

Kerr, R.A. 2005. Atlantic climate pacemaker for millennia past, decades hence. *Science* 309 (5731): 41–42.

Kogan, F.N. 2001. Operational space technology for global vegetation assessment. *Bulletin of the American Meteorological Society* 82 (9): 1949–1964.

Kogan, F., W. Guo, W. Yang, and H. Shannon. 2018. Space-based vegetation health for wheat yield modeling and prediction in Australia. *Journal of Applied Remote Sensing* 12 (2). https://doi.org/10.1117/1.JRS.12.026002.

Kogan, F., W. Guo, and W. Yang. 2017. SNPP/VIIRS vegetation health to assess 500 California drought. *Geomatics, Natural Hazards and Risk.* doi:https://doi.org/10.1080/19475705.2017.1337654.

Kogan, F., and W. Guo. 2017. Strong 2015-2016 El Niño and implications to global ecosystems from space. *International Journal of Remote Sensing* 38 (1): 161–178. https://doi.org/10.1080/01431161.2016.1259679.

Kogan, F., Z. Popova, and P. Alexandrov. 2016. Early forecasting corn yield using field experiment dataset and Vegetation health indices in Pleven region, north Bulgaria. *Ecologia i Industria (Ecology and Industry)* 9 (1): 76–80.

Kogan, F., and W. Guo. 2015. 2006–2015 mega-drought in the western US and its monitoring from space data. *Geomatic, Natural Hazards and Risk*. doi:https://doi.org/10.1080/19475705.2015.1079265.

Kogan, F., M. Goldberg, T. Schott, and W. Guo. 2015a. SUOMI NPP/VIIRS: improve drought watch, crop losses prediction and food security. *International Journal Remote Sensing*. https://doi.org/10.1080/01431161.2015.1095370.

Kogan, F., W. Guo, A. Strashnaia, A. Kleshenko, O. Chub, and O. Virchenko. 2015b. Modelling and prediction of crop losses from NOAA polar-orbiting operational satellites. *Geomatics, Natural Hazards and Risk.* https://doi.org/10.1080/19475705.2015.1009178.

Kogan, F., and W. Guo. 2014. Early Twenty-First-Century Droughts during the Warmest Climate. *Geomatics, Natural Hazards and Risk*: 1–11. https://doi.org/10.1080/19475705.2013.878399.

Kogan, F., T. Adamenko, and W. Guo. 2013. Global and Regional Drought Dynamics in the Climate Warming Era. *Remote Sensing Letters* 4: 364–372. https://doi.org/10.1080/2150704X.2012.736033.

Kogan, F.N. 1983. Perspectives for grain production in the USSR. *Journal of Agricultural Meteorology* 28 (3): 213–227. https://doi.org/10.1016/0002-1571(83)90027-4.

Levitus, S.J. Antonov, T. Boyer, O. Baranova, H. Garcia, R. Locarnini, A. Mishonov, J. Reagan, D. Seidov, and E. Yarosh. 2012. World ocean heat content and thermosteric sea level change (0-2000 m), 1955–2010. *Geophysical Research Letters* 39 (10): L10603.

Lucht, W., I.C. Prentice, R.B. Myneni, S. Sitch, P. Friedlingstein, W. Cramer, P. Bousquet, W. Buermann, and B. Smith. 2002. Climate Control of the High-Latitude Vegetation Greening Trend and Pinatubo Effect. *Science* 296: 1687–1688.

GWCh. 2018. Global warming challenge:3 Months into second decade (TA–deviation from 1981–2010). http://www.theclimatebet.com/graph-full.png.

Miller, J. 2017.Coral isotopes track the rise of anthropogenic nitrogen. *Physics Today*, July, 19–21.

Murph, B.F., and B. Timbal. 2008. A review of recent climate variability and climate change in southeastern Australia. *International Journal of Climatology* 28: 859–879.

Myneni, R.B. 1997. Increased plant growth in the northern high latitudes from 1981–1991. *Nature* 386: 698–702.

NASA. 2018. Ozone Watch. https://ozonewatch.gsfc.nasa.gov/.

Nemani, R.R., C.D. Keeling, H. Hashimoto, W.M. Jolly, S.C. Piper, C.J. Tucker, R.B. Myneni, and S.W. Running. 2003. Climate-driven increases in global terrestrial net primary production from 1982 to 1999. *Science* 300 (5625): 1560–1563.

NOAA/NESDIS. 2018. Vegetation health indices and products. https://www.star.nesdis.noaa.gov/smcd/emb/vci/VH/index.php.

NOAA. 2016. Annual mean CO_2 at Mauna Loa. ftp://aftp.cmdl.noaa.gov/products/trends/co2/co2_annmean_mlo.txt.

NOAA/NCDC. 2017. Global Climate Report – November. December. https://www.ncdc.noaa.gov/sotc/global/201711.

———. 2016. Global Climate Report. www.ncdc.noaa.gov/cag/time-series/global.

NOAA/NCEI (National Centers for Environmental Information). 2017. State of the Climate: Global

Climate Report for March 2017, April. https://www.ncdc.noaa.gov/sotc/global/201703.

Polonskii, A.B., and E.N. Voskresenskaya. 2004. On the statistical structure of hydrometeorologycal fields in the North Atlantic. *Morsk Gidrofiz Journal* 1: 14–25.

Raa, L., H.A. Dijkstra, and J. Gerrits. 2004. Identification of the mechanism of inter-decadal variability in the North Atlantic Ocean. *Journal of Physical Oceanography* 34 (12): 2792–2807.

Ren, H., Y.-C. Chen, X.T. Wang, G.T.F. Wong, A.L. Cohen, T.M. DeCarlo, and M.A. Weigand. 2017. 21st-century rise in anthropogenic nitrogen deposition on a remote coral reef. *Science* 356 (6339): 749–752. https://doi.org/10.1126/science.aal3869.

Rice, D. 2018. A major climate boundary in the central U.S. has shifted 140 miles due to global warming. *USA Today*, April 13.

Samset, B.H. 2018. How clean air changes the climate. *Science* 360 (6385): 148–150.

Schlossberg, T. 2016. 12,000 years ago, humans and climate change made a deadly team. *The New York Times*, Science, June 17. https://www.nytimes.com/2016/06/18/science/patagonia-extinctions-global-warming.html.

Seager, R. 2018. Whither the 100th meridian? The once and future physical and human geography of America's arid–humid divide. Part I: The story so far. *Journals Online*, March 21. doi: https://doi.org/10.1175/EI-D-17-0011.1.

Serreze, M.C. 2018. *Brave new arctic the untold story of melting north*, 269. Princeton: Princeton University Press.

Solomon, S. 1999. Stratospheric ozone depletion: A review of concepts and history. *Reviews of Geophysics* 37 (3): 275–316.

Staehelin, J., A. Renaud, J. Bader, R. McPeters, P. Viatte, B. Hoegger, V. Bugnion, M. Giroud, and H. Schill. 1998. Total ozone series at Arosa (Switzerland): Homogenization and data comparison. *Journal of Geophysical Research* 103 (D5): 5827–5841.

UCS (Union of Concerned Scientist). 2017. Is there a connection between the ozone hole and climate warming? July 17. https://www.ucsUS.org/global-warming/science-and-impacts/science/ozone-hole-and-gw-faq.html#.Wrgcs4jwaUl.

UNESCO. 2018. Climate Change and Water Security. https://en.unesco.org/themes/addressing-climate-change/climate-change-and-water-security.

UNFCCC. 2014. Kyoto Protocol. http://unfccc.int/kyoto_protocol/items/2830.php.

———. 2015. Adaptation Fund. http://unfccc.int/cooperation_and_support/financial_mechanizm/adaptation_fund/items/3659.php.

UN (United Nations). 2016. Paris Agreement. http://unfccc.int/paris_agreement/items/9485.php.

Ward, P.L. 2016a. *What Really Causes Global Warming?* New York: M & J Publishing Co. pp 235.

———. 2016b. Ozone depletion explains global warming? *Current Physical Chemistry* 6: 275–296 https://whyclimatechanges.com/pdf/Papers/Ward2016OzoneDepletionExplains.pdf.

Watts, J. 2018. Water shortages could affect 5bn people by 2050, UN report warns. *The Guardian*, March 19. https://www.theguardian.com/environment/2018/mar/19/water-shortages-could-affect-5bn-people-by-2050-un-report-warns.

WB. 2017. Agriculture & Rural Development. https://data.worldbank.org/topic/agriculture-and-rural-development.

Weisberger, M. 2018. *Life Science*, April. https://www.livescience.com/62283-weakest-atlantic-ocean-circulation.html.

WEF (World Economic Forum). 2016. What is Paris Agreement on climate change? https://www.weforum.org/events/world-economic-forum-annual-meeting-2016.

WMO (World Meteorological Organization). 2014. Global surface temperature data sets. http://www.wmo.int/pages/prog/wcp/wcdmp/GCDS_3.php.

———. 2016. 2015 is the hottest year on record. https://public.wmo.int/en/media/press-release/2015-hottest-year-record.

10 Application of Vegetation Health Data and Products for Monitoring Food Security

10.1 Why and How to Use Vegetation Health?

The year 2018. Almost one-fifth of the twenty-first century has already past and the Earth has still been continuing the previous tendencies for a rapid population growth, declining stock of natural resources, climate warming, land cover changes, increasing natural disasters, etc., which have intensified considerably world's concerns about the future food supply/demand and global food security (USDA 2017; FAO 2017, 1999; Heibuch 2011). Most of the indicated problems are related to a deterioration of environmental conditions. As has never been before, decision makers of the world, countries, communities, international organizations, and businesses need reliable and timely information to understand, monitor, and predict impacts of weather/climate and environmentally based Earth's changes on global food security (FS).

Unfortunately, weather, climate, and Earth changes have already affected not only global food security but also earth environment and surface. In efforts to produce more food, agriculture put considerable constraint on the environment and earth cover in overexploiting soil fertility, causing land degradation, diminishing fresh water, and deteriorating ecosystems and climate. Following some publications, the current climate has already started to limit staple crop production and fresh water, increased soil degradation, and slowed down the rate of agricultural output (IPCC5 2014; IPCC4 2007; Kerr 2005). There is currently an opinion that warmer climate is likely to constrain much stronger agricultural production due to an expected increase in severity and frequency of large-scale weather extremes (ClimateBet 2018; IPCC5 2014; Alexandratos and Bruinsma 2012; IPCC4 2007). This situation is further complicated by anticipated drought intensification and expansion, leading to a further reduction of agricultural production and FS deterioration (Schmidhuber and Tubiello 2007). If moderate-to-intensive droughts cover more than 20% of the world's main agricultural area, food production becomes less than what the world needs for consumption. This situation has already deteriorated in the twenty-first century, when in the first 17 years, world grain production (the principal staple food) was below consumption in half of those years (Kogan et al. 2018; WB 2018). Drought was the major environmental disaster affecting FS and world's sustainability in all these years (Kogan et al. 2009, 2013a, b, 2015a, b 2017; Kogan 2002). In order to predict drought damages to agriculture and developing advanced measures to improve food security, the world, countries, communities, and farmers need information on early drought start, intensity, duration, affected regions, and impacts. They have to know if vegetation is under stress or healthy, if soil moisture is limited or soil is saturated, if fire risk

F. Kogan, *Remote Sensing for Food Security*, Sustainable Development Goals Series,
https://doi.org/10.1007/978-3-319-96256-6_10

is high, if mosquitoes' activity is intensive to spread malaria, etc. These goals require high-resolution (0.5, 1 and 4 km^2), temporal (at least weekly) and multi-year data to monitor weather conditions, including extreme events, especially droughts, to assess their impacts on agriculture and prediction of food security.

In the 1990s, National Oceanic and Atmospheric Administration (NOAA) has developed new satellite-based vegetation health method (Kogan 1987, 1989, 1998, 2002; Kogan et al. 2009; Kogan and Wei 2009, 2017) with high spatial and temporal resolution to monitor operationally environmental impacts on agriculture and other weather- and climate-related activities. This chapter is written for users who are already applying and would like to apply vegetation health (VH) data and products in their work to monitor and predict weather and climate impacts on vegetation health and through VH on crops/pasture production and food security. Applying VH data and products, the users would be able to monitor effectively first of all agricultural production, secondly, climate, land cover and weather changes, and finally, food security. Vegetation health is a very powerful tool, available on WEB since the early twenty-first century (NOAA/NESDIS 2018) to: (a) assess moisture and temperature condition of crops and pasture, (b) derive area and intensity of moisture and thermal stress, (c) detect drought early enough (up to 4 weeks ahead of its start) in order to plan irrigation and other measures, (d) monitor drought area, intensity, duration, and impacts on agriculture for identifying food insecure areas and intensity of the problem, (e) model agricultural production (crops and pasture), (f) predict crop yield for an early assessment of food availability, (g) assess soil moisture (both deficit and saturation), (h) identify areas with potential weather-related threat of fire, (i) detect areas with elevated activity of mosquitoes and malaria spreading, (j) estimate malaria intensity and predict the number of affected people, (k) identify climate warming impacts on land cover, and (l) assess multiyear trends in drought, soil excessive wetness, crop and pasture production losses, spread of malaria, and food security problems. This short list of applications would help users to provide advises to farmers, agricultural producers, business community, international organizations, and policy developers for making decisions in food security enhancement. Moreover, following our activity and communication with users, VH is also useful for scientific, educational activity and development of new applications.

Vegetation health data and products are currently provided through the following WEB site https://www.star.nesdis.noaa.gov/smcd/emb/vci/VH/index.php (NOAA/NESDIS 2018; Kogan et al. 2013a, b, 2015a, b; Kogan 2002, 2009), where data and products are presented in the form for immediate applications and also explains how to use the information. The WEB displays real-time and historical (since 1981) weekly 16, 4, 1, and 0.5 km^2 resolution for the entire world and also mean country and administrative regions data in the form of images of moisture (VCI), thermal (TCI), and moisture–thermal combine (VHI) conditions, drought, vegetation stress, fire risk, vegetation greennes and temperature etc. and their changes between weeks and years (Kogan et al. 2013a, b, 2015a, b, 2017; Kogan and Guo 2017; Kogan 1987, 1989, 1998). The 38-year data and products are presented in the form of images, time series, and digital values for each pixel of the world, for administration regions, countries, continents, and the entire world.

10.2 Vegetation Health Users

Over the time of VH-WEB functionality, the users from nearly 200 world countries have accessed the images and digital data and used them for monitoring weather-climate impacts on crops. Moreover, the users are able to ftp area mean data and product for countries to develop models for prediction of crop and pasture yield and production. In addition to countries, for the first-order administration divisions in each country (for example, states in the USA, oblast in Russia, etc.), 4 km^2 area mean values of 38-year (1981–2018) weekly time series of VH data and products can be displayed for the years and weeks

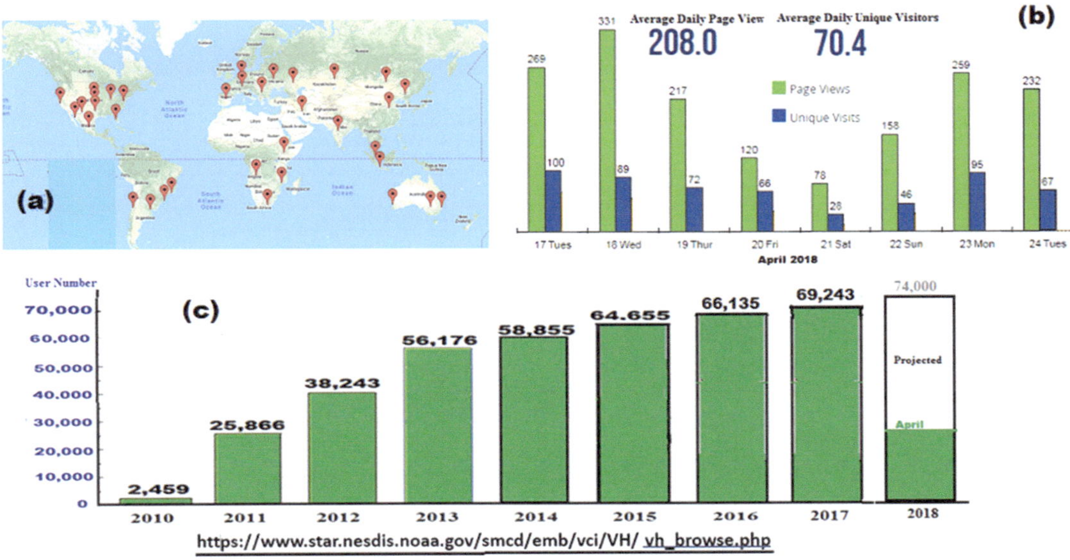

Fig. 10.1 Visits to Vegetation Health WEB page (**a**) 1-day map in May 2018, (**b**) 1 week in April 2018, (**c**) annual during 2010 through April 30, 2018

selected by users. The users have access to around 4000 administrative regions from nearly 200 world countries. The users of VH-based WEB represent a wide spectrum of interests, including agriculturalists, traders, commodity dealers, bankers, stock Co., farmers, researchers, media, and others.

This WEB page was first presented in 2010 and attended by 2459 users (Fig. 10.1). Since then, the number of users have increased considerably. Figure 10.1 displays these visits (StarStat 2018). The map (Fig. 10.1a) displays 1-day visit. Since May, which is the beginning of the growing season north of the Equator, most visitors are from the Northern Hemisphere assessing the conditions ahead of the start of spring crop planting. Meanwhile, 20–30% visitors are from the Southern Hemisphere, finishing agricultural year. The number of visitors during April 17–24 are shown in Fig. 10.1b. As seen, most visitors attend the page during Monday and Tuesday, from Thursday the number is reducing almost in half and from Sunday start to increase again. The Vegetation Health WEB page become very popular since the number of users is increasing every year. During the last 3 full years (2015–2017), the number of users attending the WEB page were slightly over 64,000, 66,000 and 69,000, respectively. In the first 6 months (January–June) of 2018 the VH-WEB page was attended by slightly over 37,000 users. So the prediction for the entire 2018 is 74,000–75,000 visits, the number is again increasing from the last year.

10.3 Vegetation Health WEB Pages

Figure 10.2a displays (a) the front page of Vegetation Health WEB (https://www.star.nesdis.noaa.gov/smcd/emb/vci/VH/index.php) after opening and (b) how to access to the data and (c) available indices and products. The main menu for accessing to images and data is on the left side of Fig. 10.2a ("Vegetation Health Home", (VHH) in red). The principal items to access the data are in the first few lines up to the "Background and Explanation." These few lines indicate the available data and images with the corresponding resolution 16, 4, and 1 km^2. The 16 and 4 km^2 data are weekly data from 1981 through present. They are blended combining AVHRR-based VH (NOAA satellites) from 1981 through 2012 and VIIRS-based (adjusted to AVHRR) from SNPP

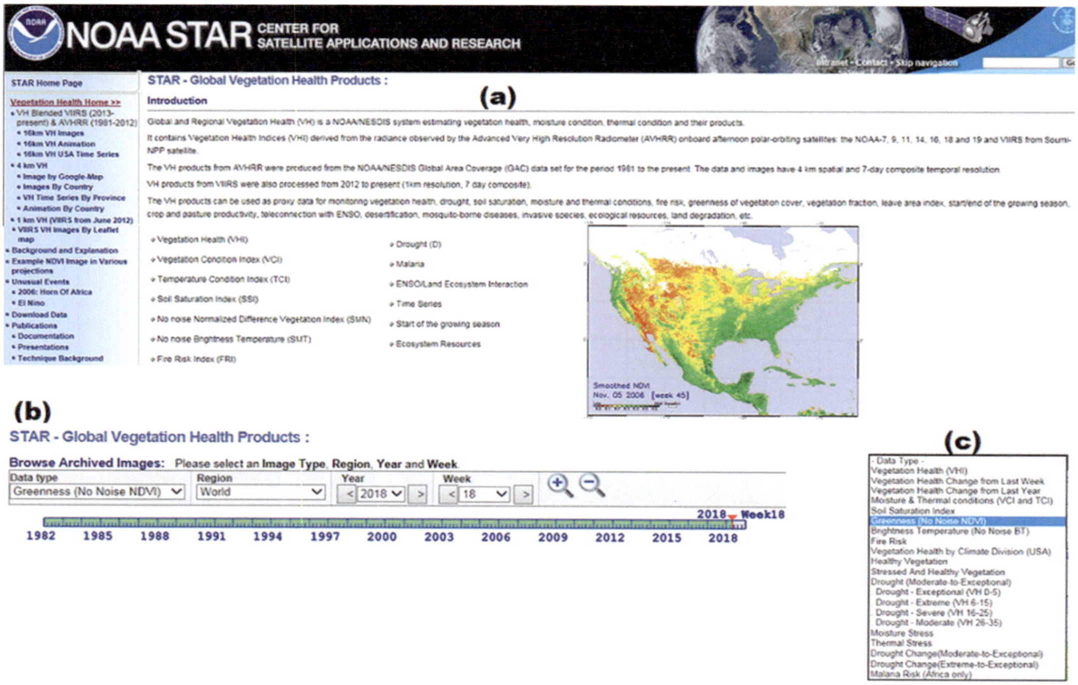

Fig. 10.2 General view of Vegetation Health WEB (https://www.star.nesdis.noaa.gov/smcd/emb/vci/VH/index.php) beginning pages (**a**) first one after opening the WEB, (**b**) the page showing how to access the data and (**c**) available indices and products for display and getting data (for details open the WEB (NOAA/NESDIS 2018) and follow the instructions)

satellite from January 2013 through 2018 and the new NOAA-20 satellite started from mid-2018. From VHH, the users can access to 16 and 4 km^2 blended data and images. There are three 16 km^2 items to display "16 km VH images," "16 km VH Animation," and "16 km VH USA Time Series." The 4 km^2 resolution data contain four items to display "Images by Google Map," "Images by Country," "VH Time Series by Province," and "Animation by Country." Further discussion presents the main points some of the important pages.

Clicking on the "16 km VH images," the user is coming to the upper menu (Fig. 10.2b), where it is possible to select the "Data Type," "Region," "Year," and "Week" to display images of indices and products (Fig. 10.2c), which can be selected clicking on the symbol "v" from "Data Type." The images are displaying vegetation greenness (SMN), brightness temperature (SMT), moisture conditions (VCI), thermal conditions (TCI), vegetation health (VHI), drought (area and intensity), moisture/thermal stress, fire risk, etc. and changes of some parameters from last week and last year. The region includes the world, continents, and some important countries (the USA, China, India, etc.). Years can be selected from 1981 through the present (currently 2018) and weeks selection cover from week 1 (the 1st week in January) through week 52 (the last one in December). Although the 4 km^2 resolution data are represented by four items to display, the explanation is focus only on the "Images by Country" and "VH Time Series by Province." The "Images by Country" show the same data type indicated in Fig. 10.2c for 191 countries (in the "country/region" menu). These countries were selected following the GIS system. The "VH Time Series by Province" displays time series of provincial mean value of indices and products for the 191 countries and provinces, which are the first-order administrative regions in each country (for example, states in the USA, oblast' in Russia, etc.). The display material is weekly indices/products from 1981 through the current (based on users' selection). The products include moisture, thermal, and moisture–thermal conditions (VCI, TCI, and VHI, respectively), vegetation greenness and temperature, drought area and favorable vegetation conditions (by two intensities). The SNPP/VIIRS 1 km^2 images are available from the "1 km VH (VIIRS from June 2012)" on the left front page menu. There are two items "VIIRS VH images browser" and "VIIRS VH images by Leaflet map." These data are from VIIRS sensor only. Therefore, they are available from mid-2012. It is important to emphasize that in addition to the products indicated in Fig. 10.2(c), there are regional (country) mean values for the following products: regional mean vegetation greenness (SMN) and temperature (SMT), moisture (VCI), thermal (TCI) and vegetation health (VHI) conditions, percent drought area (from the total administration region area) with severe-to-exceptional (VHI < 15) and moderate-to-exceptional (VHI < 35) intensities. In addition to displayed images and products, some of the digital data can be transferred through the *ftp* procedure.

10.4 Vegetation Health Applications

Since the development of vegetation health methodology, numerous applications have been developed and used in agriculture, water resources management, drought detection and monitoring, prediction of weather impacts, climate studies, soil moisture and fire risk assessments, vector-borne diseases monitoring and others. These applications were validated thoroughly against in situ data over many years. The most important applications for food security predictions were related to modeling crop and pasture yield and production, prediction of drought start, development and impacts assessments on agriculture, climate changes, weather-related water management, long-term prediction ocean impacts (ENSO) on vegetation conditions, and others (Adamenco and Prokopenko 2009; Kogan et al. 2013a, b, 2015a, b, 2018; MinHao et al. 2009). These applications using the data and products from Vegetation Health WEB site are discussed further.

Vegetation Health is estimated through combined moisture–thermal conditions assessed from VH-based vegetation health index (VHI). As was described in Chap. 4, VHI is an index combining deviation from climatology of no noise moisture and temperature indices. The VHI provides estimates on a scale from 0 (extreme vegetation stress) to 100 (extremely favorable condition). Such extreme cases (0 and 100) have occurred one to two times since 1981, when most vegetation might be dead due to the exceptional stress case or extremely healthy (very green and vigorous) under favorable conditions (100 index value). An example of VHI for May 14, 2018, is shown in Fig. 10.3a. Mid-May is the harvest time of winter-based crops (for example, winter wheat) and the beginning of the growing season in Northern Hemisphere. Following the VHI map, the USA is showing moderate-to-extreme vegetation stress (VHI is between 36 and 12) in the principal grain area (central part), which indicates unfavorable conditions for planting 2018-season crops. Northern part of Sub-Sahara Africa has also indicated unfavorable moisture–thermal conditions, especially in southern Sudan (extreme-to-exceptional vegetation stress, VHI < 12). Grain-planted area in the central India has also experienced vegetation stress. Moderately stressed vegetation conditions have also observed in Ukraine (VHI is between 48 and 25). In Southern Hemisphere, Australia's wheat area (far west and south) is under strong-to-extreme vegetation stress (VHI is between 24 and 6) and Argentina's grain area also experiences light-to-moderate stress. At the same time the major grain areas (see Fig. 9.9) of Europe, Russia, China, Brazil, and South Africa have experienced favorable conditions (VHI > 55). The 2018 crop conditions will depend much on summer precipitation and temperature. However, the beginning of Northern Hemisphere's growing season has started with some negative contribution to food security.

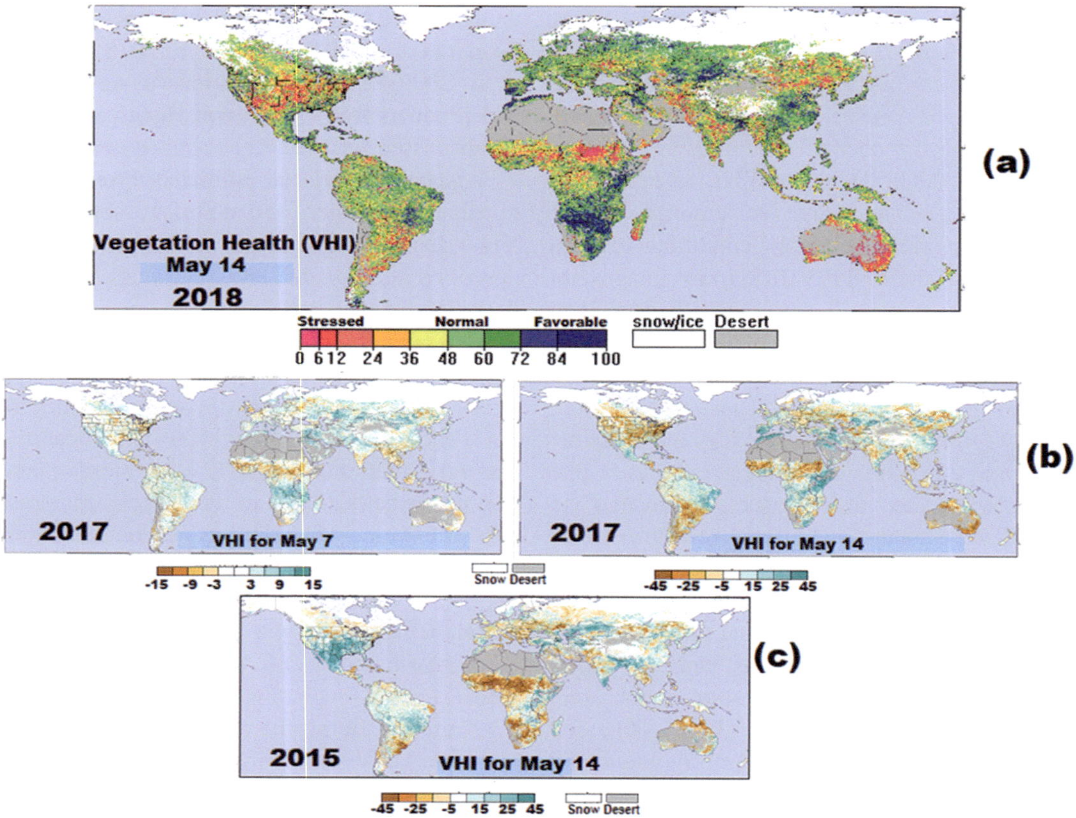

Fig. 10.3 VH-based vegetation health index (VHI): (**a**) May 14, 2018 (**b**) VHI difference between May 14 and May 7 in 2018 (left image) and between May 14, 2018 and 2017 (right image) and (**c**) VHI difference between May 14, 2018 and 2015 (NOAA/NESDIS 2018)

Additional contribution to grain crop analysis and food security can be obtained from a comparison of May 14, 2018, VHI index with its value for the previous week (May 7, 2018) and especially with previous year, show in Fig. 10.3b. Weekly comparisons indicate that all areas showing VHI-based vegetation stress on May 14 were continuing such conditions from previous week (Fig. 10.3b left), which adds credibility to the analysis and conclusion. Moreover, some deterioration of moisture–thermal conditions is observed in the areas, where previously VHI was greater than 55 (for example Europe). Comparison of May 14, 2018, VHI with the same time last year indicates that the current growing season has started with a worth conditions compared to 2017. This situation is observed in the USA, northern Sub-Sahara, Argentina, Ukraine, most of east Europe, India, and Australia. Since the 2017 grain crop production is already known, this comparison provides additional prediction for potentially unfavorable food security in 2018. For example, following the spring 2018 grain crop analysis in the USA (Fig. 10.3b), it is possible to summarize that the weather-based winter wheat yield will be below the 2017. This conclusion was supported by the recent USDA report (USDA/NASS 2018), showing that on May 1, 2018, mean USA winter wheat yield is predicted to be 4% down compared to 2017. State wise, in Kansas and Oklahoma (the largest winter wheat grower), 2018 yield is expected to drop 23% compared to 2017; in Taxes and Colorado (two other big winter wheat grower) the drop will be 7 and 4%, respectively. It is interesting that in 2015, the USA's winter wheat yield was nearly

6% below the projected 2018 yield, which is supported by much better 2018 moisture–thermal conditions (VHI in 2018 is 25–45 larger than in 2015, blue area) in the main winter wheat area of the central USA (Fig. 10.3c). The VH products used for this analysis are items 2 and 3 in the table (c) on Fig. 10.2. Comparison of the current year's VHI conditions can be done with any week and year since 1981.

Moisture and Thermal Conditions. In extreme environmental situations the combined moisture–thermal index (VHI) is not sufficient to estimate correctly the specificity of crop conditions. For example, in Chap. 5, we described fall 2017 extreme vegetation stress form thermal-based drought (air temperature exceeded 40 °C) in Portugal, Spain, France, and Italy, which almost killed grape plantations, which in turn affected wine production. Such extreme stress, either thermal or moisture (especially thermal), might decimate vegetation in a matter of days. Therefore, in addition to the described VHI-based assessments, it is worth to estimate moisture and thermal conditions separately. The VH method and the VH WEB provide weekly moisture (VCI) and thermal (TCI) conditions from 1981 through present for each 16 and 4 km^2 pixels and also area-mean VCI and TCI values for nearly 200 countries and 4000 first-order administrative divisions.

The example below covers global moisture (VCI) and thermal (TCI) conditions for May 14, 2018 (Fig. 10.4). The worse environmental situation for crops is normally developing when both moisture and thermal conditions are stressful. As seen in the figure, such conditions affected Australia wheat-growing area in May 2018, with much stronger thermal stress (TCI below 15). Similar to Australia vegetation condition affected northern Sub-Sahara Africa, with very strong thermal stress (TCI below 15) and also moderate moisture stress (except western part, where moisture conditions were favorable). In the USA, the environmental conditions were opposite to the previous two regions: moisture conditions in the main grain-growing area were much more stressful (VCI below 15) than thermal (TCI between 36 and 24). From the other world grain regions, Europe had very good moisture conditions (VCI greater than 60), while elevated temperatures produced mild level of thermal stress (TCI between 36 and 24). This analysis indicates that for the purposes of food security, it is important to estimate moisture-based and thermal-based vegetation stress first, shown in Fig. 10.4b, in order to understand which regions are going to

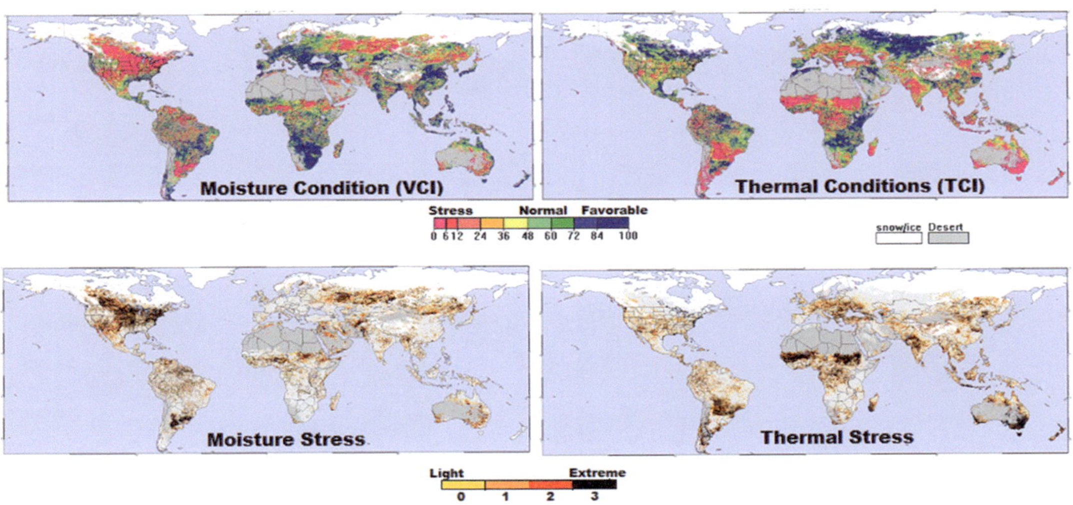

Fig. 10.4 May 14, 2018, global VH-based (**a**) moisture conditions from VCI and temperature conditions from TCI, (**b**) vegetation stress due to a lack of moisture (VCI < 40) and excessive temperature (TCI < 40)

have losses in agricultural production and needed food assistance. Opposite to moisture and thermal stress, assessments of favorable conditions are also important to indicate which regions might produce above normal grain production and might offset food insecurity situation global or regional.

Drought is the major environmental disaster affecting food security. Since droughts are global phenomenon, they normally occupy large portion of the world and occurred frequently. Moreover, they might continue from a few days to a few years, reduce agricultural production and in extreme cases affect strongly food security. An example of global and regional droughts during 12-year (2001–2012) period, estimated by moisture–thermal vegetation health index (VHI), is shown in Fig. 10.5. In some important agricultural regions, droughts might develop every year as it is indicated in Australia during 2002 through 2010 (Fig. 10.5). Among the most recent examples (not shown), it should be mentioned 12-year drought, which affected the USA's California (one of the major producers of fruits, vegetables, nuts, etc.) from 2006 through 2018 (Kogan et al. 2017, 2018). Therefore, VH-based drought detection

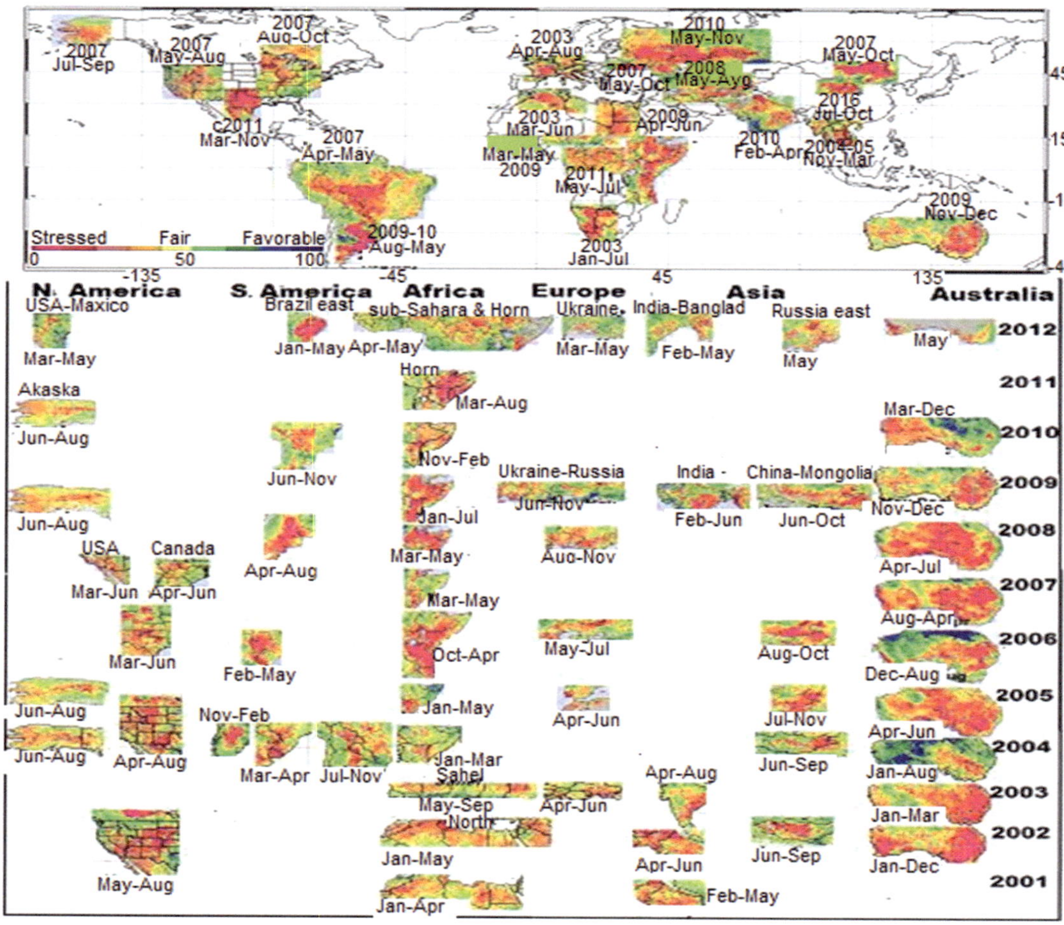

Fig. 10.5 Mosaic of global and regional moisture–thermal conditions (VHI) focusing on droughts during 2001–2012. The global map on the top was composited of regional images (under the global map). The regional images were selected for the week of the strongest drought (the lowest VHI). The months under the images indicate drought duration; years are shown on the right side

and monitoring is used widely for estimation of grain losses and food security prediction.

The VH-WEB (NOAA/NESDIS 2018) provides several drought products designed for an early drought detection, estimation of its intensity, area, duration, and impacts. One of the most important drought products estimates drought area and intensity from the VH-based moisture (VCI), thermal (TCI), and moisture–thermal (VHI) condition indices. Drought intensity is assessed in five categories: light (VH = 45–36), moderate (VH = 35–26), severe (VH = 25–16), extreme (VH = 15–6), and exceptional (VH < 6) or in a sum of the indicated categories. An example of global and Africa drought on July 1, 2015, is shown in Fig. 10.6. As seen, in mid-summer 2015, quite strong drought affected large area of Sub-Sahara Africa (northern, central, and southern), wheat area in northern Australia and pastures in northern Mongolian. Light-to-severe drought affected western USA and Europe (global image). Following individual drought intensities (bottom images), the entire Africa was affected by moderated drought (left Africa image). The exceptional (the strongest) drought affected limited areas in northern Nigeria, central Sudan, eastern Tanzania and southern Angola (second bottom image from the right). These countries supposed to experience substantial losses in agriculture and might need food assistance following food insecurity situation.

In some circumstances, the users might experience the need for more accurate assessment of agricultural production in advance of harvest. If they do not have crop-VH or crop-weather models for more accurate predictions, preliminary assessments might be done comparing the current-year VH with previous year or with VH of several weeks earlier, to identify the area where drought conditions worsened. Figure 10.7 demonstrates such comparison, indicating the areas with deteriorating or improving VH conditions. The areas with deteriorating condition indicate potential agricultural losses. For example, knowing agricultural production obtained in 2014, the July 1, 2015, VH-based assessment (Fig. 10.7a) indicate deterioration of moisture–thermal conditions in the most of Sub-Sahara Africa compared to 2014. The countries experiencing worse drought conditions in 2015 might produce smaller agricultural production depending on the intensity of deteriorated vegetation conditions.

In addition to occupied area and intensity, drought impacts on agriculture depend on drought duration. Longer droughts normally lead to stronger losses in agriculture. Such example is shown for Kenya, when more than 70% of the country was affected by severe drought in summer of 2009 and 2011 (Fig. 10.8b). The most affected grain (mostly corn)-producing area in the south (GOOGLE 2018) was in 2009, when drought continued for 2.5–4.5 months and was extreme and exceptional, which was confirmed by Fig. 10.8a. Moreover, drought situation in 2009 was complicated by severe thermal stress (bottom image). Cereal production in 2009 was reduced 23% (below multiyear trend level (WB 2018)). Although Kenya was also affected by drought in 2011, grain production was slightly reduced, but still was at the trend level, because the major grain area experienced only moderate drought (Fig. 10.8b). A very important drought parameter is drought start. It is difficult to estimate this parameter from weather data. For example, a lack of rainfall for 1 or 3 weeks does not indicate drought start if soil moisture is sufficient from earlier rains or another example, extreme temperature might decimate vegetation very quickly. Vegetation health indices reflect today's situation with moisture and thermal conditions, which are created over a short or long period of time. Figure 10.8c displays how 2007 drought has started. Following the time series, moisture–thermal (VHI) and thermal (TCI) indices started to drop from healthy vegetation conditions (both indices 70–80) during the first one half (March–July) of 2006. When these indices crossed down 45 level threshold, the drought (both thermal and moisture–thermal) has started (in the second one half of 2006). This 2006–2007 drought was triggered by a very strong thermal stress (TCI = 5–20) resulted in a strong soil moisture depletion and strong reduction in moisture–thermal (VHI) index.

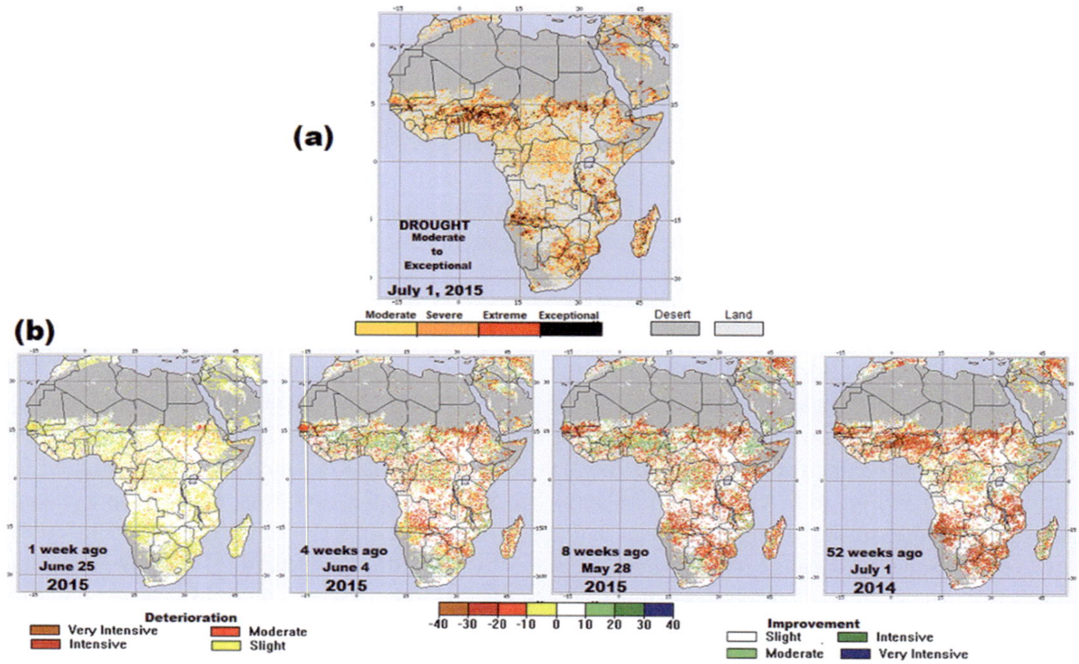

Fig. 10.6 Drought area and intensity on July 1, 2015: (**a**) Global, moderate-to-exceptional intensity; (**b**) Africa, moderate, severe, extreme, exceptional and moderate-to-exceptional intensity

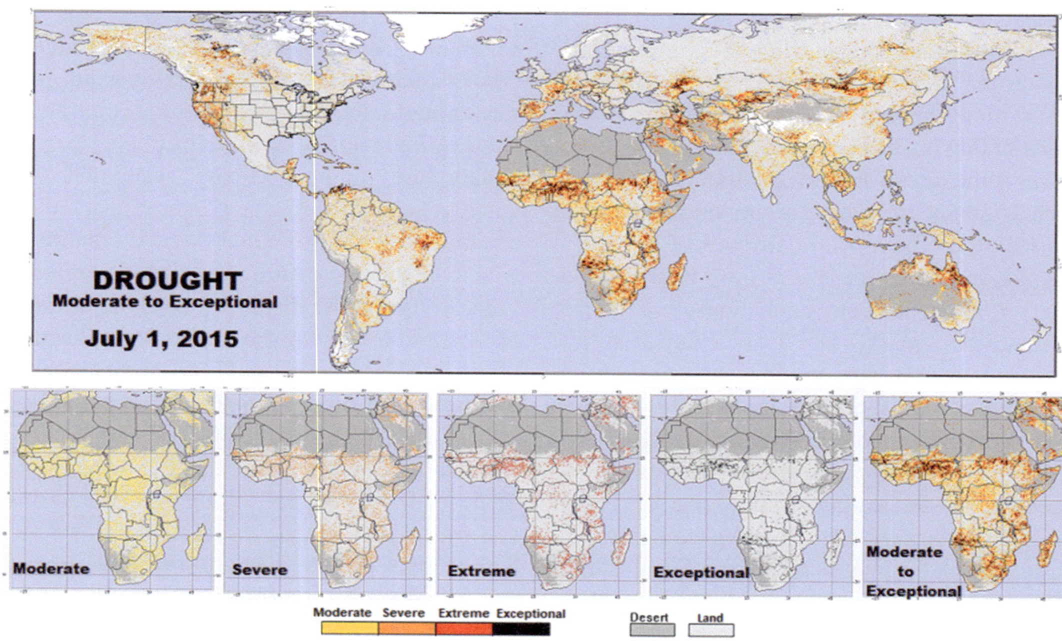

Fig. 10.7 Drought moderate through exceptional in Africa (**a**) July 1, 2015 and (**b**) Drought changes from week 1 (June 25), 4 (June 4), 8 (May 28), and 52 (July 1, 2014) before July 1, 2015

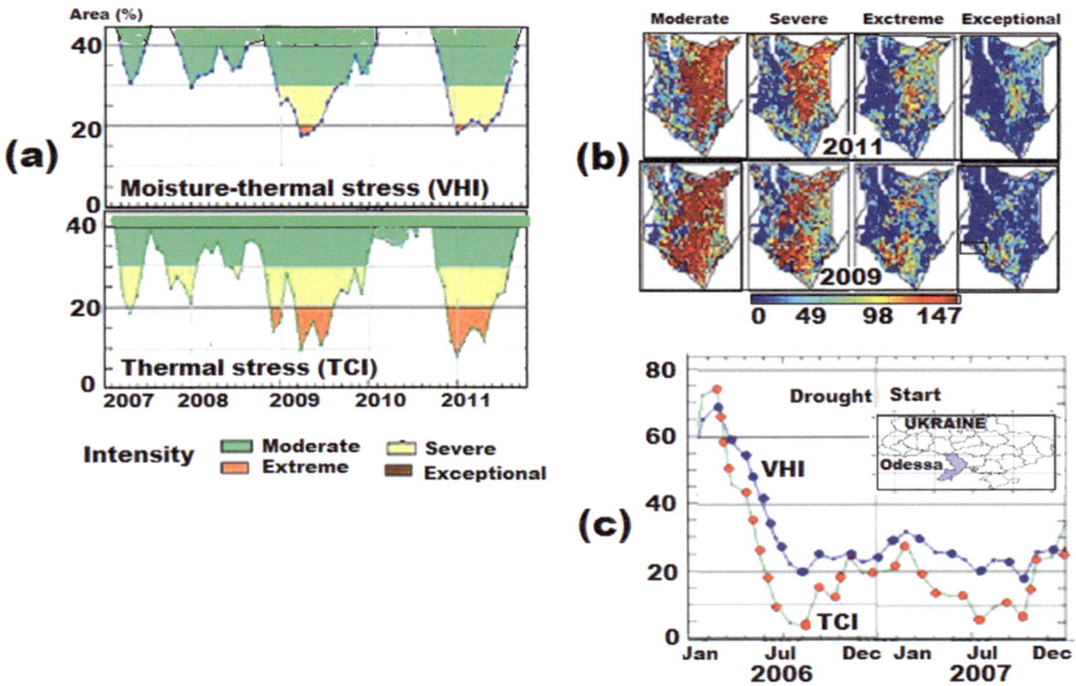

Fig. 10.8 Weekly drought in Kenya (**a**) Area by intensities and duration in 2007–2011 from moisture–thermal (VHI) and thermal (TCI) stress and (**b**) Number of drought days for areas with different intensities from moisture–thermal (VHI) stress; (**c**) Drought start in Ukraine's Odessa oblast (administrative region)

Other VH-based Products. Among other products vegetation health indices are used for global monitoring excessive soil moisture, identifying fire risk area, modeling and estimation of crop yield and monitoring malaria in Africa (MinHao et al. 2009; Rahman et al. 2016). Figure 10.9 displays and discusses these applications in some specific cases validated by ground data.

Fire risk area is assessed based on VH-estimated (moisture, thermal, and moisture–thermal) drought intensity and duration. Areas where VH-based drought intensity is estimated stronger than severe (indices below 35) from both moisture (VCI) and thermal (TCI) vegetation stress conditions and such drought is continuing for 2–3 weeks, then the environmental conditions are potentially favorable for triggering open vegetation fire (in case of ignition). Fire risk is estimated in six categories from low risk (or no risk) to light, moderate, severe, extreme, and exceptional (in the worse situation). An example in Fig. 10.9a showed area of fire risk conditions developed in Ukraine and during exceptional drought in summer 2007. Following Kogan and Guo 2009, that drought started in the southern part of Ukraine at the end of 2006 with extremely warm weather. Although that area experienced favorable moisture conditions (VCI around 90) in early 2007, extreme-to-exceptional thermal stress (TCI <10) deteriorated moisture conditions to the level of moisture-based stress (VCI < 30) by the end of May 2007. As the result of both moisture–thermal severe vegetation stress, their total contribution during June–August resulted in extreme-to-exceptional vegetation decimation and development of conditions for extreme fire risk. In addition to fire an intensive risk, winter wheat yield was reduced 14% compared to 2006 (also unfavorable crop year) and 35% compared to favorable crop year in 2008 (Adamenko and Prokopenko 2009). The 2007 fire risk area and intensity were extreme compared to 2010, a year of minor drought and only a few fire risk spots in northeastern Ukraine (Fig. 10.9a).

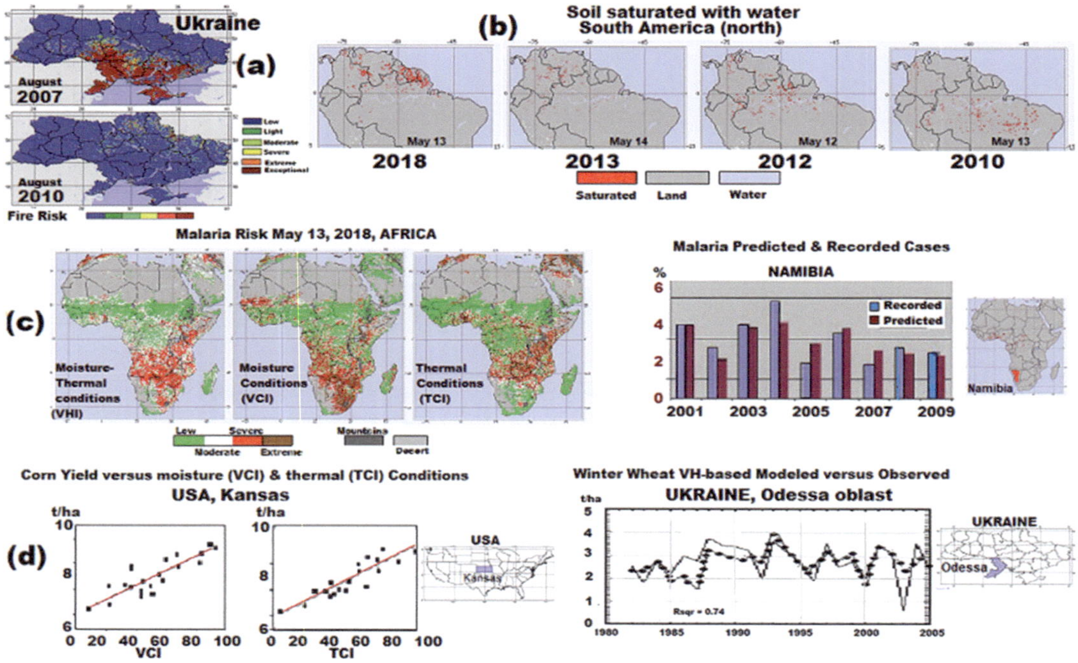

Fig. 10.9 VH-based environmental products: (**a**) Fire risk area in Ukraine and Russia in 2007 and 2010; (**b**) Soil saturation with water in northern South America in May 2018, 2013, 2012, and 2010; (**c**) Malaria risk area and intensity in Sub-Sahara Africa on May 13, 2018, and malaria cases predicted (from malaria-VH model) and recorded in Namibia during 2001–2009; and (**d**) Correlation of Kansas state (USA) corn yield with mean-state moisture (VCI) and thermal (TCI) conditions in mid-summer during 1996–2014 and comparison of VH-based modeled Odessa oblast (administrative region), Ukraine winter wheat yield with collected yield during 1982–2005

Soil saturation or soil extreme wetness can appear in case of excessive rainfall (much above normal), which fill completely soil pores to such level that water even stays on soil surface. If soil saturation continues for 1–3 weeks, vegetation becomes depressed, losing turgor and greenness, becoming unhealthy. Such vegetation conditions are normally characterized by a reduced moisture (VCI) and thermal (TCI) values (below 40), which usually happens in tropical and other wet climate areas. The VH criteria for soil saturation are low VCI and TCI (each below 40). The process of soil saturation becomes more intensive when both indices are approaching to zero. As seen in Fig. 10.9b, a large area of soil saturation was developed in May 2018 in northern South America. In the other three presented years (2010, 2012, and 2013) soil saturation spread around northern South America. Soil saturation is a very local phenomenon and has to be evaluated on a high-resolution (least 1 km^2) pixel-based level.

Another important VH-based product was developed to identify a start of increased malaria cases and monitor malaria intensity in Africa. Malaria is transmitted by infected mosquitoes. Their activity is intensified in a moderately warm and moist environment. If the environmental conditions are dry and hot mosquitoes become less active and the number of infected people is reduced. These considerations were set up in the development of VH-based criteria for reduced and increased mosquito's activity. In general, drought-based conditions reduce the number of malaria-infected people and wet and cool conditions stimulate an increase in the number of people (Rahman et al. 2006). The values of moisture (VCI), thermal (TCI), and moisture–thermal (VHI) indices for derivation of malaria reduction and intensification were established correlating the number of infected people during several-year period with weekly values of the indices and development of models. Malaria conditions are estimated every week in Africa (NOAA/NESDIS 2018). Figure 10.9 (c, left image) shows that in mid-May 2018, moisture (VCI), thermal (TCI), and moisture–thermal (VHI) conditions were favorable for an increased number of malaria cases in Sub-Sahara Africa (mostly between 3 and 16°S and also in eastern Africa, north of 3°S and south of 16°S). The figure on the right displays independently modeling results in the number of malaria cases compared the recorded number in Namibia during 2001–2009. In six cases the difference (recorded minus predicted) was very small (<7%) and in the 3 years, larger number of cases was predicted in 2004 and smaller number in 2005 and 2007.

Finally, two examples display application of VH indices for modeling crop yield (Fig. 10.9d). Kansas is one of the major producers of corn in the USA. The left diagram shows correlation between Kansas-mean corn yield and state's area-mean moisture (VCI) and thermal (TCI) condition indices for mid-summer. As seen, the correlation is very strong (correlation coefficient 0.81 for VCI and 0.85 for TCI). Following these results, models were developed to predict corn production in Kansas (Kogan et al. 2015a, b). Another example (right diagram) is showing independent validation of VH-based model for winter wheat in Ukraine's Odessa oblast (administrative region), which is the main producer of this crop in the country. Following Kogan et al. (2009), the model was developed regressing 1982–2005 winter wheat deviation from statistically-derived multiyear technological trend with 2-week VCI in February and 2-week TCI in May. Independent validation indicates that VH-simulated winter wheat yield correlates strongly ($R^2 = 0.74$) with the recorded yield. Only in 3 years, 1986, 1987, and 2003, (from 24) the difference between predicted and recorded yield was 1–2 t/ha. For the rest of the years the difference was below 0.75 t/ha.

Climate and Weather Products for Food Security Prediction. Vegetation health indices are available since 1981 through present. This data set started from the old generation of operational afternoon polar-orbiting satellites with AVHRR sensor (from 1981 through 2012: NOAA-7, -9, -11, -14, -16, -18, and -19), continued with intermediate satellites generation, designed to test VIIRS sensor (2013 through present: S-NPP) and the new generation of NOAA series satellites with VIIRS sensor, started in 2018 form NOAA-20 satellite, which will continue until the mid-century with NOAA-21, -22, -23, and -24. Thirty-eight years VH indices can be currently used to identify tendencies and changes in vegetation health, weather and climate dynamics. This information can be presented in a form of products to characterize short-term changes and long-term tendencies in vegetation health, including greenness, vegetation cover temperature, moisture, thermal, and combined conditions, vegetation stress, and healthy conditions. These products are regularly used for assessment of crop yield, losses in pasture and crop production, monitoring drought impacts, and prediction of food security. In addition, 38-year data with sensors adjustment and high- and low-frequency noise removal can be used for analysis of climate and land cover change and prediction of future global, country and regional food security. The examples of these applications are shown in Figs. 10.10 and 10.11.

In some circumstances, there is a need for preliminary prediction (2–3 months in advance of harvest) of food security in a region. In such situation, it is reasonable (1) to assess percent area affected by vegetation stress and/or with healthy vegetation, (2) compare the area size with the areas in previous years with known crop and pasture production, and (3) provide preliminary prediction of food security. Figure 10.10 displays such situation for India's Madhya Pradesh state (one of the major corn producer) in May 13, 2018. As seen in the figure (far right end of lower diagram), 50 and 24% of the state were affected by stronger than severe and extreme drought, respectively, by mid-May; healthy vegetation area was very small, less than 10% (far right end of upper diagram). In the past 3 years (2015–2017) such situation was observed during the end of 2016 through the beginning of 2017; however,

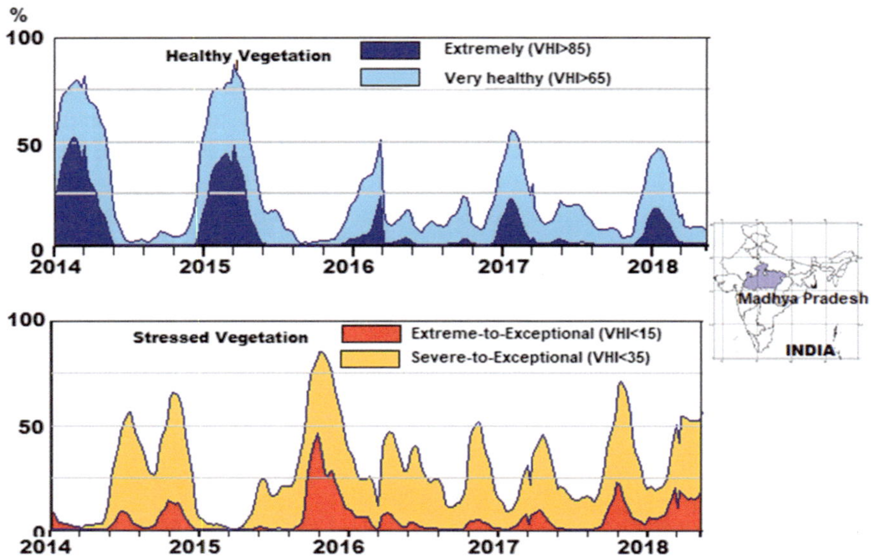

Fig. 10.10 Percent Madhya Pradesh state (India) with healthy (upper diagram) and stress (lower diagram) vegetation estimated from moisture–thermal vegetation conditions (VHI) during 2014 through May 13, 2018

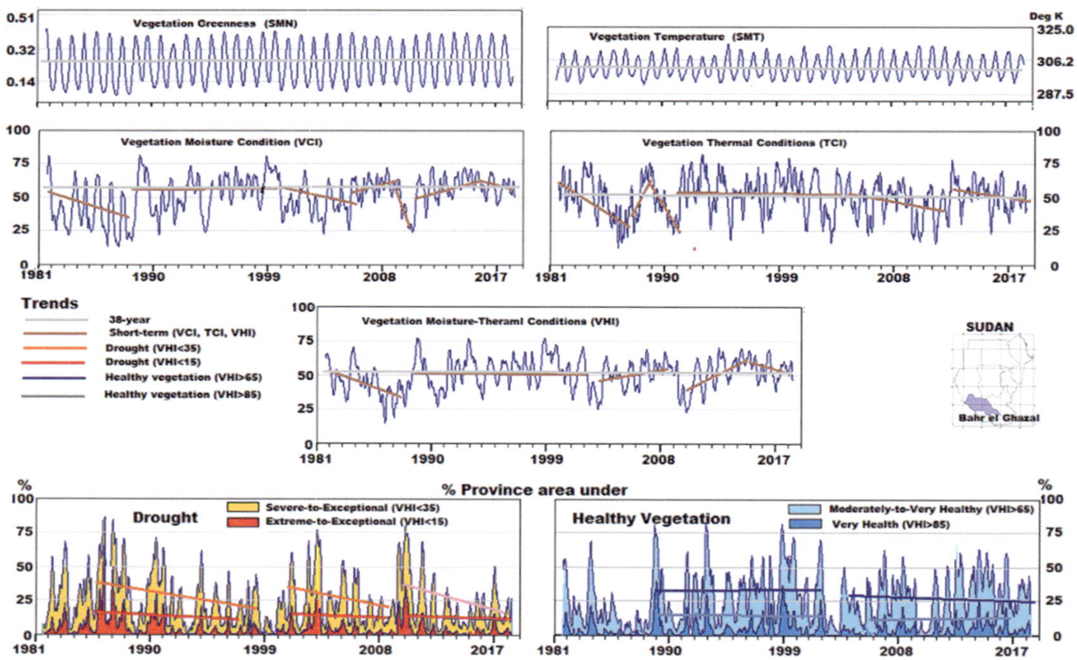

Fig. 10.11 Climate-based dynamics of vegetation greenness (SMN), surface temperature (SMT), moisture (VCI), thermal (TCI), and moisture–thermal (VHI) conditions, drought, and vegetation health areas during 1981–2018 in Sudan's Bahr el Ghazal province

the strongest drought occupied smaller area (<5%). Knowing corn production in 2017, it is possible to predict potential corn losses in 2018. In addition, the diagram in Fig. 10.10 shows that in summer 2014 and 2016 large area of the sate experienced vegetation stress and crops losses, while winter (December–February) crops and pasture had favorable conditions.

The current climate publications indicate drought intensification and expansion due to climate warming tendency (Forzieri et al. 2017; IPCC5 2014; IPCC4 2007. Hansen et al. 2000). Following such drought trends, vegetation surface temperature supposed to increase and vegetation greenness supposed to decline. Moreover, moisture and thermal vegetation conditions supposed to deteriorate reducing in grain production. These conditions would provide negative contribution to food security. Therefore, some of the important advantages of 38-year VH indices and products would be their application in analysis of multiyear trends in these characteristics and their potential tendencies in the future grain production and food security. Chapter 9 has discussed large-scale trend analysis. This Chapter demonstrates how to use historical data (NOAA/NESDIS 2018) for smaller regions. Figure 10.11 shows time series and trends for Bahr el Ghazal, one of the important agricultural provinces of Sudan. The upper diagram provides mean-area 38-year weekly time series of vegetation greenness (SMN) and surface temperature (SMT). Since the trend for both characteristics is flat, there is currently no evidence in the deterioration of these conditions due to global warming. Similar to SMN and SMT, 38-year flat trend is observed in moisture (VCI), thermal (TCI, second from the top line diagrams) and moisture–thermal (VHI, third from the top line diagram) conditions. Meanwhile, in some periods, there are short-term (6–20 years) trends of deterioration/improvement in these conditions. In the latest 4–6 years, moisture and thermal conditions in the province slightly deteriorated, although in the 1980s deterioration was much stronger. These trends should be verified against in situ (rainfall, temperature, soil moisture, etc.) data. Drought time series (lower line, left diagram) are showing three short-term declining trends (1985–1999, 2000–2009 and 2010–2018), which indicate that, at the background of global warming, there are regions with some improvements in drought conditions (reduction of drought area and intensity). Therefore, food security assessments should be done regionally for both evaluation of real-time situation and for verification purposes.

10.5 Conclusion

In the past four-five decades, climate and weather affected global and regional food security (FS), limiting staple crop production and fresh water, increasing soil degradation and slowing down the rate of agricultural output. Therefore, world's decision makers need reliable and timely environmental information to understand, monitor, and predict impacts of weather, climate and long-term Earth changes on global FS. In the 1990s, the National Oceanic and Atmospheric Administration (NOAA) has developed new operational satellite-based vegetation health method for high spatial and temporal monitoring environmental impacts on agriculture and other weather- and climate-related problems. These data and products are presented on the following NOAA WEB https://www.star.nesdis.noaa.gov/smcd/emb/vci/VH/index.php where data and products are presented for making environmental and economic decisions. The WEB displays real-time and historical weekly data since 1981 for each 16, 4, 1, and 0.5 km^2 resolution pixels for the entire world and also mean country and administrative regions data in the form of images and digital values of moisture (VCI), thermal (TCI), and moisture–thermal combine (VHI) conditions, drought, vegetation stress, fire risk, etc. and their changes between weeks and years. This chapter explains how to apply vegetation health (VH) data and products to monitor environmental impacts on vegetation health and through VH estimate crops/pasture production and predict food security. Applying VH data and products, the users would be able to monitor effectively first of all agricultural production, secondly, climate and weather changes and finally, food security.

Vegetation health is a very powerful tool, available on WEB since the early twenty-first century to: (a) assess moisture and temperature condition of crops and pasture, (b) derive area and intensity of stressful environment conditions (mostly moisture and thermal), (c) detect drought early enough (up to 4 weeks ahead of its start) in order to plan irrigation and other measures to offset the damage, (d) monitor drought area, intensity, duration and impacts for identifying food insecure areas and intensity of the problem, (e) model agricultural production (crops and pasture), (f) predict crop yield for an early assessment of food availability, (g) assess soil moisture (both deficit and saturation), (h) identify areas with potential weather-related threat of fire, (i) detect areas with elevated activity of mosquitoes and malaria spreading, (j) estimate malaria intensity and predict the number of affected people, (k) identify climate warming impacts on land cover changes, and (l) assess multiyear trends in such natural disasters as drought, landslides, soil excessive wetness, crop and pasture production losses, and food security situation. This short list of applications would help users to provide advises to farmers, agricultural producers, business community, international organizations and policy developers for making decisions in food security enhancement. Moreover, following our activity and communication with users, VH is also useful for scientific, educational activity and development of new applications.

Over the time of VH-WEB functionality, the users from nearly 200 world countries have accessed the images and digital data, applied them for monitoring weather impacts on crops/pasture and predicting food security. Moreover, the users are able to ftp area mean data and product for countries to develop yield-VH models and use them for prediction of crop/pasture yield and production. In addition to countries, for the first-order administration divisions in each country (for example, states in the USA, oblast in Russia, etc.), 4 km^2 area mean values from 38-year (1981–2018) weekly time series of VH data and products can be displayed for the years and weeks selected by users. The users have access to around 4000 administrative regions from nearly 200 world countries. The users of VH-based WEB represent a wide spectrum of interests, including agriculturalists, traders, commodity dealers, bankers, stock Co., farmers, researchers, media, and others. Between 200 and 400 users (depending on a season) attend the VH-WEB every day; during 2015–2017, nearly 200,000 users applied VH data form the WEB page.

This VH-WEB is popular because every week we present data and images of vegetation health, moisture and thermal condition, scaled from extreme stress-to-healthy, which characterize weather-base moisture and temperature impacts on vegetation, the time of drought start, its area, intensity and potential impacts on crops and pasture. In addition, VH-WEB displays fire risk area (where conditions are favorable for fire development), malaria risk area in Africa, moisture and thermal stress of crops and pasture, changes of the indicated characteristics from previous week and from last year. For the users, who need country and first-order administrative division information, they can collect and/or look at the time series (from 1981) of regional mean VH, percent area under drought or no drought of different intensities, estimate land cover changes in nearly four decades and monitor global and regional food security. The users can also *ftp* yearly/weekly data and products by their choice since 1981. This Chapter provides many examples on data use for food security assessment.

References

Adamenko, T., and A. Prokopenko. 2009. Monitoring droughts and impacts on crop yield in Ukraine from weather and satellite data. In *Use of Satellite and In-Situ Data to Improve Sustainability*, ed. F. Kogan, A. Powell, and O. Fedorov, 1–11. Dordrecht: Springer.

Alexandratos, N., and J. Bruinsma. 2012. World Agriculture towards 2030/2050: The 2012 Revision. ESA Working Paper No 12-03, June, FAO, Viale delle Terme di Caracalla 00153 Rome, Italy, pp 147. http://www.fao.org/docrep/016/ap106e/ap106e.pdf.

ClimateBet. 2018. Global Warming Challenge. http://www.theclimatebet.com.

FAO. 2017. How Close We Are to Zero Hunger. http://www.fao.org/state-of-food-security-nutrition/en/.

———. 1999. Food Insecurity: When people must live with hunger and fear of starvation. *The State of Food Insecurity in the World Report*, Rome, 76 pp.

Forzieri, G., R. Alkama, D.G. Miralles, and A. Cescatti. 2017. Satellites reveal contrasting responses of regional climate to the widespread greening of Earth. *Science* 356 (6343): 1180–1184. https://doi.org/10.1126/science.aal1727.

GOOGLE. 2018. Major grain growing area in Kenya. https://www.google.com/search?q=map+of+grain+area+in+kenya&rlz=1C1GGRV_enUS752US752&tbm=isch&source=iu&ictx=1&fir=ez0oelxf91D4SM%253A%252CbgHyrk0GffatXM%252C_&usg=__N5bNDGKPncPp9dVnsJjdmcz_qxg%3D&sa=X&ved=0ahUKEwji_f75io3bAhUHzlkKHTwkAPQQ9QEIKTAA#imgdii=l5whdhKlREDhDM:&imgrc=GbzxrF8SWHDucM.

Heibuch, J. 2011. Food insecurity on the rise. *Treehugger*, October 31. https://www.treehugger.com/health/food-insecurity-rise-infographic.html.

Hansen, J., M. Sato, R. Ruedy, A. Lacis, and V. Oinas. 2000. Global warming in the twenty-first century: An alternative scenario. *Proceedings of the National Academy of Sciences* 97: 9875–9880.

IPCC5. 2014. Climate Change 2014: Synthesis Report. 5th Assessment, eds R.K. Pachauri and L. Meyer, pp 151. Geneva: IPCC. https://www.ipcc.ch/pdf/assessment-report/ar5/syr/SYR_AR5_FINAL_full_wcover.pdf.

IPCC4. 2007. Climate Change 2007: Synthesis Report. Geneva: IPCC. pp 104. https://www.ipcc.ch/publications_and_data/publications_ipcc_fourth_assessment_report_synthesis_report.htm.

Kerr, R.A. 2005. Atlantic climate pacemaker for millennia past, decades hence. *Science* 309 (5731): 41–42.

Kogan, F., Wei Guo, Wenze Yang, and Shannon Harlan. 2018. Space-based vegetation health for wheat yield modeling and prediction in Australia. *Journal of Applied Remote Sensing* 12 (2): 026002. https://doi.org/10.1117/1.JRS.12.026002.

Kogan, F., W. Guo, and W. Yang. 2017. SNPP/VIIRS vegetation health to assess 500 California drought. *Geomatics, Natural Hazards and Risk*. https://doi.org/10.1080/19475705.2017.1337654.

Kogan, F., and W. Guo. 2017. Strong 2015-2016 El Niño and implications to global ecosystems from space. *International Journal of Remote Sensing* 38 (1): 161–178. https://doi.org/10.1080/01431161.2016.1259679.

Kogan, F., M. Goldberg, T. Schott, and W. Guo. 2015a. SUOMI NPP/VIIRS: Improve drought watch, crop losses prediction and food security. *International Journal of Remote Sensing*. https://doi.org/10.1080/01431161.2015.1095370.

Kogan, F., W. Guo, A. Strashnaia, A. Kleshenko, O. Chub, and O. Virchenko. 2015b. Modelling and prediction of crop losses from NOAA polar-orbiting operational satellites. *Geomatics, Natural Hazards and Risk*. https://doi.org/10.1080/19475705.2015.1009178.

Kogan, F., T. Adamenko, and W. Guo. 2013a. Global and regional drought dynamics in the climate warming era. *Remote Sensing Letters* 4: 364–372. https://doi.org/10.1080/2150704X.2012.736033.

Kogan, F., N. Kussul, T. Adamenko, S. Skakun, O. Kravchenko, O. Kryvobok, A. Shelestov, A. Kolotii, O. Kussul, and A. Lavrenyuk. 2013b. Crop modeling in Ukraine based on earth observation, meteorological data and biophysical models. *International Journal of Applied Earth Observation and Geoinformation* 23: 192–203. https://doi.org/10.1016/j.jag.2013.01.002.

Kogan, F., and W. Guo. 2009. Early detection and monitoring droughts from NOAA environmental satellites. In *Use of Satellite and In-situ Data to Improve Sustainability*, ed. F. Kogan, A. Powell, and O. Fedorov, 11–18. Dordrecht: Springer.

Kogan, F., T. Adamenko, and M. Kulbida. 2009. Satellite-based crop production monitoring in Ukraine and regional food security. In *Use of Satellite and In-situ Data to Improve Sustainability*, ed. F. Kogan, A. Powell, and O. Fedorov, 99–104. Dordrecht: Springer.

Kogan, F. N. 2002. World droughts in the new millennium from AVHRR-based Vegetation Health Indices. *Eos, Transactions of the American Geophysical Union* 83 (48): 562–563.

———. 1998. A typical pattern of vegetation conditions in southern Africa during El Nino years detected from AVHRR data using three-channel numerical index. *International Journal of Remote Sensing* 19 (18): 3689–3695.

Kogan, F. 1989. Remote sensing of weather impacts on vegetation in non-homogeneous areas. *International Journal of Remote Sensing* 11 (8): 1405–1419.

Kogan, F.N. 1987. Vegetation Health index for areal analysis of NDVI in monitoring crop conditions. *Preprint 18th Conference on Agricultural and Forest Meteorology*. AMS, Boston, 103–114.

MinHao, W., S.P. Urbanski, B. Nordgren, and A. Petkov. 2009. Daily fire occurrence in Ukraine from 2002 to 2008. In *Use of Satellite and In-situ Data to Improve Sustainability*, ed. F. Kogan, A. Powell, and O. Fedorov, 49–60. Dordrecht: Springer.

NOAA/NESDIS. 2018. Vegetation Health Indices and Products. https://www.star.nesdis.noaa.gov/smcd/emb/vci/VH/vh_browse.php.

Rahman, A., F. Kogan, and L. Roytman. 2006. Analysis of malaria cases in Bangladesh with remote sensing data. *American Journal of Tropical Medicine and Hygiene* 74: 17–19.

Schmidhuber, J., and F.N. Tubiello. 2007. Global food security under climate change. *Proceedings of the National Academy of Sciences of the United States of America* 104: 19703–19708.

StarStat. 2018. Visits to the VH WEB Page. https://statcounter.com/p6394160/visitor_map/?start=60.

USDA. 2017. Food Security. https://www.usda.gov/topics/food-and-nutrition/food-security.

USDA/NASS. 2018. Crop Report 1936 3737. http://usda.mannlib.cornell.edu/usda/current/CropProd/CropProd-05-10-2018.pdf.

WB (The World Bank). 2018. Agriculture and Rural Development: Cereal Yield. https://data.worldbank.org/topic/agriculture-and-rural-development.

Short Summary

More than one billion people, nearly a sixth of the world's population, are currently suffering from chronic hunger and malnutrition due to a lack of food. This creates food security (FS) problems. Our Earth is very big. We need strong technology to monitor environment and its impacts on food production to predict FS ahead of time when a lack of food problem appears. Satellite technology has already proved to be successful in this endeavor. This book describes new NOAA operational satellite-based method and its use to enhance food security prediction.

Among economic, political, population, agriculture, and environmental factors affecting food security, the world population consuming food and annual agricultural production mainly determines food supplies and demand and through their balance FS situation. Since grain is the principal product used by the entire world for both food and feed, the amount of grain produced by agriculture annually can be considered as an indicator of global food supply. Food demands can be estimated by food consumption. Food security-insecurity can be considered as a multiyear and single-year issue. The first one is mostly controlled by economic, political, administrative, etc. resources and the second one primarily by environmental resources. Annual FS can be assessed predicting grain production from operational satellites and also estimate if the predicted grain amount is sufficient for consumption. FS prediction can be successfully controlled by monitoring drought (area, duration, intensity) and estimating grain production losses on every 0.5, 1, 4 and 16 km^2 of land surface, aggregated to administration region, country and even the world.

Using satellite data is not straightforward process since atmosphere is on a way of satellite beam, contaminating its signal (for example, clouds or aerosol, etc.) and land cover plus weather and natural disasters (volcanoes, etc.) are experiencing considerable changes over space and time, complicating the development of reliable VH products. Therefore, we discuss satellite-processing data in detail that the users' community was absolutely sure that this tool is very reliable. National Oceanic and Atmospheric Administration (NOAA) has started regular satellite-based monitoring land surface, ocean and atmosphere serving global community since 1981. By now, these services have been continuing for 38 years, with AVHRR sensor at the beginning and new VIIRS sensor (considerably improve satellite system) from 2011. From 2018, the new generation of NOAA satellites (starting with NOAA-20) with VIIRS sensor is planning to be operational up to the mid-twenty-first century. Vegetation health (VH) technology will continue to monitor weather impacts, especially drought on agriculture, predict crop and pasture production, and assess in advance food security situation in the world. An important consideration is given to noise detection and elimination

F. Kogan, *Remote Sensing for Food Security*, Sustainable Development Goals Series,
https://doi.org/10.1007/978-3-319-96256-6

in the data (volcano-based aerosol, satellites' orbit degradation, sensor deterioration and change, satellite malfunctioning, and other problems) and, what is the most important, verification of the results.

Since vegetation health (VH) method is absolutely new and very powerful tool used for drought detection and monitoring and assessment of crop and pasture production losses, which are the main indicators of food security, the entire chapter 4 is devoted to its description. The VH method is very unique because: (a) operational satellites proved to be a very good single source of information (without weather and other *in situ* data) for environmental monitoring and prediction of such natural disaster as drought; (b) the VH method is a new biophysically-based powerful tool, which comprehensively validated on many crops, in different world climates and ecosystems; (c) many developed VH products were successfully used for prediction of crop and pasture production, including losses, helping to improve food security and sustainability; (d) VH data and products have been extremely popular since from 2015 nearly 70,000 world users are applying this information in their work; (e) users represent a wide spectrum of interests, including agriculturalists, traders, commodity dealers, bankers, stock Co., farmers, researchers, and others; and (f) for the first time all of the VH-based applications dealing with global FS are presented in one publication.

Among all natural disasters, drought is the most damaging environmental phenomenon. In the twenty-first century, nearly 20% of global lands were drought-stricken almost every year, but in some years this number increases considerably. Drought is a typical phenomenon of the Earth's climate. Losses from droughts, especially in agriculture, are staggering. Developing countries of Africa, South America and Asia have been the most drought-affected, facing food security problem. Reducing global grain production, droughts have affected food security, in half of the years of the twenty-first century. Therefore, the entire chapter 5 is devoted to VH-based drought detection and monitoring its impacts on grain production in order to predict food security situation. Traditionally, drought was monitored with weather and soil moisture-measured data. With the introduction of vegetation health method, drought was monitored by a combination of moisture- and temperature-based indices, which were well grounded by biological laws. Drought was VH-based classified following yield reduction and validated in 36 countries of the world. A lot of drought examples are presented, especially in the twenty-first century up to spring 2018.

Among many tasks, an advanced estimation of crop yield and production well ahead of harvest is the most important helping to improve food security assessment. There are currently two methods for modeling an advanced estimation of crop yield. They are based on weather data and satellite indices. Satellite-based crop yield modeling advanced considerably in the 1990s with the introduction of vegetation health (VH) technology. Regional and country-based yield-VH models (Chapter 6) have been developed for such grain crops as spring and winter wheat, corn, rice, barley, sorghum, some non-grain crops (potatoes), and for pastures in the USA, Brazil, Argentina, Ukraine, Greece, Morocco, Zimbabwe, China, Mongolia, India, Poland, Bulgaria, and other countries. They showed not only good results of independent verification and 2-month advance assessment of crop production, but also potential for 3–4 months advanced prediction of crop yield. The VH method is working well because the parameters the models are composited on are biologically oriented and the data is distributed frequently over space (0.5, 1, 4 and 16 km^2 resolution) and time (weekly) compared to weather data. We introduced in the book a method for yield modeling with satellite-based VH indices and provided an independent modeling results for major grain crops in the principal grain-producing countries such as the USA, China, Russia, India, Argentina, Brazil, Ukraine, and some countries of Africa.

Agricultural production is a risky economic activity. In addition to market, institutional, political, and farmers' individual risks, strong production reduction in some years due to adverse weather, especially drought, is frequently expected. In such situations, farmers can effectively replace the damage with insurance payments having a crop insurance contract, based on so-called index-insurance. The effectiveness of index-based insurance products is, however, highly dependent on the availability of reliable and cost-effective damage-assessed sources. Compared to weather-based insurance with a sparse meteorological network, satellite-based insurance has very effective advantages. Experiments in Kazakhstan (Chapter 7) have been proven that satellite-based VH indices estimating moisture (VCI) and thermal (TCI) conditions can be used effectively to insure yield losses. The copula approach was successfully applied for estimating dependence between crop yields and VH indices, and designing and rating insurance contract. Our empirical results for wheat-producing farms in Kazakhstan showed that survival Gumbel copula provided an appropriate fit when modeling joint distributions of farm wheat yields and VCI, while Frank copulas showed a better performance in describing dependence of farm yield on TCI. However, their performance strongly depends on quality of yield and resolution of satellite data, as well as regional and local specifics in climate and wheat growing.

Drought has been proven deteriorate food security situation. Therefore, an advance drought prediction can provide valuable information to decision maker about the countries and regions of FS problem and required food assistance. This part of the book (Chapter 8) provides the results of testing VH indices for 2–6 months advanced drought prediction and impact assessment on crop losses using global phenomena, called El Niño Southern Oscillation (ENSO). VH-ENSO teleconnection was derived by correlating pixel-based monthly VH indices in boreal winter, spring and summer with monthly mean SST anomaly in Nino-3.4 area during November–February for all and the strongest ENSO events in the past 36-year. The results indicate that there are areas on all continents where vegetation is sensitive to ENSO during boreal winter. In ENSO years, vegetation of northern Brazil, southern Africa, islands of Southeast Asia, eastern Australia, western part of Africa' Sahel, and central Canada experiences moderate-to-strong stress during El Niño and healthy condition during La Niña. The opposite situation (healthy conditions in El Niño and unhealthy in La Niña) occurred in central and northern Argentina, southern Brazil, Horn of Africa, eastern Asia, and some parts of southwestern USA (especially California). Since the 2015–2016 strong El Niño has triggered drought in Brazil, southern Africa, southeastern Asia islands, and eastern Australia during December–February, agricultural crops and pastures were predicted to experience some losses of production, deteriorating food situation.

It is known that the Earth climate is warming up since 1850s. Such climate changes have recently triggered many unusual environmental, economic, and social events. Since the mid-1940s, climate warming progressed more intensively along with carbon dioxide (CO_2) increase. Experts from the United Nations and some countries are warning that climate warming has already stimulated a reduction of moisture for crops, intensified thermal stress, enlarged drought area and increased drought duration, intensity and impacts, resulted in a reduction of grain production and deterioration of food security. Having 38-year of satellite-based vegetation health data and products, such as drought, vegetation stress, crops' damages, and others, we investigated if vegetation conditions, droughts, crops losses intensified following climate warming tendency and if food security situation deteriorate as well in the near future (Chapter 9). We investigated drought tendency in the entire world, grain-producing countries, and the major grain areas. What we found is that drought trends are principally mismatching with global warming trends: instead of drought intensification and area expansion due to global warming, drought area and intensity are reducing during 1981–2015. Moreover, in a few countries drought impacts on grain crops do not change. It is important to emphasize that during 1981–2015, (a) global and hemispheric droughts have been frequent; (b) on the average, global and hemispheric droughts in severe-to-exceptional (S-to-E) category covered nearly 15–20% of the entire world, 7–8 and 2–3%

in extreme-to-exceptional (E-to-E) and exceptional (E) intensity, respectively; (c) in the extreme years, drought area in the indicated three intensities might jump up to 45, 25, and 10%, respectively; (d) during the period of intensive global warming (1981–1997), drought areas in all categories have surprisingly not increased and not intensified, which was claimed by many publications; (e) during the period of stable global temperature anomaly (1998–2014), drought area mismatched with global temperature trend, increasing for two categories (S-to-E and E-to-E) for the globe and both hemispheres (especially southern) and not changing for the extreme category. Finally, climatic aspects of food security supposed to continue the past 30-year tendencies since no major changes in long-term environmental impacts on grain production is expected.

Finally, vegetation health data and products are currently provided through the following WEB site https://www.star.nesdis.noaa.gov/smcd/emb/vci/VH/index.php, where they are presented in the form of immediate applications which explains how to use the information (Chapter 10). The WEB displays real-time and historical (since 1981) weekly 16, 4, 1, and 0.5 km^2 resolution data for the entire world and also mean country and administrative regions data in the form of images of moisture (VCI), thermal (TCI), and moisture–thermal combine (VHI) conditions, drought, vegetation stress, fire risk, etc. and their changes between weeks and years. The 38-year data and products are presented in the form of images, time series, and digital values for each pixel of the world, for administration regions, countries, continents, and the world.

Index

A

Advanced Technology Microwave Sounder (ATMS), 42
Advanced Very High Resolution Radiometer (AVHRR), 24, 53, 82, 192
  channels, 24
  characterization, 24
  data, vegetation monitoring, 25–28
  NOAA (*see* NOAA/AVHRR)
  polar-orbiting satellites, 24
  sensor, 24–25
African Development Bank, 10
Aggregate yield index, 163
Agricultural production, 1
Anomaly vegetation index (AVI), 84
Applications, VH
  climate and weather products, 237–239
  drought, 232–235
  environmental products, 236
  fire risk area, 235
  grain crop analysis and FS, 230
  in situ data, 229
  Kansas, 237
  malaria intensity, 236, 237
  methodology, 229
  moisture-thermal conditions, 230–232
  soil saturation/soil extreme wetness, 236
  VHI, 229, 230
  VH-WEB, 233
  winter wheat in Ukraine's Odessa oblast, 237
Archimedean copulas, 165
Arctic Oscillation (AO), 194
Atlantic Meridional Overturning Circulation (AMOC), 194
Atlantic Multidecadal Oscillation (AMO), 194
Average risk reductions, 170

B

Biophysical laws, 2
Brazil (South America), 157, 158
Brightness temperature (BT), 37, 201
Bulgaria (Europe), 153, 154
Burkina Faso (Africa), 152

C

California drought, 102, 104
California's Central Valley, 102, 103, 105
Cereals in Russia, 140–145
Chlorofluorocarbon (CFC) gases, 198, 199
Clayton copula, 165
Climate change, 2, 3
  See also Global warming
Climate warming, 201–219
  ambient temperature, 220
  AMO mechanism, 194
  application of 38-year satellite-based VH data, 192
  CO_2 emission, 193
  CO_2 emitters, 193
  CO_2 trend, 195–197
  coal-mining place, 194
  description, 191
  deterioration of FS, 220
  developing and developed countries, 191
  drought intensification, 221
  earth vegetation, 192
  Earth's history, 194
  GDP, 194
  global land cover changes (*see* Land cover changes)
  global temperature, 192, 195–197
  global vegetation greenness, 221
  greenhouse gases, 193
  ice age, 194
  IPCC Report, 194
  IPCC4 and IPCC5, 192
  IPSS4 and IPCC5, 193
  issue, 192
  Kyoto Protocol, 193
  Paris Agreement, 193
  quasi-periodic oscillations of MHT, 194
  satellite-based VH, 220
  the United Nations, 191
  VCI and TCI conditions, 221
  vegetation surface temperature, 221
  VH, 221
Cloud detection methods, 31
Clouds and Earth's Radiant Energy System (CERES), 42
Coal-mining place, 194

F. Kogan, *Remote Sensing for Food Security*, Sustainable Development Goals Series,
https://doi.org/10.1007/978-3-319-96256-6

Copula approach, 165, 166
Corn
  in Argentina, 130–133
  in China, 119–122
  in USA, 122–128
  in Zimbabwe, 148–151
Cramer-von-Mises criterion, 169
Crop Reporting Districts (CRD) mean corn yield anomaly, 67, 69

D
Democratic People's Republic of Korea (DPRK), 9
Devastating droughts
  in 2017 and early 2018, 106–109
  vegetation stress, 62
  VH, 62
Drought prediction, 175
  data and method, 177
  economic and societal activities, 175
  ENSO (*see* El Niño southern oscillation (ENSO))
  GFS, 175
  intensification, expansion and penetration, 175
  SE intensity, 175
  21st century, 175
Droughts, 79, 163
  advanced detection, 2
  in Africa, 191
  agriculture, 2
  applications, VH, 232–235
  area and intensity, 15, 16, 215–218
  and Argentina, 16
  average annual cost, 75
  cost, 76
  Countries and parts, 15
  definition, 76–78
  detection and monitoring (*see* Monitoring drought from space)
  devastating, 106–109
  Earth's climate, 1
  economic damages, 76
  expansion, 192
  features, 78
  FS problems, 14
  global, 2
  grain losses, 15, 20
  intensification, 1, 20, 192
  losses, 75
  measuring and estimating, 78
  MMT, 15
  moderate-to-intensive, 1
  monitoring, 2
  natural disasters, 75–77
  negative P–C balance, 15
  reduces grain production, 19
  satellite technology, 2
  type, 79
  U.S. Billion-dollar, 77
  USDM, 75
  VH, 16, 20, 192
  weather-based drought detection, 2
  weather data, 75
  world, 217–219
  yield losses (*see* Yield losses)
Drought start, 70, 78, 81, 84, 85, 87, 88, 98, 109, 177, 225, 229, 233, 235, 240

E
Earth's intercepting solar lights, 28
El Niño southern oscillation (ENSO), 175, 177–183, 197
  characteristics, 177
  climate forcing phenomenon, 175
  ENSO-VH teleconnection analysis, 176
  environmental impacts, 176
  events, 175
  prediction of 2015–2016, 185–186
  research and sea surface temperature, 176
  societal impact, 175
  SST (*see* Sea surface temperature (SST))
  SSTa, 177, 178
  VH during 2015–2016, 183–185
  warm with positive and cold with negative SST anomaly in TP, 177
  weather and ecosystem patterns
    boreal summer, 177
    boreal winter, 177
    cases, 178
    colour coded maps of PCC, 180
    during 1981–2016, 182
    ecosystems, 178
    monthly SSTa, 178, 179
    monthly VHI *vs.* monthly SSTa distribution, 180, 181
    PCC, 183
    precipitation and temperature patterns, 177
    VHI *vs.* SSTa, 182
    years and VH-SSTa correlation, 181
  weather and ecosystems response, 176
  weather types, 176
Empirical distribution function (EDF), 37–39
Enhanced vegetation index (EVI), 83
Environmental laws, 54
Environmentally-based FS, 2
Equator crossing time (ECT), 34, 35
Equatorial Africa NDVI, 32–34
Evaporative drought index (EDI), 83, 84

F
Fallow-wheat-wheat-barley-wheat, 167
Farmer insurance
  contract parameters, 165, 166
  contracts, 163, 164, 169, 171
  index-based insurance, 163–165, 170
  payments, 163, 164
  premium, 165, 166
  remote-sensing data, 164

satellite-based VH index, 172
TCI-based insurance, 169
VCI-based insurance, 169, 170
weather-based insurance, 163
Fire risk area, 235
Fires disasters, 77
Food and Agriculture Organization, 10
Food security (FS), 1, 10–20, 115, 164, 191
agricultural production, 191
and climate change (*see* Climate change)
climate and weather products, 237–239
climatic aspects, 222
deterioration of, 220
developed countries, 9
developing countries, 9
DPRK, 9
drought (*see* Droughts)
drought-related yield losses (*see* Yield losses)
during 1999–2009, 9
earth changes, 1
environmentally based, 2
by financial, economic, political and environmental factors, 9
frequent monitoring situation, 10
global, 1, 9
goals, 9
grain production, 220, 221
and insecurity (*see* FS/insecurity)
malnutrition, 9
New Alliance for Food Security and Nutrition, 9
organizations, 10
performance, 10
prediction and VH-crop modeling (*see* VH modeling crop production)
satellite-based models, 10
satellite-based predictions, 2
satellite data for assessment, 10
twentieth century, 9
warmer climate, 1
WFP, 9
Forest drought-stress severity index (FDSI), 80
Frank copula, 165, 170
FS/insecurity, 10–12, 14–20
annual grain losses, 20
chronic hunger and malnutrition, 20
famine, 10
grain production and balance, 20
long-term, 12–14
measure (*see* Measure FS/insecurity)
short-term (*see* Short-term FS/insecurity)
United Nations, 10
world population, 20

G
Gaussian copula, 165
Geostationary (GEO)
global coverage, 23
global weather monitoring, 23
GOES, 37
space to frequently monitor, 23
weather monitoring and prediction, 23
Global area coverage (GAC, 24
Global droughts, 89–92
Global food security (GFS), 89–92, 110, 175
Global temperature measurements, 192
Global vegetation index (GVI), 25, 84
Global warming
CFC gases, 198
chlorine-induced ozone depletion, 198
ClimateBet report, 200
CO_2 and O_3 trends, 199
CO_2-triggered TA upward-going (general) trend, 197
GHG, 197, 198
mean land temperature anomaly, 200
Ozone (O_3), 198
parameters, 199
temperature anomaly, 198
trend analysis, 199
ultraviolet radiations, 198
Global weather monitoring, 23
Grain losses
assessments, 20
drought-related, 15
weather-related annual, 14
Grain yield
agricultural technology, 117
prediction, 130
and production, 158
Greenhouse gases (GHG), 198
Green Revolution, 11
Gross domestic product (GDP), 194
Gumbel copula, 165, 170

H
High-resolution drought monitoring, 102, 104
Hydrothermal coefficient (GTK), 80

I
Index-based insurance, 163, 164
India (Asia), 155–157
Insurance contract parameters, 166
International Decade for Natural Disaster Reduction (IDNDR), 76
International Fund for Agricultural Development, 10
IPCC Report, 194

J
Joint Polar Satellite System (JPSS), 42

K
Kolmogorov–Smirnov test, 168
Kyoto Protocol, 193

L

Land cover changes
  continents, 204
  drought area and intensity, 215–218
  global warming and flat TA trend, 201
  grain crop area, 210, 211
  grain-producing countries, 204–207, 210
  greenness and temperature, 201, 211–215
  linear equation's trend, 201
  moisture and thermal conditions, 211–215
  moisture, and thermal conditions, 211
  TA and CO_2 *vs*. TA and O_3depletion trends, 201
  VH, 201
  world, 201–203
  world drought, 217–219
Law of Minimum (LOM), 84
Law of Tolerance (LOT), 84
Liebig's Law-of-minimum (LOM), 54
Local area coverage (LAC), 24
Long-term FS/insecurity, 12–14
  and expectations
    farm income in USA, 13, 14
    forecasts, 13
    grain production and consumption, 14
    grain stocks, 13
    P–C balance, 14
    population growth, 12
    world population, 13

M

Malaria intensity, 236, 237
Malnutrition, 9
Measure FS/insecurity
  annual food demand/consumption, 10
  annual food supply/production from agriculture, 10
  assessment, 10
  global food demand (consumption), 10
  global food supply, 10
  grain production *vs*. consumption, 11, 12
  MMT, 12
  world population and global grain production, 10, 11
Mega-drought (US-WMD), 101
Meteorological methods, 80
  indices
    FDSI, 80
    GTK, 80
    PDSI, 80
    Ped, 80
    SPEI, 80
    SPI, 80
    USDM, 80
  parameters, 79
Meteosat Second Generation (MSG-2) European satellites, 34
Million metric tons (MMT), 12
Moderate Resolution Imaging Spectra-Radiometer (MODIS), 43
Moisture-thermal (VCI-TCI) stress
  chlorophyll, 53
  definition, 59
  in July 2012, 60
  mid-summer, 60
  SMN and SMT, 58
  space-based monitoring, 52
  VH, 53
  weather-related conditions, 60
Mongolia (Asia), 153–155
Monitoring drought from space, 101–106
  0.5 and 1 km^2 resolution from VIIRS
    California, 101
    California's Central Valley, 101–103, 105
    in FS issues, 101
    high resolution, 101, 102, 104
    in irrigated and non-irrigated areas, 104, 105
    operational activities, 101
    recovery, 105, 106
    USDM, 101, 103
    US-WMD, 101
    and VH-500, 102, 103, 105
  meteorological methods, 79–81
  and operational satellite-based VH, 84–88
  satellite-based methods, 81–84
  soil moisture and vegetation methods, 81

N

NASA's Pathfinder, 25
National Agricultural Statistical Services (NASS), 64
National Oceanic and Atmospheric Administration (NOAA), 23, 226
  and NESDIS 2018, 15
  operational afternoon polar-orbiting satellites, 53, 60
  POES (*see* POES NOAA)
Natural disasters, 75–77
New Alliance for Food Security and Nutrition, 9
NOAA/AVHRR, 25–27, 45–48
  cross-track scanning system, 24
  GAC and LAC resolution, 24
  morning and afternoon operational satellites, 24, 25
  and SNPP/VIIRS data records (*see* SNPP/VIIRS)
  vegetation monitoring
    application of infrared channels, 25
    channels' calibration, 25, 26
    chlorophyll-a and carotenoids, 26
    CLAVR-x processing system, 25
    GIMMS, 25
    global data sets, 25
    global VH, 25
    GVI, 25
    Landsat 1 satellite VIS and NIR data, 26, 27
    NASA's Pathfinder, 25
    Plate Carree map projection, 25
    radiometric calibration, 25
    solar radiation, 27
    thermal channel data, 26
    VIS and NIR channels, 25, 26
    wheat and bare soil reflectance, 27
  wavelengths, solar spectrum, 24
NOAA-1 to 5, 24
NOAA-6, 24

NOAA operational polar-orbiting satellites, 2
Normalized biomass (NorBio), 81
Normalized difference drought index (NDDI), 83
Normalized difference vegetation index (NDVI), 29–41, 164, 192, 201, 202
  and AVHRR-based reflectance, 28, 29
  characterization, 28
  and chlorophyll content, 28
  chlorophyll monitoring, 28
  Earth's intercepting solar lights, 28
  growing season, 29
  properties, 28, 29
  removing noise from
    aerosol-based distortion, 37
    afternoon NOAA satellites, 34, 35
    average and CDF-corrected, 39, 40
    AVHRR-based NDVI week, 38, 39
    in AVHRR data, 29
    AVHRR sensors, 35
    benchmark statistical sample, 36
    BT, 32, 37
    and BT weekly composite time series, 30, 31
    cloud detection methods, 31
    clouds reflections, 30
    compound median filter, 31, 32
    daily and seven-day composite, 30
    ECT, 34, 35
    EDF, 37–39
    El Chichon eruption, 32, 34
    equatorial Africa, 32–34
    estimate distortion, 36
    Illinois's areal mean raw, 32
    indices, 31
    latitudinal profiles, 38, 39
    least squares statistical technique, 31, 32
    noise-detected and noise-applied corrections, 37
    normalization procedure, 37, 38
    operational satellites, 34
    optimal time-based approach, 30
    orbit-degradation errors, 36
    pixel-level, 36, 37
    pixels, 30
    pre-Pinatubo time, 38
    satellite orbital drift and AVHRR sensor degradation, 34
    SEVIRI, 34
    SMN, 32, 39–41
    SMT, 32, 41
    SRF, 35
    stratospheric aerosol, volcanoes, 32, 33
    and temperature annual time series, 31
    volcanic aerosols, 34
    weekly composite technique, 30
  satellite-based methods, 82
  vegetative ecosystems, 28
  VIS channel performance, 28
  water near the shore, 28
  weather variation, 29
Normalized multiband drought index (NMDI), 82
North Atlantic Oscillation (NAO), 194

O
Ocean–atmosphere system oscillations, 175
Operational polar-orbiting environmental satellites, 81
Operational satellite-based VH and drought monitoring
  classification, 85, 86
  environmental laws, 84
  GAC, 84
  global moisture–thermal index, 85
  GVI, 84
  indices characterizing, 84
  NOAA/AVHRR satellite data, 84
  NOAA/SNPP/VIIRS data sets, 84
  products and validation, 86–88
Operational satellites, 23
  economic and cultural development, 23
  GEO (*see* Geostationary (GEO))
  POES (*see* Polar-orbiting environmental satellites (POES))
Optimal time-based approach, 30
Ozone (O_3), 198
Ozone Mapper Profiler Suite (OMPS), 42

P
Pacific Decadal Oscillation (PDO), 194
Palmer drought severity index (PDSI), 80, 101
Paris Agreement, 193
Partial correlation coefficients (pCC), 118
Pearson correlation coefficients (PCC), 168, 183
Ped index, 80
Perpendicular drought index (PDI), 84
Plate Carree map projection, 25
POES NOAA, 24, 28
  assessment of environmental impacts, 24
  AVHRR (*see* Advanced Very High Resolution Radiometer (AVHRR))
  meteorological satellites, 24
  NDVI (*see* Normalized difference vegetation index (NDVI))
  NOAA-1 to 5, 24
  NOAA-6, 24
  observations, 24
  parameters, 24
  performance and applications, 24
  TIROS-1, 24
  TIROS-2 to 10, 24
  vegetation monitoring, 42–45
Polar-orbiting environmental satellites (POES), 23
  daily observations, 23
  data, 23
  measurements, 23
  NOAA (*see* National Oceanic and Atmospheric Administration (NOAA))
Population and grain production trend, 12
Precipitation
  annual, 217, 218
  reduction, 217
Principle of Carrying Capacity (PCC), 54, 84

Q

Quasi-periodic oscillations, MHT, 194

R

Reclamation drought index (RDI), 84
Regional droughts
  grain-producing countries, 91–94
  severity indicator, 80
  warmest global temperatures, 89
  watchers, 81
Relative risk reduction (RRR), 171
Rice in Bangladesh, 137–140
Russia's Saratov administrative region, 142

S

Satellite-based crop yield modeling, 116
Satellite-based methods
  advantages, 82
  air temperature, 81
  AVHRR, 82
  AVI, 84
  cloud monitoring, 81
  CMI, 84
  drought monitoring, 82
  EDI, 83, 84
  EVI, 83
  impacts assessment gap, 82
  in situ measurements, 84
  measuring precipitation, 81
  NDDI, 83
  NDVI, 82
  NMDI, 82
  operational polar-orbiting environmental satellites, 81
  PDI, 84
  RDI, 84
  remote-sensing, 82
  scale drought condition index, 84
  simple difference index, 84
  simple ration index, 84
  SNPP, 82
  SVI, 84
  SWSI, 84
  TIROS-N, 82
  TVDI, 84
  VegDRI, 83
  VIIRS, 82
  VSDI, 83
  VSWI, 83
  WDRVI, 83
Satellite-based VH index, 164, 165
Satellite technology, 2
Scale drought condition index, 84
Sea surface temperature (SST), 175–177
Second Green Revolution, 13
Severe-to-exceptional (SE) droughts, 88
Severe-to-exceptional (SE) intensity, 175
Shefield's Law-of-tolerance (LOT), 54
Short-term FS/insecurity
  annual changes, 14
  annual major grain production *vs.* consumption, 14, 15
  area and intensity, 16
  droughts, 14–16
  global wheat, corn and rice individual production *vs.* consumption, 17, 18
  P–C balance, 14–19
  rice consumption in Sub-Sahara, 19
  satellite-based VH indices, 15
  VH-derived droughts, 19, 20
  weather variations, 14
Simple difference index, 84
Simple ration index, 84
Sklar's theorem, 165
Smoothed BT (SMT), 32, 41, 43, 47, 48, 54–58, 201–206
  and SMN, 54
Smoothed NDVI (SMN), 32, 39–41, 43, 47, 54–58, 64, 201–207
Smoothed temperature (SMT), 202, 207
SNPP/VIIRS, 45–48
  advanced observation, 42, 43
  JPSS, 42
  and MODIS/AQUA, 44
  and NOAA/AVHRR
    BT, 45–47
    J-x/VIIRS series, 45
    life expectancy, 45
    NDVI, 45, 47
    polar-orbiting operational satellites, 45
    SMN and SMT, 47, 48
    SRF, 45, 46
  NOAA's, 43
  sensors, 42
  services, 43
  SMN and SMT indices, 43
Soil Moisture anomaly (SMA), 81
Sorghum in USA, 122–128
Spectral response functions (SRF), 35, 45, 46
Spinning Enhanced Visible and Infrared Imager (SEVIRI), 34
Spring wheat in Kazakhstan, 144–148, 167, 168
SSTa in TP, 177
Standard vegetation index (SVI), 84
Standardized precipitation-evapotranspiration index (SPEI), 80
Standardized precipitation index (SPI), 80, 101
Strong droughts
  in grain-producing and trading countries, 94, 95, 97
Suomi NPOESS preparatory platform (SNPP)
  sensors, 42
Surface water supply index (SWSI), 84

T

Temperature
  ambient, 220
  anomaly, 195–200, 220
  and CO_2 Trend, 195–197

distribution, 207
global trend, 221
greenness condition, 192, 211–215
measurements, 192
moisture condition, 211–215
radiative, 201, 204
records, 193
stabilization, 199
surface, 219–221
thermal condition, 211–215
warming surface, 195, 210
Temperature condition index (TCI)
estimates, 63
exceptional VIIRS, 63
global 4 km2 resolution, 61
in mid-summer 2012, 60
in Ukraine, 60
and VCI, 59, 60, 67
VH assessments, 63
VHI, 59, 60
Temperature vegetation drought index (TVDI, 84
TIROS-1, 24
TIROS-2 to 10, 24
TIROS-N, 82
Twenty-first century, 1

U

Ultraviolet radiations, 198
The United Nations' World Food Program, 10
United States Agency for International Development (US AID), 144
United States Department of Agriculture (USDA), 64
US Drought Monitor (USDM), 64–66, 75, 80, 101, 103

V

Validation, VH
CRD mean corn yield anomaly, 67, 69
crop yield, 65, 66
drought area and intensity, 66
drought classification scheme, 67
ground measurements, 64
mean Kansas corn yield, 67
NASS, 64
P&VP conditions, 64
satellite data verification, 65
satellite data with ground observations, 64
SMN in Sudan, 64
Ukraine winter wheat data, 65
USDA, 64
USDM, 64–66
VHI, 65, 66
yield anomaly and VH indices, 67–69
Vegetation (land cover) and soil moisture (LCSM) method, 81
Vegetation condition index (VCI)
BT estimates, 54
characterization, 56
CRD-mean, 69
drought, 67
global 4 km2 resolution, 61
reduction, 60
and TCI, 59, 60, 67
thermal stress, 60
VHI, 59, 60
in wetter ecosystem, 56
Vegetation drought response index (VegDRI), 83
Vegetation health (VH), 53, 54, 60–70, 88–101, 115, 163, 175, 229–239
agricultural output, 225
agricultural production, 226
algorithm (*see* VH algorithm)
applications (*see* Applications, VH)
assessment principals, 52
biophysical
AVHRR, 53
BT, 53, 54
green vegetation, 53
IR emission, 53
NDVI, 53, 54
NIR, 53
NOAA satellites, 53
S-NPP, 53
solar radiation, 53
unhealthy, 53
VIIRS, 53
VIS, 53
BT, 54
climate, 51
climate warming, 226
components, 51
condition, 52
crop and pasture production, 51
drought prediction (*see* Drought prediction)
drought-related yield losses (*see* Yield losses)
droughts (*see* VH droughts)
Earth's system, 51
ecosystems, 51
environmental factors, 52
fire risk, 225, 226, 229
healthy and stresses, 52
land cover changes, 225
malaria intensity, 226
modeling crop (*see* VH modeling crop production)
moisture–thermal combine, 226
moisture–thermal conditions, 229
NDVI, 54
NOAA, 226
physiological process, 52
precipitation and air temperature, 52
principal component of land and biosphere, 52
products, 226
quantity and quality of biomass, 52
scientific satellites, 53
short-term weather components, 52
space-based monitoring moisture, 52
thermal conditions, 52
users, 226–227
validation (*see* Validation, VH)

Vegetation health (VH) (*cont.*)
VCI-TCI stress, 61
very green and vigorous, 52
weather, 51
weather, climate/Earth's changes, 225
WEB pages, 226–229
at work
devastating drought, 62
global resolution VCI, TCI images, 60
thermal droughts, 62
VCI-TCI stress, 60
VHI, 60–62
VIIRS/TCI-based thermal stress, 63
world grain production, 225
Vegetation health home (VHH), 227
Vegetation health index (VHI)
calculation, 55
colour coded map, 184
crop production, 117, 118
and crop yield, 66
design, 59
global pixel, 178
in mid-July 2012 and 2017, 62
moisture-thermal conditions, 60
moisture-thermal drought index, 64
vs. monthly SSTa, 180
satellite-based data, 65
severe-to-extreme drought, 64
and soil moisture deficit, 64, 65
vs. SSTa, 178, 180, 182
time series, 65
and USDM, 64–66
and VCI and TCI, 59, 60
Vegetation index (VI), 28
NDVI (*see* Normalized difference vegetation index (NDVI))
Vegetation monitoring
VIIRS data, 42–45
Vegetation (land cover) and soil moisture (LCSM) method, 81
Vegetation supply drought index (VSDI), 83
Vegetation supply water index (VSWI), 83
VH algorithm, 55, 59
calculation
of NDVI and BT, 55
of SMN and SMT climatology, 55
of VHI, 55
data collection, 54
data processing, 54
high and low frequency noise, 55
SMN and SMT, 55–58
smoothed annual weekly, 55
TCI (*see* Temperature condition index (TCI))
VCI (*see* Vegetation condition index (VCI))
VHI (*see* Vegetation health index (VHI))
weekly compositing of NDVI and BT, 55
VH droughts
in Africa and FS, 96–100
agricultural regions, 88
GFS, 89–92
global grain production, 88–92
regional, 91–94
SE, 88
societal sustainability, 88
strong, 94, 95, 97
VH indices
and constructing, 172
and crop yields, 164, 165
dependence parameters, 169
global, 166
insurance purposes, 164
insuring wheat yield losses, 164
PCC, 168
satellite-based, 164, 166, 171
spatial variation, 167
and TCI, 166
wheat yield, 166
VH modeling crop production, 118
application, 158
Earth population, 158
in situ observations, 115
large-area assessment, 115
satellite-based crop yield modeling, 116
types of agricultural production, 115
VHI, 117, 118
weather-based models, 115
weather disasters, 115
weather variables, 158
yield time series, 116, 117
yield-VH models (*see* Yield-VH models)
VH products for users
from NOAA operational polar-orbiting satellites, 2
VH Time Series by Province, 228
VH-500/VIIRS drought, 102, 103, 105
VH-WEB pages, 227–229
Visible Infrared Imager Radiometer Suite (VIIRS), 42
algorithm from data collecting/reading, 43
ATMS, 42
CERES, 42
design, 42
environmental and socioeconomic services, 42
global weekly-composite NDVI, 44
I-band, 43
in situ observations, 42
JPSS, 42
M-bands, 43
NASA's MODIS/AQUA scientific satellite, 43
NDVI and BT, 43
observations, 43
OMPS, 42
operational polar-orbiting satellites, 42
role of satellite observations, 42
and SNPP (*see* SNPP/VIIRS)
Suomi NPOESS, 42
technical and scientific features, 42
world community, 42

W

Weather-based index, 101, 164
Weather-based insurance, 163
Wheat in Australia, 133–137

Wide dynamic range vegetation index (WDRVI), 83
Winter wheat
 in Ukraine, 128–130
 in USA, 122–128
World Bank, 10
World Meteorological Organization (WMO), 76

Y

Yield losses
 aggregate yield index, 163
 agricultural production, 163
 average risk reductions, 170
 copula approach, 165, 166
 copula dependence parameters, 169
 Cramer-von-Mises criterion, 169
 critical periods, 170
 empirical procedure, 166
 extreme weather events, 163
 farmers, 163
 Frank copula, 170
 Gumbel copula, 170
 index-based insurance, 163, 164
 in Kazakhstan, 164, 165
 NDVI, 164
 PCC, 168
 quality of satellite data, 164
 remote-sensing data, 164
 RRR, 171
 satellite-based VH indices, 164
 satellite data, 166–168
 type of crop insurance, 163
 VCI- and TCI-based insurance contracts, 169–171
 weather-based index, 164
 weather-based insurance, 163
Yield modeling
 VH-based, 4
Yield-VH models, 119–133, 148–151
 Brazil (South America), 157, 158
 Bulgaria (Europe), 153, 154
 Burkina Faso (Africa), 152
 cereals in Russia, 140–145
 corn
  in Argentina, 130–133
  in China, 119–122
  in USA, 122–128
  in Zimbabwe, 148–151
 correlation and regression analysis, 118
 critical period, 118
 data analysis, 118
 global grain and FS, 119
 highly-correlation, 118
 India (Asia), 155–157
 Mongolia (Asia), 153–155
 multiyear linear trend, 118
 pCC, 118
 rice in Bangladesh, 137–140
 sorghum in USA, 122–128
 spring wheat in Kazakhstan, 144–148
 stability and performance, 118
 types, 118
 wheat in Australia, 133–137
 winter wheat
  in Ukraine, 128–130
  in USA, 122–128

MIX
Papier aus verantwortungsvollen Quellen
Paper from responsible sources
FSC® C105338

If you have any concerns about our products,
you can contact us on
ProductSafety@springernature.com

In case Publisher is established outside the EU,
the EU authorized representative is:
Springer Nature Customer Service Center GmbH
Europaplatz 3, 69115 Heidelberg, Germany

Printed by Libri Plureos GmbH
in Hamburg, Germany